Circular Economy and Sustainability

This book series aims at exploring the rising field of Circular Economy (CE) which is rapidly gaining interest and merit from scholars, decision makers and practitioners as the global economic model to decouple economic growth and development from the consumption of finite natural resources. This field suggests that global sustainability can be achieved by adopting a set of CE principles and strategies such as design out waste, systems thinking, adoption of nature-based approaches, shift to renewable energy and materials, reclaim, retain, and restore the health of ecosystems, return recovered biological resources to the biosphere, remanufacture products or components, among others.

However, the increasing complexity of sustainability challenges has made traditional engineering, business models, economics and existing social approaches unable to successfully adopt such principles and strategies. In fact, the CE field is often viewed as a simple evolution of the concept of sustainability or as a revisiting of an old discussion on recycling and reuse of waste materials. However, a modern perception of CE at different levels (micro, meso, and macro) indicates that CE is rather a systemic tool to achieve sustainability and a new eco-effective approach of returning and maintaining waste in the production processes by closing the loop of materials. In this frame, CE and sustainability can be seen as a multidimensional concept based on a variety of scientific disciplines (e.g., engineering, economics, environmental sciences, social sciences). Nevertheless, the interconnections and synergies among the scientific disciplines have been rarely investigated in depth. One significant goal of the book series is to study and highlight the growing theoretical links of CE and sustainability at different scales and levels, to investigate the synergies between the two concepts and to analyze and present its realization through strategies, policies, business models, entrepreneurship, financial instruments and technologies. Thus, the book series provides a new platform for CE and sustainability research and case studies and relevant scientific discussion towards new system-wide solutions. Specific topics that fall within the scope of the series include, but are not limited to, studies that investigate the systemic, integrated approach of CE and sustainability across different levels and its expression and realization in different disciplines and fields such as business models, economics, consumer services and behaviour, the Internet of Things, product design, sustainable consumption & production, bio-economy, environmental accounting, industrial ecology, industrial symbiosis, resource recovery, ecosystem services, circular water economy, circular cities, nature-based solutions, waste management, renewable energy, circular materials, life cycle assessment, strong sustainability, and environmental education, among others.

Vinay Kandpal • Anshuman Jaswal
Ernesto D. R. Santibanez Gonzalez
Naveen Agarwal

Sustainable Energy Transition

Circular Economy and Sustainable Financing for Environmental, Social and Governance (ESG) Practices

Springer

Vinay Kandpal
Department of Management Studies
Graphic Era Deemed to be University
Dehradun, Uttarakhand, India

Ernesto D. R. Santibanez Gonzalez
CES 4.0 and Industrial Engineering
Department, Faculty of Engineering
Universidad de Talca
Curicó, Chile

Anshuman Jaswal
Director, SVKM's NMIMS
Indore, Madhya Pradesh, India

Naveen Agarwal
Centre of Continuing Education (CCE),
Knowledge Acres, Kandoli
University of Petroleum and Energy
Studies (UPES)
Dehradun, Uttarakhand, India

ISSN 2731-5509 ISSN 2731-5517 (electronic)
Circular Economy and Sustainability
ISBN 978-3-031-52942-9 ISBN 978-3-031-52943-6 (eBook)
https://doi.org/10.1007/978-3-031-52943-6

This Springer imprint is published by the registered company Springer Nature Switzerland AG
The registered company address is: Gewerbestrasse 11, 6330 Cham, Switzerland

Paper in this product is recyclable.

Foreword

डॉ. वी.के. सारस्वत
Dr. V.K. Saraswat
सदस्य
Member
Tele : 23096566, 23096567
Fax : 23096603
E-mail : vk.saraswat@gov.in

भारत सरकार
नीति आयोग, संसद मार्ग
नई दिल्ली - 110 001
Government of India
National Institution for Transforming India
NITI Aayog, Parliament Street,
New Delhi - 110 001

Amidst the prevailing period of notable environmental and socioeconomic issues, the demand for sustainable solutions is becoming urgent. The book titled *Sustainable Energy Transition: Circular Economy and Sustainable Financing for Environmental, Social, and Governance (ESG) Practices* is an invaluable resource that provides knowledge and guidance throughout this critical moment. This book is not only timely, but it also calls for a radical shift in how we perceive and engage with the Earth's resources and the financial organizations that oversee their allocation.

Environmental, social, and economic challenges are a complex web in the modern world. The issues of climate change, resource depletion, and socioeconomic inequality are closely linked within the framework of our global society. This work is pertinent, given the growing recognition of the imperative for a thorough overhaul that encompasses energy systems, economic models, and governance structures.

The authors examine the intricate relationship between sustainable energy practices, the circular economy, and funding that prioritizes environmental, social, and governance (ESG) considerations. Individually, each of these variables is crucial, but when they are combined, they possess the ability to catalyze an enduring and beneficial transformation in our global society. This book provides a comprehensive analysis of these interconnected areas, giving readers a nuanced understanding of each and illustrating how, when integrated, they can drive us towards a more sustainable future.

The book's prospective perspective is what renders it particularly attractive. It not only highlights the challenges and the imperative for change, but also presents a well-defined strategy for achieving this transition. The book equips readers with the essential knowledge and resources to actively participate in this transition by integrating theoretical insights, practical examples, and case studies. The book's exploration of the circular economy is both current and revolutionary. This undermines the conventional linear economic models of consumption and waste, providing a persuasive argument for a systemic transition towards reuse, recycling, and regeneration. This section of the book is not solely theoretical; it is replete with practical illustrations and success narratives that showcase the practicability and advantages of circular economic activities.

An outstanding aspect of this book is its thorough examination of sustainable finance for ESG practices. Redirecting resources towards sustainable businesses is crucial in a society where financial incentives and investments heavily influence decision-making. The book provides a comprehensive analysis of financial instruments and techniques that can effectively support ESG activities, serving as a guide for both investors and regulators.

This work possesses a remarkable characteristic in the form of its exceptional accessibility. Irrespective of your role as a student, professional, policymaker, or curious citizen, you will discover the content to be engaging and educational. The author exhibits a unique ability to articulate complex ideas in a manner that is comprehensive and accessible, making this book a valuable resource for a diverse audience. As we advance into the twenty-first century, the decisions we make now will have substantial repercussions for future generations. This book serves as a persuasive plea, urging us to reevaluate our relationship with the planet and with each other. This compels us to consider not only the monetary aspects of our decisions but also their environmental and social consequences.

It has been a delight to write this foreword for a book that I believe will greatly contribute to the discourse on sustainable development. The quest for a sustainable future is a collective undertaking, and this book symbolizes a noteworthy advancement in that goal. I desire to inspire the readers of this work to employ the knowledge obtained from these pages with the aim of generating a positive impact on the world.

Best Wishes to Dr. Vinay Kandpal and his team for the success of this book.

V. K. Saraswat

Preface

In an era of fast global development, we must undertake a profound paradigm shift in our approach to energy, resources, and finance. The central theme of this book centers on a compelling vision for the future of our planet. This vision entails an energy transition that is both inclusive and sustainable, the implementation of strong circular economic models, and the establishment of resilient financing mechanisms that are supported by environmental, social, and governance (ESG) practices.

The journey of transitioning to sustainable energy involves more than just technology advancements. It encompasses a multidimensional transformation that encompasses social, economic, and environmental paradigms. This book examines how the concepts of a circular economy, which seeks to eliminate waste and promote the ongoing utilization of resources, might serve as a driving force in facilitating this shift. We tried to explore many models and practices that illustrate the practicality and advantages of adopting a circular strategy in the energy industry.

Furthermore, there is a strong emphasis on the importance of sustainable funding in promoting and sustaining environmental, social, and governance (ESG) activities. Sustainable financing is more than simply allocating investments towards environmentally friendly initiatives; it involves changing the core principles that determine the value of an investment. The interdependence between energy and economy is of significant magnitude, particularly in an era characterized by growing populations and escalating urbanization. It is evident that our prevailing linear paradigm of 'extraction, production, and disposal' is unsustainable. We find ourselves at the precipice of a period that necessitates a re-evaluation of growth, the generation of value, and the ability to endure over the long term. The circular economy emerges as a solution that challenges conventional paradigms by prioritizing waste reduction, promoting the continuous use of products and materials, and facilitating the restoration of natural systems.

The journey of the circular economy extends beyond its initial implementation. In order to effectively implement and support various sustainable initiatives, it is imperative to have a robust and resilient financial infrastructure. Sustainable financing, which is focused on environmental, social, and governance (ESG) practices, serves as a means to reconcile the divergence between financial profitability and the promotion of beneficial societal outcomes. By prioritizing investments in environmentally conscientious, socially accountable, and ethically regulated enterprises, we can foster a future that achieves both economic prosperity and sustainability.

This manuscript provides an in-depth exploration of the interconnected domains, utilizing case studies, pioneering research, and insightful narratives as sources of information. We navigate through the complex systems of renewable energy, cutting-edge recycling technologies, environmentally friendly supply chain frameworks, green bonds, impact investment, and other related areas. For some who harbor skepticism, the journey towards achieving a sustainable future may seem fraught with challenges, encompassing potential financial obstacles and technological hindrances. However, history provides ample evidence of the success of human ingenuity and cooperation in overcoming barriers that were previously considered insurmountable. It is argued that by establishing appropriate frameworks and a collective vision for the future, the shift towards a sustainable energy paradigm, bolstered by circular economic practices and finance that prioritizes environmental, social, and governance (ESG) factors, is not only feasible but also unavoidable.

This book highlights the necessity of adopting a comprehensive approach that takes into account the interaction between energy, economy, and funding in order to accomplish sustainable development. The book culminates with an exhortation, imploring readers to actively participate in crafting a sustainable future. The collective endeavor necessitates the collaboration of several stakeholders, including policymakers, enterprises, financiers, communities, and individuals. The subsequent chapters encourage readers to not only comprehend the intricacies of this transition but also to actively engage in it. Ultimately, this transformation encompasses more than just ensuring the long-term sustainability of our planet; it involves revisioning the fundamental nature of our collective human existence.

In the face of ecological, economic, and societal concerns, there is an urgent demand for sustainable solutions. We earnestly aspire for this book to enhance the expanding reservoir of knowledge in this domain and act as a catalyst for constructive transformation.

Welcome to a narrative that aims to inspire, educate, and stimulate critical thinking. Welcome to the era of sustainable energy transformation.

Dehradun, India Vinay Kandpal

Indore, India Anshuman Jaswal

Talca, Chile Ernesto D. R. Santibanez Gonzalez

Dehradun, India Naveen Agarwal

Contents

About the Authors

Vinay Kandpal With over 18 years of experience in academia and industry, Dr. Vinay Kandpal is a finance expert and educator who teaches and mentors students at Graphic Era University, Dehradun, India. He is a Professor in the Department of Management Studies, Graphic Era University, Dehradun, India, where he imparts his knowledge and insights on various topics in finance, management, and education. He holds a DLitt (Postdoc) and a PhD in Finance from Kumaun University, India, and is also a Postdoctorate Scholar of PG Administracao-Paulista University – UNIP, Brazil. Dr. Kandpal is also a passionate and prolific researcher and author, who has published more than 62 research papers and 6 books on diverse subjects, such as sustainability, fintech, smart cities, risk management, circular economy, and financial inclusion. He has presented his work at prestigious international conferences and institutions and serves on the Editorial and Reviewer Boards of several well-regarded journals. He is a Full-Time Member of the Association of North America Higher Education International (ANAHEI). Dr. Kandpal's impressive career highlights his commitment to advancing knowledge and fostering innovation in finance.

Anshuman Jaswal Dr. Anshuman Jaswal is currently the Director and Head of NMIMS Deemed University's Indore campus. The campus has four schools, for Business, Law, Engineering and Technology, and Commerce, respectively. Earlier, he has held Leadership and Teaching positions at the School of Business at UPES, the Indian School of Hospitality (ISH), Gurugram, and Bennett University's School of Management, Greater Noida. He has extensive research and consulting experience and has taught as visiting faculty at leading Indian business schools including IIM Indore and IIM Kozhikode. His research interests include development of the Indian economy, financial services, non-profit sector, impact of environmental change on emerging markets, and topics related to economics and law. Before joining academics, he had worked in senior roles at firms such as Primetech and Celent in the United States. He has authored more than 100 research reports covering strategy and technology topics in financial services. He also has prior work experience in corporate banking, market research and consulting, life insurance and wealth management, pension plan management, and transfer pricing. He has also been a founding partner in an education-based business. Dr. Jaswal is a Fellow in Management (equivalent of a PhD) with a specialization in Economics from India's premier B-school, IIM Ahmedabad. Earlier, he was awarded a Bachelor of Science in Economics (Honors) by the School of Oriental and African Studies, University of London, an MBA in Finance by NMIMS, Mumbai, a Bachelor of Laws (Honors) degree from the University of London, and a Master of Laws degree from the University of Pittsburgh School of Law. In his previous roles, Dr. Jaswal has been quoted regularly in media, including the *Economic Times*, *Business Standard*, *Mint*, *Financial Times*, *Wall Street Journal*, *Reuters*, *American Banker*, *Bloomberg*, *Forbes*, *New York Times*, and financial services technology trade journals. His hobbies include traveling, reading, and running.

Ernesto D. R. Santibanez Gonzalez Dr. Ernesto passion is to address and solve problems at the interface of climate change and sustainability that arise in many different fields, such as supply chain, energy efficiency, low carbon economy, carbon capture and storage, smart city, smart manufacturing, sustainable and resilience supply chain, and sustainable operations management. Current research of Dr. Santibañez-Gonzalez is characterized by integrating mathematical models, big-data, blockchain, and Internet technology to understand, model, and provide practical solutions to companies and organizations that need to address climate change and sustainability problems. His experience in Industrial Engineering, Operations Management, and Operations Research with a PhD in Production Engineering from the Federal University of Rio de Janeiro (UFRJ) in a joint program with the London School of Economics and Political Science (LSE) gives him a good understanding of how to integrate information technology with quantitative methods to solve real-life problems. He is serving as Associate Editor of *Journal of Cleaner Production* (Elsevier), *Journal of Intelligent Manufacturing* (Springer), *Modern Supply Chain Research and Applications* (Emerald, new journal), *International Journal of Big Data Mining for Global Warming*, and *Heliyon Business and Economics* (Elsevier). He is also Editor-in-Chief of *Sustainable Operations and Computers*, a new journal sponsored by Chinese Academy of Science and Elsevier. He has been the Managing Guest Editor of several special issues in top-tier journals, including *European Journal of Operational Research*, *Journal of Cleaner Production*, *International Journal of Production Research*, and *Sustainability* and Guest Editor for special issues in journals such as *International Journal of Production Economics*, *Computers and Industrial Engineering*, and *Science of Total Environment*. He has also served as Regional Editor for *International Journal of Physical Distribution and Logistics Management (Emerald)*. During the last 5 years, he has published more than 50 papers in top-tier journals. He has had visiting appointments at several universities in Brazil, China, and the United Kingdom, the last being, before the COVID-19 pandemic, for example, at the London Business School. Currently, he is supervising and co-supervising graduate students in Brazil, Chile, and China. He leads the Circular Economy and Sustainability 4.0 Initiative (CES4.0), is a member of the Board of Directors (Counsellor) – National Forestry Society of Chile (SNF), and has developed more than 80 research, innovation, and consulting projects providing solutions to real problems in different countries of the world in industrial sectors such as Energy, Retail, Mining, Agri-food, Forestry, Logistics, and Oil and Gas.

Naveen Agrawal Dr. Naveen Agrawal is Assistant Professor (Energy and Finance) at CCE, University of Petroleum and Energy Studies, Dehradun, focusing on areas pertaining to electricity market, power sector economics, power pricing, renewable energy, financing of energy sector projects, financial modelling, and electricity analytics. He was as a consultant in power distribution with focus on power distribution network, commercial operations, and financial analysis. He has worked on different power distribution projects like R-APDRP, RGGVY, feeder separation, capacity building, and distribution system automation. He obtained his doctorate degree from TERI School of Advanced Studies, Delhi, in area of energy economics. He holds an MBA (Power Management) from University of Petroleum and Energy Studies (UPES) and a Bachelor's Degree in Engineering from Rajasthan University, India.

List of Figures

Chapter 1
Sustainable Energy Transition, Circular Economy, and ESG Practices

Abstract To achieve sustainability and resilience, our economic models and practices must undergo a process of evolution. This paradigm shift covers the use of sustainable energy sources, the implementation of a circular economic model, and the integration of Environmental, Social, and Governance (ESG) practices. To transition from fossil fuels to renewable energy sources, a comprehensive approach encompassing technological advancements, policy reforms, infrastructural developments, and societal transformations is essential. Denmark, Germany, and China each possess distinct strategies for achieving sustainable energy transitions. The concept of the circular economy advocates for the reuse of resources, presenting a counterpoint to the linear resource consumption model characterized by the 'take-make-dispose' approach. The implementation of a circular economy model effectively mitigates the negative impacts of waste and pollution by promoting the reuse of resources and the regeneration of natural systems. These principles require the implementation of product-as-a-service and sharing economy business models. ESG underscores the importance of considering environmental, social, and governance factors when evaluating the enduring and societal consequences of investing in a firm or industry. The collaboration of technological progress, systemic design, and responsible governance synergistically contributes to the creation of a future that is both sustainable and resilient.

Keywords Sustainable energy transition · Circular economy · Environmental Social and Governance (ESG) · Renewable energy sources · Policy and regulation · Technological advancements · Urban planning · Extended Producer Responsibility (EPR) · Sustainable and resilient future

1.1 Introduction

In a time characterized by exceptional global issues, such as climate change, limited resources, social injustice, and corporate governance, the necessity for sustainable and responsible practices has become increasingly evident. The pressing need to

V. Kandpal et al., *Sustainable Energy Transition*, Circular Economy and Sustainability, https://doi.org/10.1007/978-3-031-52943-6_1

confront these complex issues has led to the emergence of three pivotal concepts that have become emblematic of our shared endeavor to achieve a fairer and more environmentally responsible future: Sustainable Energy Transition, Circular Economy, and Environmental, Social, and Governance (ESG) Practices.

This chapter explores the complex network formed by the three interconnected pillars of sustainability. This study examines the interplay between decision-making in the domains of energy, resource management, and corporate responsibility and its cumulative impact on the trajectory of our planet's future. While possessing unique areas of emphasis, each of these principles intersect to form a comprehensive perspective to foster a more prosperous and harmonious global society. The exploration commences with an examination of the dynamic domain of Sustainable Energy Transition, a movement aimed at transforming our energy landscape. This section examines the necessity of transitioning from fossil fuels to environmentally sustainable and renewable energy sources. The increasing prevalence of solar panels on rooftops and wind turbines on the horizon reveals the significant impact that sustainable energy may have on addressing climate change, enhancing air quality, and strengthening energy security (Hafner & Tagliapietra, 2020).

Subsequently, we delve into the complex networks of the Circular Economy, a conceptual framework that presents a counterpoint to the linear approach of production and consumption known as the "take-make-dispose" model. Within this circular domain, the focus lies on the creation of goods with extended lifespans, the reclamation and subsequent reuse of materials, and the minimization of waste generation. The economic, environmental, and social advantages of adopting a circular strategy demonstrate its capacity to preserve resources, stimulate innovation, and promote resilience.

ESG standards serve as a guiding framework for firms to uphold their dedication to environmental sustainability, social justice, and accountable governance. Through a thorough analysis of the fundamental principles and operational strategies that serve as the foundation for Environmental, Social, and Governance (ESG) frameworks, we can elucidate the processes by which firms can effectively harmonize their profit-driven objectives with the advancement of social and environmental well-being. In this chapter, the focus is placed on highlighting the interdependence of the three pillars of sustainability. The transition towards sustainable energy sources catalyzes the resource-efficient mechanisms of the circular economy, while environmental, social, and governance (ESG) practices promote ethical and responsible economic behavior. These themes provide a comprehensive depiction of a society where economic affluence aligns with ecological sustainability and societal welfare (Liu & Ramakrishna, 2021).

As we commence this endeavor, we extend an invitation to contemplate the significant impact of Sustainable Energy Transition, Circular Economy, and ESG Practices on our planet's current and future state. Collectively, these sources furnish us with a comprehensive guide to effectively address the intricate obstacles of our day, presenting the potential for a future that is both environmentally sustainable, adaptable, and just, ensuring the well-being of future generations. Investing in renewable energy technologies, improving energy efficiency in both consumption and production, promoting electrification of various sectors (primarily transportation), developing an intelligent grid system to accommodate renewable energy, and implementing supportive policies and regulations are all critical aspects of the sustainable energy transition.

The adoption of the circular economy model by industrial ecosystems has the potential to facilitate the achievement of Sustainable Development Goals (SDGs). The current dominant paradigm of economic growth generates significant quantities of trash, disproportionately benefiting a select few. In contrast, the circular economy model views waste as a design deficiency, hence promoting the development of more decentralized and inclusive systems (Doszhan et al., 2022). The circular economy concept has gained significant attention and interest globally, both in academic and practical spheres. Despite the considerable accumulation of research on the subject matter, the area currently faces a notable deficiency in terms of a structured theoretical framework (Schröder et al., 2018; van Dam et al., 2020).

A circular economy is a model that aims to reduce waste and promote the continuous use of resources. On the other hand, open innovation is the practice of collaborating with external partners to generate new ideas and bring innovative products and services to the market. Together, these two concepts can help to promote sustainability and innovation in the business world. Open innovation can contribute to the development of a circular economy by bringing together stakeholders from across industries to share knowledge, expertise, and resources. Companies can identify new opportunities for waste reduction and resource efficiency by working together and developing new business models that promote circularity (Babkin et al., 2023).

1.2 The Evolution of Energy Systems: From Fossil Fuels to Renewables

The transition from fossil fuel-based energy systems to renewable energy sources is a significant and continuous process that is motivated by the imperative to mitigate climate change, minimize environmental consequences, bolster energy security, and shift towards sustainable energy alternatives. This transition signifies a movement away from limited and carbon-intensive resources towards cleaner, more abundant, and sustainable energy alternatives.

1.2.1 Dependence on Fossil Fuels

In the nineteenth century, the usage of fossil fuels—coal, oil, and natural gas—began to dominate, coinciding with the Industrial Revolution. Coal was the first commonly utilized fossil fuel. It powered steam engines in factories and railways, accelerating the Industrial Revolution. Coal is a more concentrated and portable energy source than wood, and there are large, easily accessible coal reserves, particularly in countries such as the United States and the United Kingdom.

Oil and natural gas rose to prominence in the twentieth century. These resources are critical to our energy systems because of their availability, high energy density, and adaptability. Oil became the contemporary economy's lifeblood, providing fuel for vehicles, planes, and ships and raw materials for plastics and other synthetic goods. However, the strong reliance on fossil fuels revealed significant drawbacks. Burning fossil fuels results in the emission of greenhouse gases, with carbon dioxide (CO_2) being the primary contributor (Ma, 2021). These emissions significantly impact global warming and the subsequent alterations in climate patterns. Fossil fuel extraction and utilization give rise to various environmental repercussions, including but not limited to air and water pollution, habitat degradation, and oil spills. The aforementioned resources have been crucial in facilitating the advancement of manufacturing, transportation, and energy generation (Listyawati et al., 2014).

1.2.2 Environmental and Climate Concerns

The widespread utilization of fossil fuels has resulted in environmental issues, including the emission of pollutants into the atmosphere, the poisoning of water sources, and the destruction of natural habitats. In addition, burning fossil fuels results in the emission of greenhouse gases, with carbon dioxide (CO_2) being the primary contributor. These emissions have a significant impact on the phenomenon of global warming and the subsequent alterations in climatic patterns.

1.2.3 Rise of Renewable Energy

The current global energy environment is experiencing a significant and profound transformation, with a central storyline centered around the transfer from conventional fossil fuels to sustainable and renewable sources of energy. The transformation discussed is not only a passing trend but rather a necessary action motivated by the pressing requirement to address climate change, mitigate environmental consequences, strengthen energy resilience, and promote economic advancement. In light of the global implications of carbon emissions and limited resources, the shift towards renewable energy sources presents itself as a promising solution, offering a trajectory towards a future characterized by enhanced sustainability and resilience. The increasing acknowledgment of environmental and climate-related issues has spurred the advancement and acceptance of renewable energy sources, including solar, wind, hydro, and bioenergy. The sources mentioned above are plentiful, have minimal greenhouse gas emissions, and are a viable and sustainable substitute for fossil fuels.

1.2.4 Technological Advances

The inclusion of technological breakthroughs has significantly contributed to the facilitation of the transition towards renewable energy sources. The cost effectiveness and reliability of renewable energy sources have been considerably improved due to advancements in the efficiency of solar panels, the design of wind turbines, the storage of energy, and the management of power networks.

1.2.5 Government Policies and Incentives

Many governmental bodies have implemented regulations and incentives to promote the extensive use of renewable energy sources. The strategies discussed include feed-in tariffs, tax credits, renewable portfolio standards, and carbon pricing mechanisms. These policy instruments aim to establish an equitable competitive landscape and encourage investments in clean energy.

1.2.6 Energy Storage Solutions

The use of energy storage technology, such as advanced batteries, pumped hydro storage, and thermal storage, has enhanced the dependability and integration of renewable energy sources within the power grid.

1.2.7 Distributed Energy Resources

The emergence of distributed energy resources (DERs), such as rooftop solar panels, small-scale wind turbines, and home energy storage, has facilitated the ability of people and groups to independently produce environmentally friendly energy, thereby diminishing reliance on centralized power systems that rely on fossil fuels (FIDIC, 2023).

1.2.8 Grid Modernization

Efforts are being made to modernize the grid to absorb the inherent variability of renewable energy sources effectively. Smart grids, demand response initiatives, and grid-scale energy storage systems contribute to the real-time equilibrium of supply and demand.

1.2.9 Electrification of Transport

The incorporation of electric vehicles (EVs) into transportation systems is an additional aspect of the ongoing shift towards electrification. Electric vehicles (EVs) have the potential to decrease dependence on oil resources and can be powered by renewable energy sources, thus contributing to a reduction in greenhouse gas emissions.

1.2.10 Investment and Economic Opportunities

The renewable energy industry has emerged as a substantial catalyst for economic expansion, drawing capital investments, generating employment opportunities, and fostering advancements in sustainable energy technologies.

1.2.11 Global Renewable Energy Targets

Many countries have established ambitious objectives and obligations pertaining to renewable energy, aiming to attain predetermined proportions of their energy supply from renewable sources within specified timeframes. These aims align with global climate accords, such as the Paris Agreement.

1.2.12 Energy Transition Challenges

Despite the achievements made, there are still obstacles that persist in the field. These challenges encompass the requirement for upgrades in grid infrastructure, advancements in energy storage, tackling the issue of intermittency, and guaranteeing fair and equal access to sustainable energy solutions.

The transition from fossil fuels to renewable energy sources is a continuing process that encompasses various actors, advancements in technology, and legislative interventions. The statement encapsulates a significant transition towards a more sustainable and resilient energy trajectory, wherein the adoption of cleaner energy sources and conscientious energy governance plays a pivotal role in addressing climate change and securing a sustainable environment for forthcoming cohorts.

1.3 Challenges and Future Prospects

Despite the advancements made, the process of transitioning to renewable energy sources is accompanied by numerous challenges. Some of the challenges associated with solar and wind power include its intermittent nature, the requirement for significant upgrades to electricity infrastructure, and the environmental consequences of manufacturing renewable energy equipment.

The increasing use of renewable energy can be attributed to a combination of ongoing technological advancements, sustained legislative support, and a growing recognition of the significance of addressing climate change. Current research efforts are focused on exploring emerging renewable technologies, including tidal and wave energy, while concurrently witnessing a growing interest in green hydrogen as a viable and environmentally friendly fuel source.

The transition from the use of fossil fuels to the adoption of renewable energy sources is of utmost importance in addressing the challenges posed by climate change and in the pursuit of establishing a sustainable and resilient energy system. The process is intricate and ongoing, requiring advancements in technology and the implementation of appropriate legislation, enough financing, and active engagement from the public.

The evolution of energy systems demonstrates humanity's inventiveness and adaptability. As we enter an era in which renewables are essential to our energy systems, we aspire to create a more sustainable, cleaner, and resilient society that promotes economic growth without jeopardizing the environment or the welfare of future generations.

1.4 Role of Policy and Regulation in the Energy Transition

Policy and regulation play a vital role in guiding and supporting the transition from fossil fuels to cleaner and more sustainable energy sources. The successful execution of these policies is crucial for creating a favorable structure that encourages investments, minimizes risks, and facilitates the achievement of the overall goals of reducing carbon emissions, enhancing energy security, and fostering a sustainable energy future.

1.4.1 Setting Renewable Energy Targets

Governments frequently set targets for the integration of renewable energy sources, delineating the proportion of energy that must be derived from renewables within a specified period. The aims mentioned above offer a distinct trajectory for the transformation of energy.

1.4.2 Incentives and Subsidies

Governments have the potential to provide financial incentives, such as tax credits, grants, or subsidies, to stimulate investments in projects related to renewable energy. These incentives serve to mitigate the financial obstacles that are typically associated with the adoption of renewable energy technologies.

1.4.3 Renewable Portfolio Standards (RPS)

The policies of the RPS require a specified proportion of electricity generation to be derived from renewable sources. The adherence to these criteria by utilities necessitates the fulfillment of renewable energy requirements, hence stimulating the market for renewable energy and fostering investments in clean energy initiatives.

1.4.4 Feed-in Tariffs (FiTs)

Feed-in tariff (FiT) programmes ensure a predetermined payment rate for producers of renewable energy, frequently surpassing prevailing market rates. These programmes offer a level of financial stability and incentivize the advancement of renewable energy projects.

1.4.5 Net Metering and Grid Integration

Net metering policies enable consumers to effectively reduce their electricity expenses by using the opportunity to sell surplus energy generated from renewable sources back to the power grid. Grid integration laws are implemented to facilitate the smooth integration of renewable energy sources into the preexisting energy infrastructure.

1.4.6 Carbon Pricing Mechanisms

Carbon pricing methods, such as carbon taxes or cap-and-trade systems, impose a financial cost on the release of carbon emissions. These systems establish financial incentives for enterprises to mitigate emissions and shift towards more environmentally friendly energy sources.

1.4.7 Energy Efficiency Standards

The implementation of regulations that establish energy efficiency requirements for appliances, buildings, and vehicles serves to foster energy conservation and mitigate the overall use of energy resources.

1.4.8 Research and Development Funding

Governments allocate financial resources towards the advancement of research and development (R&D) in the realm of clean energy technology. These investments provide financial backing for the promotion of innovation and the rise of renewable energy solutions at a sophisticated level.

1.4.9 Environmental Regulations

The presence of regulations pertaining to emissions, air and water quality, and environmental safeguards imposes limitations on the generation of energy from fossil fuels, hence enhancing the appeal of renewable energy sources in terms of adhering to environmental compliance standards.

1.4.10 Market Mechanisms

Regulatory procedures, such as power purchase agreements (PPAs) and renewable energy certificates (RECs), play a crucial role in enabling the sale and exchange of renewable energy, thereby establishing a reliable market for producers of clean energy.

1.4.11 Electrification Policies

Policies aimed at fostering the electrification of transport and heating sectors serve to incentivize the utilization of clean power across diverse applications, hence contributing to a diminished reliance on fossil fuels.

1.4.12 Long-Term Planning and Roadmaps

Governments can formulate comprehensive energy transition roadmaps and plans that delineate the sequential actions and significant milestones necessary to attain renewable energy objectives and mitigate emissions.

1.4.13 International Agreements

International agreements, such as the Paris Agreement, serve as a comprehensive global framework that facilitates the coordination of countries' endeavors and fosters their commitment to mitigating the production of greenhouse gases.

1.4.14 Just Transition Initiatives

Policies centered around the concept of a "just transition" are designed with the objective of mitigating the adverse social and economic consequences experienced by communities and employees reliant on fossil fuel-based industries throughout their shift towards more environmentally sustainable energy sources.

The establishment of robust legislative and regulatory frameworks is of utmost importance in ensuring market stability, fostering investment inflow, and facilitating a well-coordinated and sustainable energy transition. They establish an enabling environment that encourages informed decision-making among industry participants, businesses, and individuals, thereby promoting the transition towards a sustainable and low-carbon energy future.

1.5 Renewable Energy Adoption Barriers and Solutions

Renewable energy provides a clean and sustainable solution to the world's energy demands. Despite substantial breakthroughs and cost reductions in renewable energy sources, a number of impediments to broader adoption remain. Here are some of the major roadblocks, as well as potential solutions:

- **Technical Difficulties:** While renewable technologies have advanced tremendously, there are still challenges connected to the intermittent nature of sources such as wind and solar energy, and current energy storage technology is not yet suitable for large-scale, long-term storage (Augustynowicz, 2022). Continuing to invest in Research and Development can lead to breakthroughs that make renewables more dependable and storage more efficient. Smart grid technologies and demand response tactics can also help control the intermittent nature of renewables.
- **Infrastructure Obstacle:** The existing energy infrastructure was designed for centralized, fossil-fuel-based energy generation. Adapting this infrastructure for decentralized, fluctuating renewable energy can be complex. It is critical to invest in grid modernization and infrastructure that can oversee and balance variable renewable energy. Policies that incentivize or mandate utility system upgrades can be effective.
- **Financial Barriers:** While the cost of renewable technology has fallen, the initial capital expenses can still be substantial, and obtaining financing can be difficult, particularly in underdeveloped countries or for newer technologies. Financial incentives such as tax credits, grants, or low-interest loans might assist in alleviating financial strain. Green banks and public-private partnerships can also aid in financial mobilization.
- **Regulatory and Policy Challenges:** Inconsistent policy signals and regulatory barriers might deter investment in renewable energy. Long-term policies that are stable can lessen investment risks. Policies that level the playing field, such as the elimination of fossil fuel subsidies or the imposition of a carbon tax, can make renewables more competitive.
- **Lack of Social Acceptance and Information:** A lack of information about the benefits of renewable energy, as well as reluctance to change, might impede the adoption of renewable technologies. Furthermore, local opposition to renewable energy projects may exist owing to perceived effects on landscapes, animals, or property values. Public education campaigns, community participation in renewable energy project development, and sharing project benefits with local communities can all assist in increasing support for renewable energy.
- **Skilled Workforce Barrier:** The move to a renewable-based system necessitates the development of new skills in the workforce. There may be a scarcity of qualified individuals to plan, implement, and maintain renewable energy installations. Renewable energy education and training programs can assist in generating a trained workforce to support the energy transition.

Numerous players, including governments, the commercial sector, research institutes, and civil society, must work together to overcome these obstacles. The rewards of such initiatives are worthwhile: cleaner air, lower greenhouse gas emissions, and new economic prospects.

1.6 Renewable Energy Technological Advancements: Solar, Wind, Hydro, and Bioenergy

The integration of renewable energy sources, such as solar, wind, hydro, and bioenergy, has played a pivotal role in facilitating the shift towards environmentally friendly and sustainable energy systems. The following are significant advancements observed within each respective renewable energy sector:

Solar energy is an environmentally friendly and enduring source of energy that is obtained from the radiant energy emitted by the sun. The incorporation of cleaner and more sustainable energy sources is an essential element in the process of transitioning towards a more environmentally, economically, and socially advantageous energy system. The following is a comprehensive review of solar energy.

(a) **Solar Energy Generation:** Solar energy is harnessed through the utilization of photovoltaic (PV) cells, commonly referred to as solar panels, which capture sunlight or through the concentration of sunlight using solar thermal technology. Photovoltaic (PV) cells are capable of directly converting solar energy into electrical energy, whereas solar thermal systems utilize sunlight to heat a fluid, commonly water or a heat transfer fluid, to generate steam that powers a generator.
(b) **Clean and Renewable:** Solar energy is considered a sustainable and environmentally friendly energy source due to its ability to generate electricity without emitting any greenhouse gases or air pollution. The utilization of alternative energy sources aids in diminishing dependence on fossil fuels, contributing to the mitigation of climate change and the reduction of air pollution.
(c) **Abundant Resource:** Solar energy is considered to be an abundant resource due to the substantial daily emission of energy by the sun. The resource in question possesses a nearly limitless supply, rendering it a dependable and sustainable option for fulfilling energy requirements over an extended period.
(d) **Distributed Energy Source:** Solar panels can be deployed in multiple locations, such as rooftops and open fields, or integrated into different structures, thereby establishing solar energy as a decentralized energy resource. The use of decentralization measures results in a reduction of transmission and distribution losses, hence improving the resilience of the grid.
(e) **Off-Grid and Remote Power:** Solar energy is highly advantageous in the provision of electricity in rural and off-grid regions characterized by limited or costly access to conventional power sources. It finds utility in a wide array of applications, encompassing the provision of energy to rural clinics, residential dwellings, and telecommunication infrastructure.
(f) **Net Metering and Grid Integration:** Net metering is a prevalent practise in numerous regions, whereby the extra electricity produced by solar panels can be seamlessly integrated into the power grid, effectively reducing the end user's electricity expenses. The integration of solar power into the electrical grid is necessary for the purpose of maintaining equilibrium between electricity supply and demand.

(g) **Scalability:** Solar energy systems can be customized to accommodate a range of scales, encompassing both modest home installations and huge utility-scale solar farms. The scalability of this system allows for the accommodation of a wide range of energy requirements and applications.
(h) **Cost Decline:** The cost of solar panels and solar energy systems has notably declined over time due to technological improvements and enhanced production capabilities. The decrease in costs has resulted in solar energy being more competitive than traditional fossil fuels.
(i) **Environmental Benefits:** Solar energy has been found to have a significant positive impact on the environment by effectively mitigating greenhouse gas emissions, reducing air pollution, and minimizing water consumption in comparison to power-generating methods reliant on fossil fuels. The product exhibits a limited ecological impact, hence promoting the preservation of natural resources.
(j) **Job Creation:** The solar energy sector presents several employment prospects across multiple domains, including production, installation, maintenance, research, and development. The aforementioned factor has a significant role in fostering local economic expansion and facilitating the generation of employment opportunities.
(k) **Energy Independence:** The use of solar energy diminishes a country's dependence on imported fossil fuels, hence bolstering energy security and mitigating susceptibility to disruptions in energy supply.
(l) **Technological Advancements:** Continuous research and development efforts in the field of solar technology persistently enhance the efficiency and storage capacities, hence further enhancing the accessibility and cost-effectiveness of solar energy.

Solar energy is of great significance in the global pursuit of sustainable and low-carbon energy solutions. Renewable energy sources provide a sustainable and versatile means of power generation that not only addresses environmental concerns but also fosters economic development and enhances energy accessibility on a global scale. The ongoing expansion of solar energy presents the potential for a future energy landscape that is both environmentally sustainable and robust in the face of challenges.

Wind Energy

Wind energy is considered an economically sustainable and environmentally friendly source of power that relies on the extraction of kinetic energy from the movement of air masses. This is predominantly achieved through the utilization of wind turbines. Renewable energy has emerged as a substantial contributor to the overall global electricity generation, assuming a crucial role in facilitating the move towards cleaner and more sustainable energy alternatives. The following are a few fundamental elements pertaining to wind energy:

(a) **How Wind Energy Works:** Wind turbines, alternatively referred to as windmills, harness the dynamic energy of the wind and transform it into electrical energy. The rotation of a turbine's blades is induced by the movement of wind, which then activates a generator responsible for generating electrical power.

(b) **Types of Wind Turbines:** Two primary categories of wind turbines exist: horizontal-axis wind turbines (HAWT) and vertical-axis wind turbines (VAWT). Horizontal axis wind turbines (HAWTs) are characterized by their rotor shaft being positioned horizontally. This configuration is widely prevalent in contemporary wind turbine technology. Vertical axis wind turbines (VAWTs) are characterized by their vertical rotor axis, which distinguishes them from the more prevalent horizontal axis wind turbines (HAWTs). While VAWTs are less commonly used, they offer distinct advantages in specific applications.

(c) **Onshore and Offshore Wind Farms:** Wind energy projects have the potential to be situated either onshore, generally in rural or remote regions, or offshore in aquatic environments. Offshore wind farms possess the capability to harness more robust and continuous wind resources; nonetheless, their implementation and upkeep necessitate a higher degree of intricacy.

(d) **Wind Energy Capacity:** The measurement of wind energy capacity is typically expressed in units of megawatts (MW) or gigawatts (GW). There has been a significant global expansion in wind power capacity. Notably, countries such as China, the United States, and Germany have emerged as prominent leaders in the installation of wind energy infrastructure.

(e) **Environmental Benefits:** Wind energy is regarded as an environmentally friendly and renewable form of energy, as it does not generate any direct emissions of greenhouse gases or air pollutants during the process of power production. This action aids in the reduction of carbon emissions and the mitigation of climate change.

(f) **Energy Independence and Security:** Wind energy plays a pivotal role in diminishing reliance on fossil fuels and bolstering energy security through the diversification of the energy portfolio. The implementation of measures aimed at reducing vulnerability to energy supply disruptions and price changes in fossil fuel markets is beneficial for governments.

(g) **Land Use and Environmental Impact:** In comparison to alternative kinds of energy generation, wind farms often necessitate very modest land usage. Nonetheless, the environmental consequences of their activities are contingent upon various circumstances, including geographical location, the implementation of wildlife conservation measures, and adherence to local regulatory frameworks.

(h) **Technological Advancements:** The progression of wind turbine design, materials, and control systems has resulted in enhanced energy efficiency and reliability. At greater elevations, increased wind energy may be harnessed through the utilization of taller towers and longer blades.

(i) **Grid Integration:** The integration of wind energy into the power grid can be effectively achieved in conjunction with other energy sources. Grid management and energy storage technologies play a crucial role in the effective management of variability and the maintenance of a reliable and stable energy supply.

(j) **Economic Benefits:** The wind energy sector facilitates employment opportunities across various domains, including manufacture, installation, operation, and maintenance. The activity in question has a significant role in bolstering local economies and presents avenues for fostering innovation and conducting research.

(k) **Future Growth and Potential:** The potential for substantial expansion in wind energy is evident due to ongoing technological advancements and the establishment of ambitious renewable energy objectives by several nations. The reliability and viability of energy storage and grid infrastructure are anticipated to be substantially improved by ongoing advancements.

Wind energy plays a crucial role in the worldwide shift towards cleaner and more sustainable forms of energy. The technology holds the potential to provide a dependable, environmentally friendly, and ample source of energy, so contributing to a more sustainable and resilient energy landscape.

Hydropower

Hydropower, alternatively referred to as hydroelectric power, is a sustainable energy resource that produces electrical energy by the use of the kinetic energy present in the movement of water, whether it be in the form of flowing or descending water. Renewable energy in the form being discussed is among the earliest and most extensively employed sources, boasting a rich historical lineage that can be traced back to ancient societies. The following are a few fundamental components of hydropower:

(a) **How Hydropower Works:** Hydropower systems commonly encompass the utilization of a dam or other infrastructure to regulate the water flow within a river or reservoir. The process involves the discharge of water from a higher elevation, facilitating its downward flow, and the subsequent conversion of the resulting kinetic energy of the water into mechanical energy through turbines. Subsequently, the turbines enable the operation of generators, thus generating electrical energy.

(b) **Types of Hydropower Systems:** Various types of hydropower systems exist, which encompass:

- **Large-Scale Hydropower:** These structures are characterized by their significant size and are commonly known as hydroelectric power plants, possessing the capacity to produce substantial quantities of electrical energy. Notable instances encompass the Hoover Dam situated in the United States and the Three Gorges Dam located in China.
- **Small-Scale Hydropower:** These compact facilities serve the purpose of generating electricity in rural or distant regions. Frequently, these systems are characterized as run-of-river installations, which have a reduced ecological impact compared to their larger dam counterparts.
- **Pumped Storage Hydropower:** These systems employ excess electricity during periods of reduced demand to transfer water from a lower reservoir to an upper reservoir. During periods of elevated electricity demand, the release of water from the higher reservoir is employed to generate electrical power.

(c) **Environmental Considerations:** Although hydropower is widely regarded as a sustainable energy source, the construction and operation of major dams and reservoirs can have significant environmental consequences. These include the modification of river ecosystems, disturbance of habitats, and adverse effects on aquatic organisms. Thorough strategic planning and implementation of environmental mitigation measures are needed to effectively minimize the aforementioned consequences.
(d) **Hydropower Capacity:** The measurement of hydropower capacity is commonly expressed in units of megawatts (MW) or gigawatts (GW). Certain nations, such as Norway and Brazil, possess a substantial proportion of their electrical generation capacity sourced from hydropower.
(e) **Global Hydropower Production:** Hydropower has developed as one of the leading sources of renewable energy globally, contributing a substantial share of the world's electricity production.
(f) **Role in Energy Transition:** Hydropower maintains its crucial significance in the worldwide shift towards cleaner and more sustainable forms of energy. The technology in question offers a consistent and dependable supply of electrical power that effectively supplements the intermittent nature of renewable energy sources such as wind and solar.
(g) **Technological Advances:** The continuous progression of technological innovations in turbine design, efficiency enhancement, and environmental impact reduction endeavors to enhance the operational effectiveness and long-term viability of hydropower systems.

Hydropower continues to be a significant constituent of the renewable energy repertoire, providing dependable and environmentally friendly electricity generation while simultaneously tackling the imperatives of energy storage and grid stability. The anticipated trajectory of its contribution to the progression towards a more sustainable energy landscape is projected to undergo further development in tandem with advancements in technology and environmental methodologies.

Bioenergy

Bioenergy is classified as a sustainable energy resource that originates from organic matter, encompassing vegetation, agricultural byproducts, and organic refuse. These materials can undergo conversion processes to yield diverse energy outputs, such as thermal energy, electrical power, and fuels for transportation purposes. Renewable energy is an adaptable and environmentally friendly energy alternative that encompasses various essential elements:

(a) **Biomass Feedstock:** Biomass feedstock for bioenergy can include a wide range of organic materials, such as:

- Energy crops like switchgrass and miscanthus.
- Agricultural residues like corn stover, rice husks, and sugarcane bagasse.
- Forest residues like wood chips and sawdust.

- Organic waste like municipal solid waste (MSW), food waste, and sewage sludge.

(b) **Policy and Regulations:** The promotion of bioenergy technology is heavily influenced by government policies, incentives, and restrictions. These regulations have the potential to impact the supply of feedstock, investment patterns, and market expansion.
(c) **Advanced Biofuels:** The manufacture of advanced biofuels, such as cellulosic ethanol and algae-based biofuels, has been achieved via extensive research and development endeavors. These advanced biofuels exhibit superior energy yields and provide a notable reduction in greenhouse gas emissions when compared to conventional biofuels.
(d) **Challenges and Sustainability Considerations:** Bioenergy poses several challenges, including the issue of competing land use, potential ramifications for food security, and environmental concerns arising from land-use alteration and extensive biomass cultivation. In order to effectively tackle these difficulties, it is imperative to prioritize the implementation of sustainable practices and adopt comprehensive land-use planning strategies.
(e) **Waste Reduction:** Bioenergy can be generated through the utilization of organic waste materials, thereby diverting these materials away from landfills and mitigating the release of methane emissions that are linked to the process of decomposition.
(f) **Economic Opportunities:** The bioenergy industry generates employment opportunities across several stages of biomass agriculture, including cultivation, harvesting, processing, and energy generation. The aforementioned initiative serves to invigorate rural economies and sustains local agricultural sectors.
(g) **Environmental Benefits:** Bioenergy is widely recognized as a renewable and carbon-neutral kind of energy due to the fact that any carbon dioxide (CO_2) emissions resulting from the process of combustion or conversion are effectively balanced by the CO_2 absorption that occurs during the growth of plants. When utilized as a substitute for fossil fuels, it has the potential to mitigate greenhouse gas emissions.
(h) **Energy Independence and Security:** The utilization of bioenergy serves to enhance energy independence through the mitigation of dependence on imported fossil fuels. Furthermore, the diversification of the energy mix reduces countries' susceptibility to supply disruptions and price changes in fossil fuel markets.

Bioenergy is a multifaceted and ecologically sustainable form of renewable energy that has the potential to make a substantial contribution towards the mitigation of climate change, waste reduction, and the diversification of energy resources. The sustainability of the subject in question relies on the implementation of responsible biomass management and the incorporation of optimal methodologies in both production and conversion processes.

1.7 Grid Infrastructure and Energy Storage for Renewable Integration

The integration of renewable energy sources such as wind and solar energy into the power grid brings about unique challenges due to their intermittent nature. The sun does not always shine, and the wind does not always blow, creating reliability and stability issues for the power grid. Hence, two key areas that help facilitate the integration of renewable energy sources are improvements in grid infrastructure and energy storage solutions.

Grid Infrastructure

Traditionally, power grids were designed for centralized power generation, primarily from fossil fuel power plants. Renewable energy generation is often decentralized and variable, requiring a more flexible and responsive grid infrastructure.

- **Smart Grids:** These use digital technology to monitor and manage the transmission of power from all generation sources to meet the changing electrical demands of end users. Smart grids make it easier to integrate renewable energy sources by providing real-time monitoring and demand response capabilities, which can help balance electricity supply and demand.
- **Grid Interconnection:** Interconnecting power systems across areas can help with renewable energy integration. When one location is not sunny or windy, power can be imported from another site where the sun is shining or the wind is blowing.
- **Microgrids:** These are smaller grids that can function independently of the primary grid. They can integrate renewable energy sources locally while also providing resilience in the event of grid failures.

Energy Storage

Energy storage is a vital answer to renewable energy's intermittent nature. Energy storage can assist in assuring a consistent electricity supply by storing the excess power generated during high renewable energy generation periods and releasing it during low generation periods.

- **Batteries:** Battery technology advancements, notably lithium-ion batteries, have resulted in major gains in energy storage. Batteries provide fast, flexible energy storage and can be used at multiple scales, from residential to grid-scale.
- **Pumped Hydro Storage:** This is the most widely used form of large-scale energy storage. It involves pumping water to a higher elevation when excess power is available and releasing it to generate power when needed.
- **Thermal Energy Storage:** This stores excess energy as heat or cold, which can be used directly for heating or cooling applications or converted back into electricity. For instance, concentrated solar power (CSP) plants often use thermal energy storage to enable power generation after sunset.

- **Power-to-X:** These technologies convert excess electricity into other forms of energy or commodities, such as hydrogen (power-to-gas) or ammonia. These can be stored longer and used in different sectors, such as transport or industry.

Both grid infrastructure and energy storage advancements are crucial for integrating renewable energy. Policies that support the development and deployment of these technologies and regulatory reforms that create a level playing field for storage and flexible resources in power markets can help accelerate this process.

1.8 Electrification of Transport and Its Implications on Energy Systems

The process of electrifying transportation, which entails the substitution of cars propelled by internal combustion engines (ICEs) with electric vehicles (EVs), carries substantial ramifications for energy systems. This transition signifies a significant milestone in the process of achieving decarbonization in the transportation industry, mitigating the release of greenhouse gas emissions, and improving energy efficiency. The following are a few significant ramifications:

(i) **Reduced Carbon Emissions:** The decrease in carbon emissions is a significant factor contributing to the promotion of electrification. Electric vehicles (EVs) are characterized by their ability to generate no tailpipe emissions, resulting in reduced levels of greenhouse gas emissions within the transportation industry. The magnitude of emissions reduction is contingent upon the specific electrical source employed for the purpose of charging.

(ii) **Increased Electricity Demand:** The widespread implementation of electric vehicles (EVs) is anticipated to result in a rise in the demand for power. The increased demand for energy during peak charging hours might exert pressure on energy systems, hence requiring enhancements to grid infrastructure.

(iii) **Grid Integration Challenges:** The incorporation of electric vehicle (EV) charging infrastructure into the electrical grid poses significant problems pertaining to load management and grid stability. To adequately tackle these difficulties, it is imperative to establish intelligent charging infrastructure and implement demand response techniques.

(iv) **Renewable Energy Synergy:** The integration of electric vehicles (EVs) into the transportation sector can exhibit a mutually beneficial relationship with the expansion of renewable energy sources. Electric vehicles (EVs) have the potential to function as a valuable grid resource through their participation in vehicle-to-grid (V2G) systems. This capability enables EVs to store surplus renewable energy and subsequently release it back into the grid during periods of high demand.

(v) **Energy Storage Potential:** Electric vehicle (EV) batteries possess the capacity to function as decentralized energy storage systems. The use of this technology has the potential to improve the stability of the electrical grid and provide assistance to intermittent renewable energy sources by effectively storing surplus energy during periods of low demand and subsequently releasing it during periods of high demand.

(vi) **Energy Efficiency:** Electric vehicles (EVs) often exhibit higher energy efficiency compared to internal combustion engine (ICE) vehicles. This implies that electric vehicles (EVs) can cover greater distances with the same amount of energy, so resulting in a reduction of overall energy consumption within the transportation industry.

(vii) **Impacts on Oil Demand:** The potential for widespread electrification of transport to diminish oil consumption has significant implications for the global oil business and the reduction of reliance on fossil fuels.

(viii) **Infrastructure Investment:** In order to facilitate the expansion of electric vehicles (EVs), it may be necessary for governments and utility companies to allocate resources towards the development and implementation of charging infrastructure. This includes the implementation of strategies such as the expansion of public charging networks, the provision of incentives for the installation of home charging stations, and the facilitation of charging access for individuals residing in apartments or lacking private garages.

(ix) **Impact on Energy Mix:** The proliferation of electric vehicles (EVs) can impact the composition of the energy mix by augmenting the overall demand for electricity. To establish a viable and environmentally friendly energy future, it is imperative to synchronize the widespread adoption of electric vehicles (EVs) with the creation of clean energy.

(x) **Economic Opportunities:** The electrification of transportation can generate economic prospects, such as the expansion of employment opportunities in electric vehicle (EV) manufacture, the establishment of charging infrastructure, and the growth of associated sectors.

(xi) **Air Quality Improvement:** Electric vehicles (EVs) make a positive impact on urban air quality due to their lack of exhaust emissions. This phenomenon has the potential to yield positive health outcomes and mitigate the financial burden of healthcare expenses linked to illnesses caused by air pollution.

(xii) **Policy and Regulation:** The involvement of government policies, incentives, and regulations is crucial in facilitating the adoption of electric vehicles (EVs), influencing the transition towards electric transportation, and establishing sustainable energy systems.

(xiii) **Equity Considerations:** It is imperative to prioritize the establishment of fair and equal access to electric vehicles (EVs) and the accompanying charging infrastructure. This is crucial in order to prevent the exacerbation of existing transportation disparities and to fully capitalize on the advantages of electrification for all societal groups.

The electrification of transport plays a vital role in addressing climate change and facilitating the move towards more sustainable energy sources. In order to fully harness its capabilities, the achievement of its complete potential necessitates meticulous strategic planning, substantial financial investment, and effective collaboration among diverse entities, encompassing governmental bodies, utility companies, automobile manufacturers, and end-users. In conclusion, an effectively executed shift towards electric transportation has the potential to generate positive outcomes in terms of the environment, economy, and society.

1.9 Role of Nuclear Power in the Sustainable Energy Transition

The role of nuclear power in the transition to a more sustainable energy system is a hotly debated topic among policymakers, scientists, and environmentalists. Nuclear energy is a low-carbon energy source that emits fewer greenhouse gases than fossil fuel power plants. However, it creates substantial obstacles, such as safety concerns, waste management, and expensive prices. Here are some key aspects to consider:

1.10 Advantages of Nuclear Power in the Energy Transition

1. **Low-Carbon Energy Source:** Nuclear energy can play a significant role in reducing carbon emissions. It can provide a considerable amount of reliable power while emitting no direct CO_2.
2. **Reliable Base Load Power:** Unlike wind and solar power, nuclear power is not affected by weather and can provide a consistent and reliable source of electricity. This is advantageous in grid balancing, particularly in a power system with a significant amount of fluctuating renewable energy.
3. **High Energy Density**: Nuclear power has a significantly higher energy density than other sources, which means it can generate a considerable amount of energy from a small amount of fuel.

1.11 Challenges of Nuclear Power in the Energy Transition

1. **Safety and Environmental Concerns:** The potential for nuclear mishaps, as demonstrated in Chernobyl and Fukushima, raises major safety concerns. Furthermore, while nuclear power produces low-carbon electricity, it also generates radioactive waste, which must be overseen and stored for extended periods of time.

2. **Economic Considerations:** Constructing new nuclear power facilities is both expensive and time-consuming. In many markets, it is currently more expensive than other low-carbon alternatives, such as wind and solar.
3. **Limited Fuel Supply:** While nuclear fuel has a high energy density, the supply of uranium, on which most nuclear reactors rely, is limited and non-renewable. Concerns have been raised concerning the possibility of using nuclear technology and materials for negative ends, such as the production of nuclear weapons.

Some regard nuclear power as a 'bridge' technology in the context of the energy transition, delivering low-carbon power while renewable energy technologies grow and storage options improve. Others advocate for the use of sophisticated nuclear technology, such as small modular reactors (SMRs) or Generation IV designs, which seek to be safer and more cost-effective than traditional nuclear power facilities.

It is also worth mentioning that the role of nuclear power may differ from country to country, depending on factors such as resource availability, energy security concerns, existing infrastructure, and public acceptance. As a result, while nuclear power can play a role in the transition to a more sustainable energy system, its exact position will depend on how these benefits and challenges are balanced, and it will be part of a broad mix of energy sources.

1.12 Case Studies of Countries Leading the Energy Transition

Several countries around the world are pioneering the move to sustainable energy. Here are a few case studies that demonstrate several approaches to this complicated task:

- **Germany: Energy Transition:** Germany's "Energiewende" energy transformation has been motivated by a desire to mitigate climate change, reduce energy imports, and boost technical innovation and the green economy. Despite these obstacles, Germany has successfully grown the proportion of renewable energy in its electrical mix from 6.3% in 2000 to approximately 50% by 2020. Policies such as the Renewable Energy Sources Act (EEG), which established a feed-in tariff system to stimulate renewable energy generation, have facilitated this transformation.
- **China: Superpower in Renewable Energy:** China has surpassed the United States as the world leader in the production and installation of solar panels and wind turbines. The country's commitment to renewable energy stems from a goal to cut carbon emissions, combat air pollution, and promote technological innovation. China invests in electric car infrastructure, in addition to enormous solar and wind projects, and boasts the world's largest fleet of electric vehicles.

- **Denmark is a pioneer in wind energy:** Denmark has effectively used its geographic advantages to position itself as a world leader in wind energy. It aims to be completely independent of fossil fuels by 2050. Wind energy accounted for over 50% of Denmark's electricity output by 2020, thanks to a robust renewable energy policy commitment, a high degree of community engagement, and the presence of wind energy firms such as Vestas.
- **Costa Rica: Towards Carbon Neutrality:** Costa Rica wants to become carbon-neutral by using a combination of hydro, wind, solar, and geothermal energy. It has already seen success with this multifaceted approach; in recent years, Costa Rica has satisfied more than 98% of its electricity needs through renewable sources. Forest protection and reforestation are essential components of their energy strategy for mitigating carbon emissions.
- **Sweden is the world leader in Bioenergy:** Sweden is a world pioneer in transitioning energy systems to renewable energy, intending to produce 100% renewable electricity by 2040. Because Sweden has extensive forest resources, bioenergy plays a significant role in the country's energy balance. This transformation has been aided by significant government assistance and legislative tools such as carbon taxes.
- **Norway: Electric Vehicle Adoption:** Norway is the undisputed leader in terms of electric vehicle (EV) adoption. Nearly 60% of all new cars sold in the country in 2020 were fully electric. Generous incentives, including tax breaks, toll exemptions, and free parking for EV owners, have driven this transition. The government aims to have all new cars be zero-emission by 2025. This widespread adoption of EVs significantly reduces the country's carbon emissions and transitions away from fossil fuels.
- **Portugal: Diverse renewable energy mix:** Portugal stands out in its diverse renewable energy mix, including hydro, wind, and solar power. The country occasionally runs for days powered entirely by renewable energy. In 2016, Portugal ran for four consecutive days on renewable energy alone. Portugal's success is attributed to its substantial investment in modernizing the grid, renewable technologies, and favorable natural conditions.
- **Scotland: Wind Energy Powerhouse:** Scotland has invested heavily in wind power and often produces an electricity surplus from wind power that can be exported to the rest of the UK. In 2019, wind generation provided enough electricity to power every home in Scotland and part of Northern England. Its success in wind power is due to strong governmental support, investment in infrastructure, and Scotland's naturally windy conditions.
- **Morocco: Solar Power Leader:** Morocco is home to one of the world's largest solar power plants, the Noor Complex, demonstrating the country's commitment to renewable energy. As a sun-rich country, solar power in Morocco has enormous potential. The government aims to have 52% of the country's energy come from renewable sources by 2030. Morocco's efforts are an excellent example of a non-OECD country making significant strides in adopting renewable energy.

- **Iceland: Geothermal energy pioneer:** Iceland is a world leader in geothermal energy, using its unique geographical position as a volcanic hotspot. Almost all the country's heating and approximately 25% of its electricity comes from geothermal sources. This reduces Iceland's greenhouse gas emissions and provides a stable and independent energy supply.

These case studies demonstrate that while the path to energy transition may look different for each country, progress is achievable depending on their unique circumstances and resources. With strategic planning, innovative policies, and technological advancements, these countries are showing the way forward towards a sustainable energy future.

1.13 Circular Economy

A circular economy is an alternative to the traditional linear economy (make, use, dispose of) in which we keep resources in use for as long as possible, extract the maximum value from them while in use, and then recover and regenerate products and materials at the end of each service life. This systemic approach aims to redefine growth, focusing on positive society-wide benefits.

Circular economy models can be broken down into three principles:

- **Design out waste and pollution**: This involves rethinking product design to eliminate waste, using materials that can be reused or recycled, and using renewable or nontoxic inputs.
- **Keep products and materials in use**: This includes creating new business models to increase product use (such as sharing platforms), repairing items instead of replacing them, and promoting reuse, recycling, and remanufacturing.
- **Regenerate natural systems**: This means creating ways to regenerate natural capital and enhance natural systems by returning valuable nutrients to the soil or using renewable energy instead of fossil fuels.

Principles and Models of the Circular Economy

The circular economy is an economic system that aims to eliminate pollution and sustainably use resources. It is a more sustainable alternative to the traditional linear economy (make, use, dispose of) in which we keep resources in use as long as possible, extract maximal value while they are in use, and then recover and regenerate products and materials at the end of their service life. Three core principles underpin the circular economy model:

- **Design Out Waste and Pollution:** Rather than managing waste and pollution after they are created, the circular economy suggests that we should design systems to prevent them from occurring in the first place. This includes designing reusable, repairable, and recyclable products and shifting towards renewable or biodegradable materials.

- **Keep Products and Materials in Use:** In the circular economy, the focus is on preserving the value of products and materials for as long as possible. This can be achieved through business models such as sharing, leasing, and remanufacturing, which extend product lifetimes and decouple economic growth from new resource extraction.
- **Regenerate Natural Systems:** The circular economy recognizes that we cannot run a thriving economy without a healthy environment. Therefore, it seeks to minimize environmental harm and restore and regenerate natural systems. For example, this could involve using agricultural waste to improve soil health or adopting regenerative agriculture practices that enhance biodiversity.

Several models or strategies have been proposed to operationalize these principles:

(a) **Circular Supply Chain:** Adopt renewable, reusable, nontoxic resources wherever possible.
(b) **Recovery and Recycling:** Waste from one product is used in another. This is also called industrial symbiosis.
(c) **Product Life Extension:** Extend the life of products through repair, maintenance, upgrades, and reselling.
(d) **Sharing Platform:** Increase product utilization by making it possible to share goods or by rethinking ownership (such as ridesharing or tool libraries).
(e) **Product as a Service:** Shift from selling products to providing services, which allows for better product utilization and easier return and reuse of products (such as leasing instead of buying).

It is important to note that transitioning to a circular economy requires changes in production and consumption behaviors and policy, finance, and business models. The goal is to design a restorative and regenerative economy, providing economic, social, and environmental benefits.

1.14 Role of Design in Achieving a Circular Economy: From Product Design to Urban Planning

Design plays a pivotal role in achieving a circular economy. From designing individual products to planning entire cities, incorporating circular economy principles at the design stage can significantly reduce waste and resource consumption while creating more sustainable, resilient systems. Here is a look at how design is integral to the circular economy in both product design and urban planning:

Product Design

- **Material Selection:** Designing for a circular economy begins with the selection of materials. Designers can select recycled, recyclable, or biobased materials to reduce the need for virgin materials and enable easier recovery and reuse at the end of the product's life.

- **Design for Longevity:** Products can be designed to last longer, reducing the need for replacement and thus lowering the demand for new materials and energy. This could involve choosing durable materials and developing easily repaired or upgraded products.
- **Design for Disassembly:** If products are designed to be easily disassembled, their components can be more effectively reused or recycled at the end of the product's life. This principle is particularly important for electronics and appliances, which often contain a mix of valuable and potentially harmful materials.
- **Design for Service:** Instead of designing to sell products, companies can design to sell services. For example, a company might sell lighting-as-a-service rather than light bulbs, retaining ownership of the bulbs and thus the responsibility for their end-of-life management.

Urban Planning

- **Design for Density:** Dense urban environments can support circular practices such as sharing and public transportation, reducing the need for resource-intensive private goods such as cars.
- **Design for Multifunctionality:** Spaces and buildings can be designed to serve multiple functions, increasing their utilization rate and reducing the need for additional resources. For example, a park might also include stormwater management features, or a building might be designed to generate its power through integrated solar panels.
- **Design for Connectivity:** A well-connected city can reduce the need for private vehicles and support the efficient movement of goods and resources, such as collecting recyclables or delivering shared goods.
- **Design for Local Loops:** Urban environments can be built to close local resource loops. For example, organic waste from households and businesses might be processed locally into compost for urban farms, which would then supply local food markets.
- **Design with Nature:** Cities may regenerate natural systems and provide various ecosystem services by incorporating natural elements and processes into urban settings. Green roofs and walls, for example, can assist in managing rainwater, cutting energy use, and boosting urban biodiversity.

Creating products, locations, and systems for a circular economy necessitates a significant shift in how we think about and construct them. It necessitates a comprehensive strategy that considers not only the initial manufacture and usage of a product but also its full life cycle. It also entails assessing both the specific product or building and the larger system within which it operates. Regardless of the limitations, such design methods provide significant opportunities for creating a more sustainable and resilient economy and society.

1.15 Business Models and Policies for a Circular Economy

Business models are critical in the shift to a circular economy. Instead of the typical linear model of 'take-make-dispose,' firms might embrace models that promote product and material consumption, reuse, and recycling. Here are a few examples of business models typically connected with the circular economy:

- **Product as a Service**: In this approach, businesses retain ownership of their products while providing the product's service. As maintenance and disposal costs remain with the company, this motivates businesses to develop goods that last longer, are easier to fix and upgrade, and can be more easily recycled at the end of their lifecycle. For example, Rolls-Royce's "Power-by-the-Hour" program allows clients to purchase the hours of flight time that a jet engine can give rather than the engine itself.
- **Sharing Economy**: The sharing economy model maximizes product utility by making it easy to distribute unwanted or underutilized commodities among individuals. This may be observed in platforms such as Airbnb (for lodging) and Uber (for transportation), which allow assets to be shared rather than held separately. This lessens the desire for new products while also fostering community and reducing overconsumption.
- **Remanufacturing and Refurbishment**: In this model, used products are restored to as good as, if not better than, new conditions. Cleaning, mending, and rebuilding are all part of the process. This concept increases the longevity of items while decreasing the demand for new materials. For example, a caterpillar's remanufacturing program remanufactures old machinery parts into new ones.
- **Renting and leasing**: Leasing and renting models, such as product-as-a-service, incentivize corporations to build durable and repairable items because they retain responsibility for the devices over their lifetime. Customers pay for the product's use rather than ownership. An example of this can be found in the fashion business, where companies such as Rent the Runway provide high-quality clothing rental services.
- **Reverse Logistics**: This concept entails transferring items from their usual destination to capture value or dispose of them properly. It is part of processing returned merchandise, reusing containers, recycling packing materials, and disposing of old goods. Dell, for example, has comprehensive reverse logistics programs to salvage valuable components from returned or recycled equipment.
- **Industrial Symbiosis:** This involves sharing resources among different industrial processes, where the waste or byproducts of one process become the inputs for another. This model has been realized on a large scale in places such as Kalundborg, Denmark, where various industrial processes coexist and share resources in a closed-loop system.

The policy also plays a critical role in encouraging and enabling the circular economy. Governments can set standards, provide incentives, and create a policy environment that supports circular business practices. Here are a few examples of policies that can help a circular economy:

1.16 Case Studies of Successful Implementation of Circular Economy Principles

A circular economy is a regenerative approach that seeks to decouple economic growth from resource consumption. Here are several case studies that highlight the successful implementation of circular economy principles:

1. **Philips Lighting-as-a-Service:** Dutch electronics company Philips provides a 'lighting-as-a-service' model, which sees them retaining ownership of their lighting products while clients pay for the light they use. When the lighting installations reach the end of their life, Philips takes responsibility for recycling or reusing them, closing the loop.
2. **Interface's Net-Works Program:** Interface, a global commercial flooring company, created Networks that enable fishing communities in developing countries to sell waste fishing nets back into a worldwide supply chain; the nets are then recycled into yarn for carpet tiles. This project tackles the environmental issue of marine plastic pollution while providing additional income to impoverished communities.
3. **Renault's Remanufacturing Program:** Renault, the French car manufacturer, has a remanufacturing program that restores used parts to like-new conditions for resale at a lower price than new parts. The program is cost-effective and reduces environmental impact. Remanufactured parts consume almost 80% less energy and 88% less water and emit 70% less CO_2 than new parts.
4. **Mud Jeans' Lease a Jeans Program:** Mud Jeans, a Dutch company, has implemented a leasing model where customers can lease a pair of jeans for a monthly fee. After a year, customers can choose to keep the jeans, switch them for a new pair, or return them to be recycled into new jeans, thus reducing waste and promoting recycling.
5. **Winnow Solutions:** Winnow Solutions is a UK-based tech company that has developed a smart meter to help commercial kitchens monitor and reduce food waste. The Winnow system helps chefs easily measure, monitor, and cut waste, saving money and reducing their environmental footprint.
6. **Unilever:** Unilever's Sustainable Living Plan showcases its commitment to reducing environmental impact and improving health and well-being. The company has made significant strides in reducing its carbon footprint and water usage. It also focuses on a circular approach to packaging, aiming to make all of its plastic packaging reusable, recyclable, or compostable by 2025. Unilever

maintains high ESG standards, regularly reporting its progress and engaging in community development projects.

7. **Tesla, Inc.:** Tesla is not just an electric vehicle manufacturer; it's also involved in clean energy generation and storage. The company embodies sustainable energy transition, aiming to accelerate the world's transition to sustainable energy with electric cars, solar products, and integrated renewable energy solutions. Tesla's business model revolves around high ESG standards, with a focus on innovation, safety, and sustainable manufacturing practices.
8. **Patagonia:** Known for its environmental activism, Patagonia implements circular economy principles by creating durable products and offering repair services, thus extending the lifespan of its products. The company invests in sustainable energy sources for its operations and has a strong focus on social responsibility, reflected in its supply chain practices and advocacy for environmental causes.
9. **Ørsted:** Formerly an oil and gas company, Ørsted has transformed itself into a pure-play renewable energy company, focusing on wind power, bioenergy, and green energy solutions. This dramatic shift highlights the company's commitment to sustainable energy transition. Ørsted operates with high ESG standards, focusing on reducing emissions and promoting biodiversity.
10. **Inditex (Zara)** Inditex, the parent company of Zara, has made significant efforts to integrate sustainable practices. It focuses on sustainable fabrics and recycling programs, aligning with circular economy principles. The company also invests in sustainable energy sources for its operations and adheres to robust ESG policies, including labor rights and ethical supply chain management.

These companies demonstrate that integrating sustainable energy transition, circular economy principles, and strong ESG practices is not only beneficial for the environment and society but also viable and profitable for businesses across various sectors. They serve as models for other companies aiming to embark on a similar path towards sustainability and corporate responsibility.

1.17 Policies and Regulations to Promote a Circular Economy

Governments and regulators play a significant role in expediting the transition to a circular economy by establishing clear policy frameworks and rules, giving incentives, and leading by example through procurement practices. The following are some laws and legislation that can help to develop the circular economy:

1. **Extended Producer Responsibility (EPR):** EPR rules oblige producers to manage their products' whole life cycle, including take-back, recycling, and final disposal. This encourages the development of items that are easier to recycle and last longer.

Extended Producer Responsibility (EPR) is an environmental protection strategy that makes the product's manufacturer responsible for the product's entire life cycle and especially for the take back, recycling, and final disposal (Khetriwal et al., 2009).

The OECD defines Extended Producer obligation (EPR) as "an environmental policy approach in which a producer's obligation for a product is extended to the post-consumer stage of the product's life cycle." An EPR policy includes the following elements:

(a) Moving responsibilities (physically and/or monetarily; wholly or partially) upstream, away from towns, and.
(b) Incentivizes businesses to consider environmental factors when creating their products.

Extended producer responsibility (EPR) is a waste management approach that adds all of a product's estimated environmental costs throughout its life cycle to its market price. The social costs of automobiles are market externalities.

Extended producer responsibility legislation promotes remanufacturing by focusing on "the end-of-life treatment of consumer products and aims to increase the amount and degree of product recovery while minimizing the environmental impact of waste materials." Passing polluter responsibility to manufacturers is both good for the environment and the most effective strategy to improve product design standards. EPR aims to incorporate signals about the environmental features of products and manufacturing processes throughout the product chain. These regulations hold manufacturers accountable for their goods' end-of-life management. EPR can motivate businesses to create items that are easier to recycle or dispose of properly.

The proliferation of plastic waste poses a significant threat to the Earth's atmosphere, hydrosphere, and lithosphere. The World Life Foundation (WWF) is leading the charge to rethink how we obtain, design, dispose of, and repurpose plastics as this global disaster unfolds. One of our significant issues is broken recycling commitments.

The US recycles 13% of its plastic packaging, compared to 8% globally. These worrying data reveal that our trash management and recycling systems cannot handle the massive volumes of rubbish we produce today, never alone in the future, when the world is anticipated to treble its plastic production by 2040 (EPA – United States Environmental Protection Agency, 2020).

Policymakers and corporate leaders must change our linear, "take–make–waste" connection with plastic to circular. Circular economies will reduce our demand for new virgin plastic by reusing and recycling daily plastic products and preventing necessary materials from contaminating the world. Many large-scale measures will support circular economies. The US's extended producer responsibility (EPR) has great promise but needs urgent political determination. EPR, a technical waste management approach, is at the forefront of the plastic waste discourse.

Waste Regulations: Policies that encourage recycling and composting and discourage landfilling can help keep resources in the economy rather than being lost as waste.

Public Procurement: As a major purchaser of goods and services, governments can favor suppliers that use circular practices.

Research and Development Support: Governments can provide funding or tax incentives for research into circular materials and manufacturing processes.

Education and Awareness: Policies can also focus on increasing public understanding of the circular economy to drive demand for circular products and services.

Concisely, in an understandable manner, the adoption of a circular economy necessitates a collaborative endeavor from both enterprises and policymakers. Business models must try to maintain the value of resources for as long as feasible, while policy should establish the enabling conditions for these business models to prosper.

2. **Green Public Procurement:** Governments are frequently the economy's top customers. They can stimulate the market by implementing green public procurement policies that create demand for circular products and services.
3. **Eco-design Regulations:** These regulations compel items to meet specified environmental performance standards, such as energy efficiency or recyclability, before being sold. They compel producers to create items that adhere to circular ideals.
4. **Waste Management Laws:** Legislation that establishes recycling targets prohibits specific types of garbage from landfills or levies waste disposal taxes, which can help to promote recycling and resource recovery.
5. **Support for Innovation and Research:** Governments can provide grants or tax breaks to assist the research and development of modern technologies and business models that support the circular economy.
6. **Education and Awareness Programmes:** Governments should also support education and awareness programs to encourage firms and consumers to understand and embrace circular practices.

The European Union's Circular Economy Effect Plan and China's Circular Economy Promotion Law are tangible examples of these policies. These comprehensive initiatives include a wide range of measures to promote circular practices, such as trash rules, eco-design mandates, and funding for innovation and research.

While policy and law are critical drivers of the circular economy, they must be supported by company innovation and changes in consumer behavior to produce a genuinely circular system.

1.18 The Circular Economy and Digital Technology Junction

The junction of the circular economy and digital technology is an intriguing and emerging subject. Digital technology may propel the circular economy by increasing resource efficiency, enabling new business models, and developing waste-reduction methods. Here are a few ways digital technology might help the circular economy:

1. **Internet of Things (IoT):** IoT devices can track product usage and condition in real-time, assisting in predictive maintenance and extending product lifespans. They can also provide useful information for developing more efficient goods and systems. IoT-enabled smart meters, for example, can assist utilities and consumers in more efficiently managing energy use.
2. **Blockchain Technology:** The utilization of blockchain technology can enhance the traceability and accountability of supply chains through its secure and transparent tracking capabilities. It has the potential to facilitate the implementation of responsible production and recycling practices, as well as to promote circular business models, including sharing and leasing.
3. **Artificial Intelligence (AI) and Big Data:** AI can mine enormous databases for patterns and efficiencies that humans may miss, allowing for increased resource utilization and waste reduction. In the energy sector, for example, AI can be used to optimize grid operation based on consumption patterns and renewable energy sources.
4. **Digital Platforms:** Digital platforms can facilitate product sharing, reuse, and recycling. ThredUP and The RealReal, for example, offer the resale of worn apparel, while ShareRing and Peerby enable the sharing of rarely used things such as tools or luggage.
5. **Augmented Reality (AR) and Virtual Reality (VR):** AR and VR can help users understand complex systems and visualize the impact of circular practices, which can reduce transportation needs and associated emissions.
6. **3D printing:** 3D printing allows for localized production, which can reduce transportation needs and associated emissions. It also enables more efficient use of materials and the development of goods that are easier to repair or recycle.

While digital technology has immense potential for fostering the circular economy, examining the lifecycle implications of these technologies, such as e-waste, energy consumption, and data privacy concerns, is critical. Balancing these concerns is vital for entirely using the benefits of digital technology in a circular and sustainable manner.

1.19 Recycling and Composting Are Examples of Circular Economies in Waste Management

One of the most crucial areas for adopting the circular economy is waste management. The circular economy philosophy aims to use resources efficiently while minimizing waste. Recycling and composting are two effective solutions in this context:

1. **Recyclables:** Recycling is the process of converting waste resources into new materials and objects. It is a critical component of the circular economy since it allows resources to be reused, reducing the need for virgin materials and reducing environmental impacts. However, recycling alone is insufficient. It must be integrated with eco-design principles to make items easier to recycle and to maintain material quality across several usage cycles. The development of new recycling technologies is also critical for expanding the spectrum of materials that can be recycled, as well as improving the process's efficiency and quality.

   The policy is critical in promoting recycling. Extended producer responsibility legislation, which holds manufacturers accountable for the end-of-life management of their products, can potentially increase recycling. Other measures, such as recycling objectives and waste disposal charges, can encourage the recycling business.
2. **Composting:** Composting represents a significant strategy in the realm of circular waste management. It is an environmentally beneficial process that involves the conversion of organic waste into a soil conditioner that is rich in nutrients. This practice not only enriches the soil with essential elements but also mitigates the accumulation of organic waste in landfills, where it emits methane, a potent greenhouse gas.

Composting can be performed on various scales, from home composting to municipal programs. Composting reduces the environmental impact of waste disposal, contributes to soil health, and reduces the need for synthetic fertilizers in agriculture. Again, policy plays a crucial role in promoting composting. Policies could include requirements for separate collection of organic waste, composting targets, or regulations to encourage the use of compost in agriculture. Both recycling and composting are essential elements of a circular approach to waste management, turning waste into a resource and keeping materials circulating in the economy.

1.20 Challenges and Solutions in Transitioning to a Circular Economy

The transition to a circular economy raises several economic, social, and environmental issues. Understanding these difficulties can aid in identifying the answers required to solve them:

1. **Linear Economic Models**: The current state of the global economy can be characterized as predominantly linear, exhibiting a structure that involves the sequential processes of extraction, production, and disposal. Addressing this deeply entrenched paradigm requires a substantial transformation in production and consumption dynamics, constituting a formidable undertaking. A comprehensive transformation at the systemic level is necessary, encompassing alterations in corporate structures, consumer behavior, and political frameworks. Education and public awareness initiatives have the potential to influence and modify prevailing societal norms and consumer expectations.
2. **Lack of Circular Design**: Many modern items are not created with their end-of-life in mind. They may be difficult to disassemble, repair, or recycle, and they may be made up of mixed components that are difficult to separate. Extended producer responsibility (EPR) policies, for example, can motivate manufacturers to design circular products. Eco-design principles can also assist businesses in developing products that are more compatible with a circular economy.
3. **Inadequate garbage management infrastructure**: Many communities lack the infrastructure required for successful garbage sorting and recycling. Infrastructure investment is necessary to guarantee that garbage is properly sorted and treated. To recycle complicated materials, new technologies may be needed.
4. **Economic Factors**: Because the price of virgin materials does not always reflect their environmental impact, they are often less expensive than recycled materials. Environmental taxes or subsidies can be used to reflect the true cost of materials and increase the competitiveness of circular choices.
5. **Technological problems**: Improving the efficiency and effectiveness of recycling operations, as well as developing novel materials optimized for circularity, remain technological problems. Investing in R&D can help propel technological developments in these areas.
6. **Regulatory Barriers**: Existing legislation may not enable circular practices, such as the reuse of certain types of garbage, in some circumstances. Regulatory reform may be required to guarantee that rules and regulations promote the circular economy rather than inhibit it. Policies can also be used to incentivize the adoption of circular practices.

Transitioning to a circular economy is a complicated process that necessitates systemic change; however, the potential benefits of resource efficiency, environmental preservation, and economic resilience make it desirable.

1.21 ESG Practices

ESG stands for Environmental, Social, and Governance. These three key variables assess an investment's long-term viability and societal impact on a firm or corporation. They assist investors in identifying dangers and opportunities that go beyond typical financial research.

1. **Environmental**: How a corporation behaves as a custodian of the natural environment. It encompasses a company's energy use, waste, pollution, natural resource conservation, and animal treatment, as well as compliance with environmental standards.
2. **Social:** This refers to how a business maintains its relationships with its employees, suppliers, customers, and the communities in which it operates. Employee relations, diversity, working conditions, health and safety, and how the organization interacts with the local community are all possible topics.
3. **Governance**: A company's leadership, executive remuneration, audits, internal controls, shareholder rights, and openness are all examples of governance. Good corporate governance ensures that businesses are administered in a fair and value-adding manner for all stakeholders.

The rise of ESG practices underscores investors' growing realization that nonfinancial factors can significantly impact a company's risk and return profiles. ESG elements are frequently included in sustainable investing strategies due to their potential to influence financial performance. Companies that adhere to strong ESG principles are expected to be more successful and profitable eventually.

1.21.1 ESG Investing Principles, Strategies, and Performance

Environmental, social, and governance (ESG) investing incorporates environmental, social, and governance aspects into investment decisions in addition to financial reasons. The following is an overview of ESG investing principles, techniques, and performance:

1. ESG Investing Principles
ESG investing principles are guidelines that investors follow when considering the environmental, social, and governance (ESG) practices of the entities in which they invest. The integration of ESG principles into investing acknowledges that these factors can affect both the risk and the potential return of an investment. Here are some core principles and considerations of ESG investing:

1. **Long-term Value Creation:** ESG investing often aligns with a long-term investment horizon, understanding that companies with robust ESG practices may be better positioned for sustainable success over time.

2. **Materiality:** Not all ESG factors will be equally relevant for every industry or company. ESG investing emphasizes understanding which factors are materially significant to financial performance.
3. **Risk Management:** Proactively identifying and managing ESG risks (like regulatory risks, reputational risks, etc.) can prevent or mitigate adverse impacts on investment returns.
4. **Stewardship and Active Ownership:** This involves investors taking an active role in engagements, proxy voting, and other actions to influence positive ESG behaviors in the companies they invest in.
5. **Transparency and Disclosure:** ESG investing principles prioritize clear reporting and transparency, allowing investors to make informed decisions.
6. **Integration with Investment Decisions:** ESG data and analysis are incorporated into traditional financial analysis, portfolio construction, and risk management processes.
7. **Collaboration:** Investors may collaborate to promote and achieve common ESG goals and initiatives.
8. **Continuous Improvement:** Given the evolving nature of ESG issues and data, it is essential to periodically review and refine ESG strategies, tools, and metrics.
9. **Focus on Real-world Impact:** More recent trends in ESG investing emphasize not just avoiding harm or managing risks but also making positive impacts on society and the environment.
10. **Avoidance or Exclusion:** Some ESG investment approaches exclude certain sectors, companies, or practices based on specific ESG criteria. For example, an investor might choose to exclude companies involved in the production of tobacco or fossil fuels.
11. **Best-in-Class Selection:** This approach involves investing in sectors, companies, or projects that are leaders in terms of ESG performance relative to their peers.
12. **Thematic Investing:** Investors might focus on specific themes or assets linked to sustainability, such as clean energy, sustainable agriculture, or water technologies.

Several global initiatives and frameworks support and promote ESG investing principles, including:

- **The UN-supported Principles for Responsible Investment (PRI):** A consortium of investors collaborating to integrate environmental, social, and governance (ESG) considerations into their decision-making frameworks.
- **The Task Force on Climate-related Financial Disclosures (TCFD):** Provides recommendations for consistent climate-related financial risk disclosures.
- **Sustainability Accounting Standards Board (SASB):** Offers industry-specific standards for corporate sustainability disclosure.

Adhering to ESG investing principles can lead to both financial and non-financial benefits. It allows investors to align their portfolios with their values, potentially reduce investment risk, and strive for competitive returns while also seeking to create positive societal impacts.

1.21.2 ESG Investment Strategies

ESG Integration: ESG integration pertains to the incorporation of environmental, social, and governance (ESG) considerations into the framework of investment decision-making. The prominence of ESG integration has grown as investors and stakeholders see the significance of these aspects in evaluating risks and opportunities.

Negative/Exclusionary Screening: The exclusion of specific industries, firms, or practices from investment consideration based on ESG criteria.

Good/best-in-class screening: Investing in sectors, companies, or projects that have demonstrated good ESG performance compared to industry peers.

Investing in themes or assets that are directly tied to ESG issues is known as thematic investing.

Impact Investing: Specific investments, often done in private markets, to have a positive ESG impact in addition to financial return.

1.21.3 ESG Investment Performance

According to numerous studies, organizations with strong ESG profiles may outperform their competitors in the long run, both in terms of financial performance and risk reduction. ESG investment is increasingly being viewed as a technique for identifying companies that are industry leaders, better managed, and more resilient to potential reputational harm. ESG investment can also help identify companies that are exposed to risks (such as regulatory or environmental dangers) that are not often considered in financial analysis.

Furthermore, ESG investing enables investors to align their portfolios with their values and societal goals, resulting in financial rewards and a positive social impact.

However, it is crucial to recognize the challenges associated with ESG investment, such as the lack of standardized ESG reporting, the difficulties in comparing ESG performance across firms and sectors, and the necessity for investors to analyze corporate ESG claims carefully.

1.21.4 Role of ESG Ratings and Benchmarks

ESG ratings and benchmarks hold significant importance within the financial and investment domain as they offer a standardized mechanism for evaluating and contrasting the ESG performance of companies and investment products. The following is a comprehensive summary of their respective duties and the significance they hold:

1.22 Investor Decision-Making

ESG ratings and benchmarks are seen as valuable instruments that enable investors to make educated decisions on the allocation of their resources. The assessment stated above provides an evaluation of the environmental, social, and governance (ESG) performance of a company or investment product, encompassing both quantitative and qualitative measures. This allows investors to accurately evaluate the potential risks and opportunities connected with the transaction.

1.23 Risk Assessment

ESG ratings and benchmarks function as instrumental instruments for evaluating risks associated with environmental, social, and governance factors. Investors possess the ability to discern organizations exhibiting subpar environmental, social, and governance (ESG) performance, rendering them potentially more vulnerable to regulatory, reputational, and operational risks that have the potential to affect their financial gains (Sarpong et al., 2023).

1.24 Performance Benchmarking

Benchmarks facilitate the evaluation of environmental, social, and governance (ESG) performance among various enterprises or investment products operating within certain industries or asset classes. This aids investors in discerning the frontrunners and stragglers in relation to sustainability and responsible governance.

1. **Portfolio Construction:** ESG ratings and benchmarks are utilized by asset managers and institutional investors to construct portfolios that are focused on environmental, social, and governance (ESG) factors. Investors have the option to connect their investments with specific environmental, social, and governance (ESG) goals. This entails actively seeking to promote environmental sustainability, social impact, and responsible governance through their investment decisions.

2. **Engagement and Advocacy:** ESG ratings and benchmarks provide a basis for shareholder engagement and advocacy. Investors can use ESG data to engage with companies and push for improvements in their sustainability practices, which can ultimately enhance long-term financial performance.

1.25 Reporting and Transparency

Organizations frequently employ environmental, social, and governance (ESG) ratings and benchmarks as a structured framework to disclose their ESG performance to various stakeholders, such as shareholders, regulators, and the general public. This promotes greater levels of transparency and accountability.

1.26 Regulatory Compliance

In some regions, regulatory bodies and stock exchanges require companies to disclose ESG-related information. ESG ratings and benchmarks can help companies assess their compliance and identify areas for improvement.

1.27 Market Recognition

High ESG ratings and inclusion in prominent ESG benchmarks can enhance a company's reputation and attract ESG-focused investors. It can also signal to the market that the company is committed to responsible and sustainable business practices.

1.28 Encouraging ESG Integration

ESG ratings and benchmarks encourage companies to integrate ESG considerations into their business strategies. Companies strive to improve their ESG scores to attract investment and remain competitive in ESG-focused markets.

1.29 Standardization and Consistency

ESG ratings and benchmarks provide a standardized framework for assessing ESG performance, which promotes consistency and comparability across different companies and industries.

1.30 Data-Driven Decision-Making

Investors and companies increasingly rely on ESG data and analytics to make data-driven decisions. ESG ratings and benchmarks provide a structured approach to ESG data analysis.

1.31 Sustainable Development Goals (SDGs) Alignment

ESG ratings and benchmarks can help investors and companies align their activities with the United Nations Sustainable Development Goals (SDGs) by identifying ESG areas that contribute to broader societal and environmental objectives.

1.31.1 Impact of ESG Factors on Financial Performance

The financial performance of companies and investments can be significantly influenced by Environmental, Social, and Governance (ESG) aspects. The correlation between environmental, social, and governance (ESG) variables and financial performance is complex and can be observed through diverse mechanisms.

(i) **Risk Mitigation**: Effective ESG practices can assist businesses in avoiding disputes and reputational damage that could impact their financial performance. Good governance, for example, can prevent business scandals, whereas strong environmental practices can prevent incidents such as oil spills or regulatory fines.

(ii) **Operational Efficiency**: Environmental, Social, and Governance (ESG) practices have the potential to facilitate enhancements in operational efficiency. Organizations that prioritize sustainability exhibit a propensity for embracing innovation and swiftly adopting novel technology, hence resulting in enhanced productivity and competitiveness.

(iii) **Access to Capital:** Investors and lenders tend to find companies with robust environmental, social, and governance (ESG) performance more appealing. It is possible that individuals may experience greater ease in obtaining loans under more advantageous conditions. Furthermore, an emerging phenomenon in the realm of investment is the increasing prevalence of ESG-focused strategies, whereby organizations that exhibit commendable environmental, social, and governance practices are more apt to garner investment funds.

(iv) **Stakeholder Engagement:** The active involvement of many stakeholders, such as shareholders, employees, consumers, and communities, in addressing environmental, social, and governance (ESG) concerns can foster improved relationships and facilitate a more comprehensive comprehension of market demands. This can potentially augment a company's capacity to effectively respond to dynamic and evolving circumstances.

(v) **Reputation and Brand Value:** Adverse ESG occurrences, such as instances of environmental mishaps or social disputes, have the potential to detrimentally impact a company's standing and diminish the value of its brand. These occurrences can lead to a decrease in client confidence, a reduction in sales, and a depreciation in market value.

(vi) **Regulatory Compliance:** Failure to adhere to environmental or governance rules can result in the imposition of legal and financial sanctions. Organizations that take proactive measures to manage environmental, social, and governance (ESG) concerns are more inclined to maintain compliance and mitigate the occurrence of expensive legal complications.

(vii) **Long-Term Growth:** ESG characteristics exhibit a strong correlation with the enduring viability and adaptability of a given entity over an extended period of time. Organizations that prioritize environmental, social, and governance (ESG) factors possess enhanced capabilities to effectively respond to evolving market dynamics, regulatory landscapes, and societal demands, hence fostering sustainable long-term expansion and resilience.

(viii) **Customer and Brand Loyalty:** There is a growing trend among consumers to make purchasing decisions that are influenced by a company's environmental, social, and governance (ESG) practices. Companies that demonstrate robust environmental, social, and governance (ESG) practices have the potential to cultivate heightened consumer loyalty, resulting in amplified sales and income.

(ix) **Talent Attraction and Retention:** The acquisition and retention of highly skilled individuals are crucial for achieving optimal performance and prosperity within an organization. Organizations that exhibit favorable social and governance practices, encompassing aspects such as diversity and inclusion, tend to possess heightened attractiveness to prospective employees. A workforce that is both motivated and diverse has the potential to enhance productivity and foster creativity.

However, it is essential to highlight that the impact of ESG issues on financial performance varies depending on the industry, geographical region, and other factors. Not all ESG investments will outperform, and they, like all investments, carry risk. As a result, while making investment decisions, it is critical to examine ESG aspects in addition to traditional financial criteria thoroughly. Despite the hurdles, the trend towards ESG investing is growing, driven by increased investor knowledge of the financial importance of ESG concerns, regulatory developments, and societal upheavals. As a result, understanding the influence of ESG variables on financial performance is becoming increasingly crucial for both organizations and investors.

1.31.2 Strategies for Integrating ESG into Business Models

Integrating Environmental, Social, and Governance (ESG) principles into business models is a strategic approach that aligns a company's operations with broader societal and environmental goals. Here are key strategies for effectively embedding ESG into business practices:

(a) **Leadership Commitment and Culture Shift:**

- **Top-Down Approach:** Leadership must fully commit to ESG principles, integrating them into the company's vision and mission. This commitment should be visibly championed by top executives and the board.
- **Organizational Culture:** Cultivate a culture where ESG values are ingrained in every aspect of the business, encouraging employees at all levels to embrace and advocate for these principles.

(b) **Stakeholder Engagement and Communication:**

- **Identify Stakeholders:** Understand who your stakeholders are (employees, customers, suppliers, communities, investors) and their concerns regarding environmental and social issues.
- **Engage and Communicate:** Regularly engage with stakeholders to gather feedback, communicate ESG goals and progress, and build trust.

(c) **ESG Assessment and Materiality Analysis:**

- **Conduct ESG Assessment:** Evaluate current practices to identify strengths, weaknesses, and areas for improvement in environmental, social, and governance aspects.
- **Materiality Analysis:** Determine which ESG issues are most material to your business and stakeholders. This helps prioritize efforts and resources.

(d) **Setting Clear ESG Goals and Objectives:**

- **Define Specific Objectives:** Set clear, measurable, and achievable ESG goals that align with the company's overall strategy and stakeholder expectations.
- **Create a Roadmap:** Develop a detailed plan for achieving these objectives, including timelines, responsibilities, and resources needed.

(e) **Integration into Core Business Operations:**

- **Operational Changes:** Integrate ESG considerations into daily operations, decision-making processes, and throughout the supply chain.
- **Product and Service Innovation:** Develop products or services that reflect ESG values, meeting consumer demand for sustainable options.

(f) **Employee Engagement and Training:**

- **Training and Awareness:** Provide regular training and resources to educate employees about ESG principles and their role in achieving these objectives.

- **Incentivize Participation:** Encourage employee involvement through incentives, recognitions, or incorporating ESG-related goals into performance evaluations.

(g) **Transparent Reporting and Accountability:**

- **ESG Reporting:** Regularly report progress on ESG initiatives, using standardized frameworks like GRI (Global Reporting Initiative) or SASB (Sustainability Accounting Standards Board) for transparency.
- **Accountability:** Implement mechanisms to ensure accountability at all levels for ESG performance.

(h) **Collaboration and Partnerships:**

- **Collaborate with Others:** Work with other businesses, NGOs, and government bodies to share best practices and engage in joint initiatives.
- **Leverage External Expertise:** Consider partnering with external consultants or experts in the ESG field for guidance and improvement.

(i) **Monitoring and Continuous Improvement:**

- **Regular Monitoring:** Continuously monitor ESG performance against set goals and industry benchmarks.
- **Adapt and Improve:** Be ready to adapt strategies in response to new challenges, regulations, and stakeholder expectations.

(j) **Risk Management and Compliance:**

- **Identify Risks:** Understand and mitigate risks related to ESG issues, such as regulatory changes, climate risks, or social unrest.
- **Ensure Compliance:** Stay compliant with local and international laws and standards related to ESG aspects.

By adopting these strategies, businesses can effectively integrate ESG into their business models, leading to improved sustainability, better stakeholder relationships, enhanced reputation, and long-term profitability.

1.31.3 ESG Reporting and Transparency

ESG (Environmental, Social, and Governance) reporting and transparency are critical components in demonstrating a company's commitment to sustainable practices and ethical operations. This process involves the disclosure of information related to a company's environmental impact, social responsibility, and governance practices to stakeholders, including investors, customers, employees, and the general public. Here are key aspects of ESG reporting and transparency:

(i) **Comprehensive ESG Framework:**

- Companies should adopt a comprehensive framework for ESG reporting, such as the Global Reporting Initiative (GRI), Sustainability Accounting Standards Board (SASB), or the Task Force on Climate-related Financial Disclosures (TCFD).
- These frameworks provide guidelines for reporting on various ESG aspects, ensuring consistency and comparability of data.

(ii) **Environmental Reporting:**

- This includes disclosure of the company's impact on the environment, such as carbon emissions, energy usage, waste management, water usage, and impact on biodiversity.
- Reporting should also cover the company's strategies and practices for reducing environmental footprints, such as investments in renewable energy and sustainable resource usage.

(iii) **Social Reporting:**

- Social reporting focuses on the company's practices and policies regarding labor rights, employee relations, diversity and inclusion, community engagement, and impact on local and global communities.
- It should also cover issues like product responsibility, including customer health and safety, data protection, and privacy.

(iv) **Governance Reporting:**

- Governance reporting involves disclosing information about the company's leadership structure, executive remuneration, board diversity, ethical standards, anti-corruption policies, and risk management processes.
- This component highlights the company's commitment to ethical operations, transparency, and accountability.

(v) **Transparency and Honesty:**

- ESG reports should be transparent and honest, presenting both successes and areas for improvement. This candor helps build trust with stakeholders.
- Avoiding "greenwashing" – making misleading claims about sustainability practices – is crucial for maintaining credibility.

(vi) **Regular and Consistent Reporting:**

- ESG reporting should be done regularly (usually annually) and be consistent in how information is presented over time to track progress and changes.
- Regular reporting also keeps stakeholders informed and engaged with the company's ongoing ESG efforts.

(vii) **Third-Party Verification:**

- Having ESG reports verified by an independent third party can enhance credibility. This could involve audits or reviews by external consultants or certification bodies.
- Third-party verification ensures that the data reported is accurate, reliable, and conforms to the reporting standards.

(viii) **Stakeholder Engagement:**

- Engage stakeholders in the reporting process by seeking their input on what ESG issues are important to them and how they would like the company to communicate its efforts.
- Effective stakeholder engagement can also help companies understand and address the ESG issues that are most relevant to their business.

(ix) **Integration with Financial Reporting:**

- Integrating ESG reporting with traditional financial reporting can provide a more holistic view of the company's performance and risks.
- This integration demonstrates how ESG factors are linked to financial outcomes and long-term value creation.

(x) **Utilizing Technology for Reporting:**

- Leverage technology to collect, manage, and report ESG data efficiently. Digital platforms can help in aggregating data across various departments and locations.
- Technology also enables more dynamic and interactive reporting, making it easier for stakeholders to access and understand ESG information.

Effective ESG reporting and transparency are not just about compliance; they are about demonstrating a genuine commitment to sustainable development and ethical practices, building trust with stakeholders, and contributing to the long-term success and resilience of the business.

1.32 Conclusion

Climate change is currently being seen as a significant and formidable obstacle to achieving sustainable development. The current environmental catastrophe can be attributed to the release of greenhouse gases (GHGs), primarily stemming from the combustion of fossil fuels (Reiche & Bechberger, 2004; Sánchez et al., 2023). The synthesis of Sustainable Energy Transition, Circular Economy, and Environmental, Social, and Governance (ESG) Practices embodies a robust framework for effectively tackling the urgent issues faced by our global community. Although these three interconnected principles have different areas of emphasis, they are inherently interlinked, providing a comprehensive approach to sustainability. As the global shift towards cleaner energy sources gains momentum, the adoption of circular economy principles and the adherence to environmental, social, and governance (ESG) standards not only contribute to the mitigation of climate change but also foster the development of more resilient businesses and communities. Despite the presence of various hurdles, the advantages of this approach are significant. These include decreased environmental impact, financial savings, promotion of social fairness, and enhancement of company reputation. The present moment necessitates immediate action, as well as the potential for favorable transformations that are readily attainable.

The varied and profoundly crucial significance of Sustainable Energy Transition, Circular Economy, and Environmental, Social, and Governance (ESG) practices in the attainment of global sustainability goals is evident. These concepts are intricately interconnected, forming a unified strategy that holds promise for a more sustainable future. The Sustainable Energy Transition plays a crucial role in addressing climate change and decreasing the worldwide carbon emissions. The transition from fossil fuels to renewable energy sources such as solar, wind, and hydro not only mitigates greenhouse gas emissions but also enhances energy security and promotes innovation. Additionally, this transformation facilitates job creation and stimulates economic growth in growing technological sectors. In addition, it plays a role in enhancing public health outcomes through the mitigation of air pollution, so supporting the pursuit of sustainability objectives pertaining to health and overall well-being.

Simultaneously, the concept of the Circular Economy redefines the notion of economic growth by prioritizing the attainment of positive societal outcomes through the deliberate separation of economic activity from the utilization of limited resources. The aforementioned statement underscores the significance of engaging in practices such as reusing, repairing, refurbishing, and recycling preexisting materials and products. By doing so, the lifespan of resources is prolonged, and waste generation is minimized. This method not only facilitates the preservation of resources and mitigation of environmental consequences, but also promotes enhanced resource efficiency, hence creating economic prospects and stimulating innovation. The adoption of a circular economy model enables businesses and societies to generate products and services that are more sustainable and durable, hence fostering sustainable patterns of consumption and production.

The integration of environmental, social, and governance (ESG) practices is of utmost importance in ensuring that business plans are in line with the overarching objectives of global sustainability. ESG standards facilitate the adoption of ethical, socially responsible, and environmentally sustainable practices by businesses. The environmental component of the Environmental, Social, and Governance (ESG) framework compels corporations to adopt strategies aimed at diminishing their carbon emissions, optimizing resource use, and mitigating pollution. Consequently, this aligns with the objectives of climate action and the preservation of terrestrial and aquatic ecosystems. The incorporation of the social dimension drives businesses to consider aspects such as human rights, labor standards, and community engagement.

This active consideration directly contributes to the pursuit of objectives related to excellent education, gender equality, decent employment, and economic growth. Governance practices have a crucial role in fostering accountability, transparency, and ethical conduct inside businesses, hence serving as a fundamental pillar for cultivating trust and promoting stability in both economies and society.

Furthermore, the integration of these principles together tackles the issue of energy poverty and has the potential to stimulate economic growth, particularly in developing areas. The adoption of sustainable energy sources facilitates the provision of clean and cost-effective energy, a pivotal factor in addressing poverty, enhancing healthcare, education, and overall well-being. The implementation of a circular economy has the potential to stimulate local economies, decrease expenses, and generate employment opportunities within the recycling and refurbishment industries. This, in turn, can contribute to overall economic expansion and the mitigation of societal disparities. ESG practices aim to foster equity and inclusivity by prioritizing the interests of all stakeholders, particularly marginalized communities, so ensuring that advancements are fair and accessible to all.

Within the framework of global concerns such as climate change, limited resources, and socioeconomic disparities, the incorporation of Sustainable Energy Transition, Circular Economy, and Environmental, socioeconomic, and Governance (ESG) practices presents a comprehensive and all-encompassing strategy towards achieving sustainability. This initiative supports the advancement of robust infrastructures, encourages the establishment of inclusive and sustainable industrial practices, and cultivates a climate of innovation. The promotion of responsible consumption and production plays a vital role in the achievement of the Sustainable Development Goals (SDGs) set forth by the United Nations. The aforementioned goals provide a comprehensive framework for attaining a more prosperous and environmentally sound future. These objectives encompass various critical areas, including accessible and environmentally friendly energy, the promotion of fair employment and economic advancement, the development of industry, innovation, and infrastructure, the mitigation of disparities, the establishment of sustainable urban environments, the cultivation of responsible patterns of consumption and production, the pursuit of climate action, and the fostering of collaborative partnerships to achieve these aims.

In summary, it is evident that the combined effect of Sustainable Energy Transition, Circular Economy, and ESG practices plays a crucial role in attaining worldwide sustainability objectives. The aforementioned interconnected principles serve as not just means to achieve environmental sustainability, but also as drivers for economic growth, social equity, and corporate responsibility. The integration of several components is crucial in the construction of a sustainable, resilient, and equitable global society. This integration involves the simultaneous pursuit of environmental preservation, economic prosperity, and social fairness, with the aim of benefiting both present and future generations. The requirement for achieving the lofty targets set by the global community for a sustainable future necessitates the widespread acceptance and rigorous implementation of these policies across many sectors and geographical regions.

1.33 Call to Action for Stakeholders

The imperative for businesses, governments, and individuals to actively engage in the promotion of global sustainability is evident. It is crucial for businesses to incorporate sustainable practices into their fundamental strategies, with a particular emphasis on renewable energy, circular economy models, and comprehensive ESG (Environmental, Social, and Governance) criteria. This transition not only fosters the development of new ideas and sustainable financial gains, but also ensures that corporate objectives are in harmony with the requirements of the world. It is recommended that governments implement favorable regulations, allocate resources towards the development of environmentally friendly infrastructure, and establish ambitious yet attainable objectives for the promotion of renewable energy and waste management. These measures will create a conducive climate whereby sustainable activities are not only promoted, but also incentivized. The imperative for individuals is to adopt a conscientious approach as consumers and engaged citizens, by adopting ecologically and socially responsible decisions in their everyday lives. This entails actions such as minimizing waste and energy consumption, endorsing sustainable enterprises, and pushing for progressive policies. The collaboration of individuals is of utmost importance in guiding our international society towards a future that is both sustainable, equitable, and affluent. The diverse and profoundly crucial significance of Sustainable Energy Transition, Circular Economy, and Environmental, Social, and Governance (ESG) practices in the achievement of global sustainability goals is evident. These concepts are interconnected and can be integrated into a cohesive strategy aimed at fostering a more sustainable future. The Sustainable Energy Transition plays a crucial role in addressing climate change and minimizing the worldwide carbon emissions. The transition from fossil fuels to renewable energy sources such as solar, wind, and hydro has multiple benefits. In addition to mitigating greenhouse gas emissions, this shift promotes energy security and fosters innovation. Furthermore, it generates employment opportunities and stimulates economic growth, particularly in developing industries. In addition, it plays a role in enhancing public health outcomes through the mitigation of air pollution, therefore supporting the pursuit of sustainability objectives pertaining to health and overall welfare.

Simultaneously, the concept of the Circular Economy redefines the notion of economic growth by prioritizing the generation of societal advantages that extend throughout the entire population, achieved through the separation of economic activity from the utilization of limited resources. The text underscores the significance of prioritizing the reuse, repair, refurbishment, and recycling of pre-existing materials and products, hence prolonging the lifespan of resources and reducing waste generation. This method not only facilitates the preservation of resources and mitigation of environmental impacts, but also stimulates enhanced resource production, hence offering economic prospects and promoting innovation. The adoption of a circular economy model enables businesses and society to generate products and services that are more sustainable and durable, hence fostering the development of sustainable patterns of consumption and production.

The environmental component of the Environmental, Social, and Governance (ESG) framework compels corporations to adopt strategies aimed at diminishing their carbon emissions, optimizing resource use, and mitigating pollution. Furthermore, the integration of these principles together tackles the issue of energy poverty and has the potential to stimulate economic growth, particularly in developing areas. The adoption of sustainable energy sources facilitates the provision of clean and cost-effective energy, a pivotal factor in addressing poverty, enhancing healthcare, education, and overall well-being. The implementation of a circular economy has the potential to stimulate local economies, mitigate expenses, and generate employment opportunities within the recycling and refurbishment industries. This, in turn, can contribute to overall economic expansion and the mitigation of socioeconomic disparities. ESG practices aim to foster equity and inclusivity by emphasizing the interests of all stakeholders, especially vulnerable populations.

Within the framework of worldwide predicaments such as climate change, limited resources, and societal disparities, the amalgamation of Sustainable Energy Transition, Circular Economy, and Environmental, Social, and Governance (ESG) practices presents a comprehensive methodology towards achieving sustainability. It serves to enable the advancement of robust infrastructures, advocate for inclusive and sustainable industrialization, and cultivate innovation. The promotion of responsible consumption and production plays a vital role in the achievement of the Sustainable Development Goals (SDGs) set forth by the United Nations. The aforementioned goals provide a comprehensive framework for attaining a more favorable and enduring future for humanity. These objectives encompass various crucial aspects, including accessible and environmentally friendly energy, the promotion of satisfactory employment opportunities and economic advancement, the development of industry, innovation, and infrastructure, the mitigation of disparities, the establishment of sustainable urban areas and communities, the cultivation of responsible consumption and production patterns, the undertaking of climate action, and the fostering of partnerships to achieve these goals.

It is evident that the interdependent relationship between Sustainable Energy Transition, Circular Economy, and Environmental, Social, and Governance (ESG) practices plays a crucial role in the attainment of worldwide sustainability objectives. The aforementioned interconnected principles serve as not just means to achieve environmental sustainability, but also as drivers for economic growth, social equity, and corporate responsibility. The integration of various components is crucial in constructing a sustainable, resilient, and equitable global society, wherein the objectives of environmental preservation, economic advancement, and social fairness are pursued collectively for the betterment of present and future generations. The requirement for achieving the lofty targets set by the global community for a sustainable future necessitates the widespread acceptance and rigorous implementation of these policies across many sectors and geographical regions. The imperative for businesses, governments, and individuals to actively address global sustainability is evident. Businesses must incorporate sustainable practices into their fundamental strategies, with a focus on renewable energy, circular economy models, and comprehensive ESG (Environmental, Social, and Governance) standards. This transition not only fosters the development of new ideas and sustainable financial success but also ensures that corporate objectives are in harmony with the requirements of the world. It is recommended that governments implement favorable regulations, allocate resources towards environmentally friendly infrastructure, and establish high yet attainable objectives for renewable energy and waste mitigation. These measures will cultivate an atmosphere in which sustainable actions are promoted and incentivized. Individuals are encouraged to adopt a conscientious consumer mindset and engage in active citizenship by making ecologically and socially responsible decisions in their everyday lives. This includes minimizing waste generation, conserving energy, endorsing sustainable enterprises, and pushing for progressive policies. The collaborative endeavor holds the utmost significance in guiding our international society toward a sustainable, equitable, and affluent future.

References

Augustynowicz, A. (2022). Tendencies in development of energy assistance systems in an electric car. *Combustion Engines, 190*, 95–103. https://doi.org/10.19206/CE-145714

Babkin, A., Shkarupeta, E., Tashenova, L., Malevskaia-Malevich, E., & Shchegoleva, T. (2023). Framework for assessing the sustainability of ESG performance in industrial cluster ecosystems in a circular economy. *Journal of Open Innovation: Technology, Market, and Complexity, 9*(2), 100071. https://doi.org/10.1016/j.joitmc.2023.100071

Doszhan, R., Zhuparova, A., Kozhahmetova, A., & Semerkova, L. (2022). Economic feasibility of sustainable innovations. *Models, Systems, Networks in Economics, Technology, Nature and Society, 3*, 42–59. https://doi.org/10.21685/2227-8486-2022-3-3

EPA – United States Environmental Protection Agency. (2020). Advancing sustainable materials management: 2018 fact sheet assessing trends in materials generation and Management in the United States. In *United States Environmental Protection Agency. Office of Resource Conservation and Recovery* (Issue December).

FIDIC. (2023). *Developing tomorrow's sustainable energy systems*. https://issuu.com/fidic/docs/sotw_2023_design_report_final?fr=xKAE9_zU1NQ

Hafner, M., & Tagliapietra, S. (2020). *The geopolitics of the global energy transition*. Springer Nature Singapore.. https://link.springer.com/book/10.1007/978-3-030-39066-2

Khetriwal, D. S., Kraeuchi, P., & Widmer, R. (2009). Producer responsibility for e-waste management: Key issues for consideration – Learning from the Swiss experience. *Journal of Environmental Management, 90*(1), 153–165. https://doi.org/10.1016/j.jenvman.2007.08.019

Listyawati, R. N., Meidiana, C., & Anggraeni, M. (2014). Evaluation of energy self-sufficient village by means of emergy indices. *Procedia Environmental Sciences, 20*, 30–39. https://doi.org/10.1016/j.proenv.2014.03.006

Liu, L., & Ramakrishna, S. (2021). *An introduction to circular economy*. Springer Nature Singapore. https://link.springer.com/book/10.1007/978-981-15-8510-4

Ma, B. (2021). China's current and future role in climate change. Harvard. https://insights.grcglobalgroup.com/chinas-current-and-future-role-in-climate-change/

Reiche, D., & Bechberger, M. (2004). Policy differences in the promotion of renewable energies in the EU member states. *Energy Policy, 32*(7), 843–849. https://doi.org/10.1016/S0301-4215(02)00343-9

Sánchez, A. S., Junior, E. P., Gontijo, B. M., de Jong, P., & dos Reis Nogueira, I. B. (2023). Replacing fossil fuels with renewable energy in islands of high ecological value: The cases of Galápagos, Fernando de Noronha, and Príncipe. *Renewable and Sustainable Energy Reviews, 183*, 113527. https://doi.org/10.1016/j.rser.2023.113527

Sarpong, F. A., Sappor, P., Nyantakyi, G., Ahakwa, I., Esther Agyeiwaa, O., & Blandful Cobbinah, B. (2023). From traditional roots to digital bytes: Can digitalizing ESG improves Ghanaian rural banks' brand equity through stakeholder engagement, and customer loyalty? *Cogent Business & Management, 10*(2), 2232159. https://doi.org/10.1080/23311975.2023.2232159

Schröder, P., Dewick, P., Kusi-Sarpong, S., & Hofstetter, J. S. (2018). Circular economy and power relations in global value chains: Tensions and trade-offs for lower income countries. *Resources Conservation and Recycling, 136*, 77–78. https://doi.org/10.1016/j.resconrec.2018.04.003

van Dam, K., Simeone, L., Keskin, D., Baldassarre, B., Niero, M., & Morelli, N. (2020). Circular economy in industrial design research: A review. *Sustainability, 12*(24). https://doi.org/10.3390/su122410279

Chapter 2
The Economics of Sustainable Energy Transition and the Circular Economy

Abstract This chapter analyzes the economic concepts and implications associated with the shift towards sustainable energy sources and the adoption of a circular economy model. Considering pressing global concerns surrounding climate change, resource scarcity, and environmental deterioration, there is a growing imperative to undertake a comprehensive overhaul of energy systems and production-consumption practices. The statement underscores the capacity of the circular economy to enhance resource efficiency, minimize waste, and stimulate innovation. Additionally, it emphasizes the potential of a sustainable energy transition to stimulate economic growth, generate employment opportunities, and reduce the effects of climate change. Policymakers, organizations, and individuals have the potential to navigate a trajectory toward a more environmentally sustainable and economically viable global economy by comprehending the economic imperatives and opportunities associated with these revolutionary approaches.

Keywords Sustainability · Energy transition · Circular economy · Energy efficiency · Green growth · Climate change · Circular cities · Circular business model

2.1 Introduction

Amidst a time characterized by increasing environmental apprehensions and the pressing requirement for sustainable progress, the shift towards sustainable energy sources and the embrace of circular economy ideas have become crucial. This chapter examines the complex interconnection between these two areas, investigating how they combine to influence a future that is both environmentally sustainable and economically robust.

Sustainable energy transition is the transition away from relying on fossil fuels and towards the adoption of renewable energy sources including solar, wind, and hydroelectric power. It plays a vital role in global efforts to tackle climate change (Nwokolo et al., 2023). This transformation addresses both the environmental

V. Kandpal et al., *Sustainable Energy Transition*, Circular Economy and Sustainability, https://doi.org/10.1007/978-3-031-52943-6_2

impacts of energy generation and has substantial implications for the global economy. This process involves reevaluating the techniques for producing, distributing, and using energy, while ensuring that they conform to the ideals of sustainability.

The circular economy represents a paradigm shift in economic ideology, focusing on resource optimization, waste minimization, and the restoration of natural systems. It challenges the traditional linear model of "take, make, dispose" and advocates for a closed-loop system that incorporates the reuse, repair, remanufacturing, and recycling of resources and products. Within the field of energy, this involves creating systems that maximize energy production and usage efficiency to minimize waste and ameliorate environmental damage, hence promoting a more sustainable economic structure.

The convergence of sustainable energy transition and the circular economy is crucial in the endeavor to achieve a more sustainable future. This convergence embodies both an imperative for the environment and a prospect for economic progress. This project ensures the promotion of innovation, creation of employment prospects, and enhancement of market reach, all while minimizing our ecological footprint. The purpose of this chapter is to clarify the economic dimensions of this transition. The chapter seeks to examine the economic mechanisms involved to achieve a comprehensive understanding of how we might sustain and accelerate this crucial shift, so ensuring a more affluent future for both the planet and its inhabitants.

The world is today at a critical juncture, grappling with numerous challenges stemming from climate change, resource depletion, and environmental degradation. To ensure a sustainable and resilient future, it is crucial to accelerate the transition to cleaner and renewable energy sources. Moreover, it is imperative to adopt a circular economy framework that effectively addresses waste management and maximizes resource consumption. The sustainable energy transition involves the substitution of fossil fuel-based energy sources with renewable alternatives (Romero-Castro et al., 2023). The process involves the adoption of sustainable energy technologies, such as solar, wind, and hydropower, to reduce the emission of greenhouse gases and mitigate the impacts of climate change. Implementing sustainable energy sources can enable communities to attain energy self-reliance and foster sustainable development in rural regions. It is essential for achieving the net-zero goals and decarbonization targets set by many advanced economies.

The motivation to achieve a sustainable energy transition stems from the necessity to decrease carbon emissions and alleviate the consequences of climate change. Renewable energy sources, including sun, wind, hydropower, and geothermal, offer a sustainable and eco-friendly substitute for fossil fuels, ensuring energy stability (Marwa et al., 2020). The economic factors associated with renewable energy, such as the decreasing costs and increasing competitiveness, present a compelling case for accelerating the shift.

The concept of the circular economy offers a groundbreaking approach to both manufacturing and consumption. The circular economy is an economic framework characterized by closed loops, wherein raw materials, components, and completed goods are preserved in terms of their quality and value over extended durations. Furthermore, this system utilizes renewable energy sources to power its many activities. The process entails the implementation of recycling, reusing, and remanufacturing technologies to preserve resources and mitigate environmental effects. The concept of the circular economy aims to separate economic growth from the linear processes associated with the extraction, utilization, and disposal of finite resources (Mayes-Ramírez et al., 2023; Schmid & Mendoza, 2022; Velasco-Muñoz et al., 2021).

The implementation of a circular economy can serve as a valuable means of obtaining low-carbon resources and effectively attaining sustainability objectives. It is supported by a shift towards renewable energy sources, which are not exhausted within appropriate timeframes and contribute to a more environmentally friendly economy. This concept offers substantial economic benefits, such as reduced expenses, improved resilience in the supply chain, and the opportunity to generate additional money through creative circular business models. The production and management of renewable energy technology must adhere to principles of environmental, economic, and social sustainability. In order to accomplish these goals, circular economy solutions can effectively reduce, decelerate, and ultimately eliminate resource cycles (Bocken et al., 2016).

According to Martínez-Alier (2018), The global community can enhance its likelihood of preventing hazardous climate change by transitioning to a circular economy, so enabling societies to achieve the objectives outlined in the Paris Agreement on Climate Action. Several nations have acknowledged the significance of the circular economy in achieving a state of net-zero emissions.

According to the World Resources Institute, around 33% of nationally defined contributions, which are climate action plans devised by each country to mitigate emissions and cope with climate-related effects, make reference to a circular economy. A study conducted by the Platform for Accelerating Circular Economy (PACE), World Resources Institute (WRI), Chatham House, and the National Renewable Energy Laboratory (NREL) reveals that implementing circular economy strategies can effectively (i) supplement decarbonization efforts in reducing greenhouse gas emissions; (ii) improve the ability to adapt to a changing climate; and (iii) facilitate the sustainable expansion of the clean energy transition (Schmid & Mendoza, 2022).

Faced with rapidly depleting natural resources, climate change, and the societal repercussions of linear economic models, the transition to sustainable energy and the adoption of the circular economy (CE) model has become essential. This chapter delves into the economics of sustainable energy transition and the circular economy, exploring their opportunities and challenges for businesses, economies, and society.

Transitioning to sustainable energy is necessary to solve climate change challenges, lower greenhouse gas emissions, and guarantee a safe and resilient energy future. This section focuses on the concept of a sustainable energy transition, its environmental benefits, and the economic opportunity it offers businesses and society worldwide.

The transition from fossil-based energy production and consumption systems, such as oil, natural gas, and coal, to renewable energy sources, such as wind, solar, and lithium-ion batteries, is called the energy transition. The rising penetration of renewable energy in the energy supply mix, the beginning of electrification, and advances in energy storage are all significant drivers of the energy transition. Regulation and decarbonization commitments have been inconsistent, but the energy transition will become more critical as investors prioritize environmental, social, and governance (ESG) issues.

Sustainable transition methods typically include three major technological changes: energy savings on the demand side, production efficiency and fossil fuel substitution with various renewable energy sources and low-carbon nuclear energy. Policy measures are required to direct the global electrical transition towards a sustainable energy and electricity system to remain technically and economically possible and advantageous. On a large scale, renewable energy implementation should include initiatives to increase the efficiency of existing nonrenewable sources, which continue to play an essential role in cost reduction and stabilization. Humanity's most significant sustainability concerns are greenhouse gas emissions and global climate change, with fossil fuels, led by coal, natural gas, and oil, accounting for 61.3% of worldwide power generation in 2020 (Kabeyi & Olanrewaju, 2022).

A focus on the electric power sector would be especially promising because there is the potential to save much more energy and reduce more greenhouse gas emissions than with end-use technologies in buildings and vehicles since enormous quantities of waste heat can be recovered cost-effectively and nonpollutingly. Energy efficiency enhancements can be strategically focused on as a component of worldwide endeavors to achieve mutually advantageous objectives through the establishment of an international accord on energy efficiency that explicitly addresses the electric power industry. Additional backing and advocacy for the smart grid would be beneficial to maximize efficiency improvements and minimize emissions in electric power systems. However, any international agreement needs to complement regional policies, which must be more robust and effective (Solomon & Krishna, 2011).

2.2 The Need for Sustainable Energy Transition

The need for a sustainable energy transition is apparent. The global energy landscape is undergoing a significant transformation as it shifts from traditional fossil fuels to renewable energy sources. A combination of environmental, economic, and technological factors drives this transition. A sustainable energy system meets our energy needs while minimizing environmental impacts. This system is based on renewable energy sources, such as solar and wind power, and energy efficiency measures. We need to transition to a sustainable energy system for many reasons. Here are some of the most important reasons:

Environmental Concerns

A. **Reducing greenhouse gas emissions**: The primary driver of this shift is the need to address climate change. Fossil fuels, such as coal, oil, and natural gas, are the largest source of greenhouse gas emissions, contributing significantly to global warming. Fossil fuels, when burned, release carbon dioxide (CO2) and other greenhouse gases into the atmosphere. These gases trap heat, leading to the greenhouse effect and global warming. The energy sector significantly contributes to greenhouse gas emissions, primarily by burning fossil fuels. Transitioning to renewable energy sources can significantly reduce these emissions and help mitigate climate change.

B. **Meeting climate goals**: Global commitments like the Paris Agreement have prompted countries to increase their investments in renewable energy to meet emission reduction targets. The Paris Agreement sets ambitious climate goals, including limiting global warming to below 2 degrees Celsius above preindustrial levels. Achieving these goals requires a rapid transition to renewable energy and a significant increase in renewable energy production. By transitioning to sustainable energy sources such as wind, solar, and hydroelectric energy, we can drastically reduce CO2 emissions, slowing the pace of global warming and the associated climate change impacts. Meeting climate goals will require a concerted effort from governments, businesses, and individuals. Some of the steps that can be taken to meet climate goals:

   a. **Reduce emissions:** This means reducing our reliance on fossil fuels and transitioning to renewable energy sources. It also means improving energy efficiency and changing our behavior to reduce energy consumption.
   b. **Protect forests:** Forests play a significant role in absorbing carbon dioxide from the atmosphere. We need to protect forests and replant trees to help mitigate climate change.
   c. **Investment in sustainable agriculture:** Agriculture is a major source of greenhouse gas emissions. We need to invest in sustainable agriculture practices, such as crop rotation and cover cropping, to reduce emissions from agriculture.
   d. **Adapt to the effects of climate change:** Climate change is already having an impact on our planet. We must adapt to these impacts by building seawalls to protect coastal communities from flooding, developing drought-resistant crops, and improving early warning systems for extreme weather events.

Meeting climate goals will require a significant investment of time, money, and effort. However, the benefits of taking action are clear. By meeting climate goals, we can protect our planet for future generations and create a more sustainable future for all. Here are some specific examples of what can be done to meet climate goals:

- **Governments can set ambitious emissions reduction targets and policies to support them.** For example, the European Union has targeted reducing greenhouse gas emissions by 55% by 2030.

- **Businesses can invest in renewable energy and energy efficiency measures.** They can also amend their operations to reduce their emissions. For example, Google has set a goal of being carbon neutral by 2025.
- **Individuals can amend their own behavior to reduce their carbon footprint.** This includes driving less, using less energy, and eating less meat.

Meeting climate goals is a challenge, but it is one that we can overcome. By working together, we can create a cleaner, healthier, and more sustainable future for all.

C. **Environmental benefits**: Renewable energy sources like wind, solar, and hydropower have lower environmental impacts than fossil fuels. They produce little to no air pollution, reduce water usage and have a smaller carbon footprint. Environmental benefits are the positive effects that actions taken to protect the environment can have on human health, the economy, and the natural world. These benefits can be both tangible and intangible. Some of the most common environmental benefits include the following:

(i) **Improved air quality:** Reducing emissions from cars, factories, and other sources can improve air quality, which can lead to a decrease in respiratory problems, heart disease, and other health problems.
(ii) **Cleaner water**: Protecting water resources and reducing pollution can lead to cleaner water for drinking, bathing, and recreation.
(iii) **Conserved wildlife**: Protecting habitats and reducing threats to wildlife can help to ensure the survival of endangered species and the overall health of ecosystems.
(iv) **Reduced climate change impacts**: Taking action to reduce greenhouse gas emissions can help to mitigate the effects of climate change, such as sea level rise, extreme weather events, and changes in agricultural yields.
(v) **Increased economic opportunities**: Investing in environmental protection can create jobs and stimulate economic growth. For example, the renewable energy industry is a growing sector that is creating new jobs and generating economic activity.
(vi) **Improved quality of life:** Spending time in nature has been shown to improve mental and physical health, and protecting the environment can help to ensure that future generations have access to clean air, water, and natural spaces.

The environmental benefits of taking action to protect the environment are numerous and far-reaching. By making changes to our individual and collective lifestyles, we can all help to create a cleaner, healthier, and more sustainable future. Here are some specific actions that can be taken to protect the environment and reap the associated benefits:

- **Reduce, reuse and recycle:** This can help to conserve resources and reduce pollution.
- **Conserve energy**: This can help to reduce greenhouse gas emissions and save money on utility bills.

- **Drive less:** This can help to reduce air pollution and improve public health.
- **Support sustainable businesses:** This can help to create jobs and invest in environmental protection.
- **Get involved in your community**: This can help to raise awareness about environmental issues and advocate for change.

D. **Energy security**: Fossil fuels are a finite resource, and their prices can be volatile. Diversifying the energy mix by incorporating more renewable energy sources enhances energy security by reducing dependence on finite fossil fuel resources and volatile energy markets. Energy security is the ability of a country to access affordable, sufficient, and safe energy supplies reliably. It is a critical issue for all countries, but it is especially important for countries that are heavily reliant on imported energy. There are a number of factors that can affect energy security, including:

  - **The availability of energy resources:** Countries with abundant energy resources are less likely to experience energy security problems.
  - **The reliability of energy supplies**: Countries with reliable energy supplies are less likely to experience disruptions in energy services.
  - **The affordability of energy prices:** Countries that can afford to pay for energy are less likely to experience energy security problems.
  - **The security of energy infrastructure**: Countries with secure energy infrastructure are less likely to experience disruptions in energy services.

There are a number of things that can be done to improve energy security, including:

- **Diversifying energy supplies**: This means reducing reliance on a lone source of energy, such as oil.
- **Investing in energy efficiency**: This means using less energy to do the same thing.
- **Developing renewable energy sources**: This means using sources of energy that are not finite, such as solar and wind power.
- **Improving energy infrastructure** means making energy systems more resilient to disruptions.

Energy security is a complex issue, but it is essential for the stability and prosperity of all countries. By taking steps to improve energy security, we can reduce our vulnerability to disruptions and ensure that we have the energy we need to power our economies and societies.

E. **Finite Nature of Fossil Fuels:** Coal, oil, and natural gas are finite resources. As they deplete, extraction becomes more challenging, leading to geopolitical tensions, price volatilities, and potential energy shortages.
F. **Job Creation:** The renewable energy sector has the potential to create numerous jobs – from manufacturing solar panels to maintaining wind turbines. These jobs can drive economic growth, especially in regions with abundant natural resources.

G. **Reliability of Renewable Sources:** The reliability of renewable energy sources is a complex issue. Some renewable energy sources, such as solar and wind, are intermittent, meaning that they do not always produce energy when needed. This can make it difficult to rely on them as a baseload power source. However, renewable energy sources can be made more dependable by using energy storage and demand response technologies. Here is a brief overview of the reliability of some of the most common renewable energy sources:

   - **Solar:** Solar energy is intermittent, but it can be made more dependable by using energy storage technologies, such as batteries. Solar energy is also becoming more affordable, and its capacity is increasing rapidly.
   - **Wind:** Wind energy is also intermittent, but it can be made more reliable by using wind farms in distinct locations. Wind energy is also becoming more affordable, and its capacity is increasing rapidly.
   - **Hydropower:** Hydropower is a reliable source of energy, but it can be affected by droughts and floods. Hydropower is also a relatively mature technology, and its capacity is not expected to grow rapidly.
   - **Geothermal:** Geothermal energy is a reliable source of energy, but it is only available in certain locations. Geothermal energy is also expensive, and its capacity is not expected to grow rapidly.
   - **Biomass:** Biomass energy is a reliable source of energy, but it can be harmful to the environment if not managed properly. Biomass energy is also an expensive technology, and its capacity is not expected to grow rapidly.

The reliability of renewable energy sources is improving. However, there are still some challenges that need to be addressed. By using technologies such as energy storage and demand response, renewable energy can be made more dependable and can play a greater role in meeting our energy needs. Here are some of the ways to make renewable energy more dependable:

- **Energy storage:** Energy storage technologies, such as batteries, can store energy from renewable sources when it is abundant and then release it when needed. This can help to smooth out the output of renewable sources and make them more dependable.
- **Demand response:** Demand response programs can encourage consumers to reduce their energy use during peak demand periods. This can help to reduce the need for conventional power plants and make the grid more dependable.
- **Smart grids:** Smart grids can use information and communication technologies to improve the efficiency and reliability of the grid. This includes things such as monitoring the grid in real time and adjusting the flow of electricity to meet demand.
- **Diversifying renewable sources:** By using various renewable sources, such as solar, wind, and hydro, we can reduce the risk of disruptions from any source.

By investing in these technologies and strategies, we can make renewable energy more dependable and a more critical part of our energy mix.

H. **Economic opportunities**: The transition to renewable energy can create new job opportunities, stimulate economic growth, and attract investment in clean energy technologies and infrastructure.
I. **Health benefits**: Shifting to cleaner energy sources can improve air quality and reduce the negative health impacts of fossil fuel combustion pollution.
J. **Technological advancements**: The push for sustainable energy has spurred technological advancements in energy storage, grid infrastructure, and energy efficiency. Such innovations present lucrative business opportunities and can drive economic progress.
K. **Reducing Pollution:** Beyond CO2, burning fossil fuels releases pollutants that degrade air quality. Transitioning away from these sources can lead to cleaner air, reduce respiratory diseases, and improve public health.
L. **Ecosystem Preservation**: Fossil fuel extraction, whether mining or drilling, often disrupts local ecosystems. When implemented responsibly, sustainable energy sources have a smaller ecological footprint, helping preserve biodiversity.
M. **Distributed Energy Systems**: Renewable energy sources often allow for decentralized, distributed energy systems. Such systems can be more resilient to disruptions, whether natural disasters or human-made conflicts.
N. **Meeting Future Energy Demands**: As the global population grows and developing nations industrialize, energy demand will surge. Renewable energy sources can scale up to meet these demands without exacerbating environmental challenges.
O. **Economic growth:** The renewable energy industry is a growing sector that is creating new jobs and generating economic activity. By investing in renewable energy, we can create a clean energy economy that is good for both the environment and the economy.

It is important to note that a sustainable energy transition involves not only generating power from renewable sources but also addressing other aspects, such as energy storage, electric transportation infrastructure, and land-use planning. A sustainable energy transition is crucial for addressing climate change, reducing greenhouse gas emissions, promoting environmental sustainability, enhancing energy security, and creating economic opportunities. It requires policy support, technological innovation, and collaborative efforts from governments, businesses, and individuals to achieve a cleaner, more sustainable energy future.

The need for sustainable energy transition is evident, driven by the imperatives of mitigating climate change, reducing environmental impacts, enhancing energy security, and fostering socioeconomic resilience. The transition to sustainable energy is not just a matter of ecological responsibility but a multifaceted solution addressing economic, health, security, and social challenges. As the implications of continued fossil fuel reliance become increasingly dire, the shift towards a sustainable energy paradigm becomes desirable and imperative for our global community's continued prosperity and survival. Accelerating the adoption of sustainable energy practices is an environmental imperative and a strategic and ethical obligation to safeguard the planet's well-being and its inhabitants.

The need for a sustainable energy transition is apparent. The world is facing a number of challenges, including climate change, air pollution, and energy security. These challenges can only be addressed by moving to a more sustainable energy system.

A sustainable energy system meets our energy needs while minimizing environmental impacts. This system is based on renewable energy sources, such as solar and wind power, and energy efficiency measures. The transition to a sustainable energy system will not be easy. It will require a coordinated effort from governments, businesses, and individuals. However, the benefits of a sustainable energy system are clear, and the need for this transition is urgent.

Here are some of the things that can be done to accelerate the transition to a sustainable energy system:

- **Investment in renewable energy:** Governments can invest in renewable energy projects like solar and wind farms. They can also provide tax breaks and other incentives to businesses that invest in renewable energy.
- **Promote energy efficiency:** Governments can promote energy efficiency measures, such as insulation and energy-efficient appliances. They can also set standards for energy efficiency in buildings and appliances.
- **Develop innovative technologies:** Researchers are developing modern technologies that can help to make renewable energy more affordable and dependable. Governments can fund research and development in these technologies.
- **Change our behavior:** Individuals can make changes to their behavior to reduce their energy consumption. This includes driving less, using less electricity, and recycling more.

The transition to a sustainable energy system is a complex challenge, but it must be addressed. By working together, we can create a clean energy future that is good for the environment and the economy.

2.3 Circular Economy

A circular economy is an economic system that aims to eliminate waste and pollution by keeping products, materials, and resources in use for as long as possible. This can be done through various means, such as recycling, repair, and remanufacturing. Its primary objective is to redefine growth, focusing on benefits for society, not just profit-driven motives. The circular economy is a more sustainable alternative to the traditional linear economy based on the "take, make, dispose" model. The linear economy is responsible for a significant amount of waste and pollution, and it is not sustainable in the long term.

The circular economy has the potential to provide a number of benefits, including:

- Reduced waste and pollution
- Improved resource efficiency

- Increased economic growth.
- Creation of new jobs
- Improved environmental quality.
- Increased resilience to shocks and stresses.

The circular economy is still a developing concept, but it is gaining traction in many countries worldwide. Several initiatives are underway to promote the circular economy, and there is a growing body of evidence that it can be a viable and sustainable way to organize our economy. Here are some examples of how the circular economy is being applied in practice:

- **Recycling:** Recycling is a crucial part of the circular economy. It helps to keep materials in use and prevent them from being sent to landfills.
- **Repair:** Repairing products instead of replacing them can help to extend their lifespan and reduce waste.
- **Remanufacturing:** Remanufacturing is restoring used products to their original condition or better. It can help to create jobs and reduce the need for new products.
- **Product sharing:** Product-sharing services allow people to borrow or rent products instead of buying them. This can help to reduce the number of products that are produced and consumed.
- **Design for sustainability:** Designing products with the circular economy in mind can help to reduce waste and pollution. This includes designing products that are easy to repair, disassemble, and recycle.

The circular economy is a complex and evolving concept, but it has the potential to be a major force for good in the world. By applying the circular economy principles, we can create a more sustainable and prosperous future for all.

2.4 Core Principles

A. Design out Waste and Pollution
"Design out Waste and Pollution" is one of the core principles of the circular economy. It emphasizes the need to tackle waste and pollution at the source rather than dealing with their consequences. This approach underscores that prevention is better than cure, especially regarding environmental degradation. Let us delve deeper into this principle:

Instead of just managing waste after it is produced, the idea is to eliminate waste from the outset by integrating sustainable and thoughtful design practices. This involves creating products, processes, and systems that either do not produce waste or turn the waste they do produce into resources.

Strategies to Design Out Waste and Pollution

a. **Eco-Design:** Eco-design, also known as environmentally conscious or sustainable design, refers to the practice of designing products with particular consideration for their environmental impacts throughout their life cycle. This comprehensive approach encompasses everything from the extraction of raw materials to production, distribution, use, and end-of-life management. The goal is to minimize adverse environmental impacts and create sustainable and functional products.

- **Material Selection**: Material selection is a fundamental aspect of traditional and eco-design processes. It involves choosing the appropriate materials for a product or application based on specific criteria, such as functionality, cost, aesthetic appeal, and environmental impact in the case of sustainable design. The chosen materials can significantly influence a product's performance, durability, and environmental footprint.
- **Modularity**: Modularity refers to the design and organization of a system, product, or service in which individual components can be independently created, replaced, or altered without affecting the entirety of the system. It is a strategy in which a system is broken down into smaller, interchangeable parts (modules) that can function independently but collectively form a whole.

b. **Efficient Production Processes**: Efficient production processes are crucial for companies seeking to minimize waste, reduce costs, and increase productivity. They involve optimizing every aspect of the manufacturing process to use resources more effectively and generate fewer unwanted byproducts. With an increased focus on sustainability and the circular economy, these efficiencies can also contribute significantly to reducing a company's environmental footprint.

Elements of Efficient Production Processes

1. **Lean Manufacturing**: An approach that aims to reduce waste in the production process without compromising productivity. It focuses on value creation for the end customer and involves tools and strategies to identify and eliminate non-value-added activities.
2. **Process Automation**: Using technology and machinery to automate repetitive tasks can significantly boost efficiency, reduce human error, and ensure consistency in product quality.
3. **Continuous Improvement (Kaizen)**: A philosophy that emphasizes ongoing, incremental improvements in the production process. It involves every level of the organization and focuses on enhancing productivity and quality.
4. **Six Sigma**: A set of techniques and tools aimed at process improvement by identifying and removing the causes of defects and reducing variability in manufacturing.
5. **Just-in-Time (JIT) Production**: A strategy that aligns raw material orders from suppliers directly with production schedules, reducing inventory costs and waste.
6. **Total Quality Management (TQM)**: An approach that seeks to improve product quality and output by involving all organization levels in finding and eliminating reasons for production defects.
7. **Resource Efficiency**: Using advanced technologies and strategies to ensure that materials and energy are used as effectively as possible.

Benefits

1. **Cost Reduction**: Renewable energy technologies have seen dramatic reductions in costs over the past decade. Solar and wind energy, in particular, are now often cheaper than new fossil fuel capacity in many regions.
2. **Improved Product Quality**: Product quality and consistency are enhanced by reducing defects and variability.
3. **Higher Productivity**: Streamlined processes can lead to faster production cycles.
4. **Reduced Environmental Impact**: Using fewer resources and generating less waste can reduce the environmental footprint of production.
5. **Competitive Advantage**: Companies that implement efficient production processes can often deliver products faster and at a lower cost, giving them an edge in the market.

Challenges

1. **Initial Costs**: Implementing innovative technologies or training staff in new methodologies can be expensive initially.
2. **Change Management**: Employees might resist changes to the production process, especially if they feel their roles are threatened.
3. **Complex Implementation**: Some strategies, such as Six Sigma, require in-depth statistical knowledge and can be complex to implement.

Real-World Applications

1. **Automotive Industry**: Car manufacturers, led by pioneers such as Toyota with its Toyota Production System, have embraced lean manufacturing, JIT, and other efficient production methodologies to drastically reduce waste and improve output.
2. **Electronics**: Companies such as Apple and Samsung have optimized their supply chains and production processes to meet the high demand and ensure the consistent quality of their products.
3. **Clothing and Textiles**: With the move towards more sustainable practices, some companies in this sector are exploring efficient production processes to reduce waste, especially given the industry's historically elevated levels of waste.
4. **Food and Agriculture**: Process improvements are being applied to reduce waste in food production, given the significant global concern about food waste.

Efficient production processes are not just about cutting costs but are crucial for sustainability, quality, and competitiveness in the modern market. As consumer awareness of environmental issues grows, companies that invest in efficient production are likely to be viewed more favorably and achieve better market positions.

c. **Reduce Toxicity**

- **Green Chemistry**: Design chemical products and processes to reduce or eliminate the use and generation of hazardous substances.
- **Nontoxic Materials**: Avoid using harmful substances in products, making them safer for use and easier to recycle.

d. **Waste as a Resource**

- **Upcycling**: Transform waste into higher-value products.
- **Industrial Symbiosis**: Interindustry collaborations where waste from one industry becomes a resource for another.

Business and Economic Implications

- **Cost Savings**: Efficient resource use and waste reduction can lead to significant savings in production costs.
- **New Market Opportunities**: Green and sustainable products can cater to an increasingly eco-conscious consumer base, opening new markets.

Challenges

- **Research and Development**: Designing waste might require substantial investment in R&D.
- **Consumer Expectations**: Some consumers might be accustomed to disposable products or might equate durability with quality.

Examples

- **Packaging**: Companies design packaging that is either reusable or made from materials that are fully recyclable or compostable.
- **Electronics**: Design products without harmful chemicals, making them safer to recycle and creating modular devices that can be easily upgraded.
- **Fashion**: Producing clothing from sustainable materials and designing them for longevity rather than short-lived trends.

Broader Environmental and Social Impacts

- **Biodiversity Protection**: By reducing pollution and waste, ecosystems face fewer threats, leading to the protection of biodiversity.
- **Health benefits**: Reducing pollution and toxic materials can lead to fewer health issues in communities, particularly in areas close to production sites or waste disposal areas.

"Design out Waste and Pollution" is a forward-thinking approach that challenges businesses, designers, and innovators to rethink how products and processes are conceived. By integrating circular principles from the outset, we can transition towards a more sustainable and regenerative economic model.

Rethink Packaging

"Rethink packaging" is an essential component of the "design out waste and pollution" principle of the circular economy. Since packaging waste, mainly plastic, has emerged as a significant environmental challenge, reimagining packaging is pivotal. This means questioning current packaging practices and redesigning them to minimize waste, maximize resource efficiency, and reduce environmental impact.

The Issue with Current Packaging

- **Overuse**: Many products are overpackaged, using more material than necessary.
- **Nonrecyclable Materials**: A considerable proportion of packaging materials, especially mixed materials and certain plastics, are not recyclable.
- **Single-Use**: Packaging, especially in the food industry, is often designed for sole use, leading to an enormous waste problem.

Strategies to Rethink Packaging

a. **Reduce**
    - **Minimalistic Design**: Use the minimum amount of packaging necessary for product protection and information.
    - **Concentrated Products**: For instance, concentrated detergent requires less packaging per wash.

b. **Reuse**
    - **Returnable Packaging**: Implement systems where consumers can return packaging for reuse, such as glass milk bottles.
    - **Reusable Containers**: Encourage customers to bring their containers for products such as groceries or takeaway food.

c. **Recycle**
    - **Single Material Packaging**: Design packaging with a single material to make recycling easier.
    - **Compostable packaging**: For organic products, packaging that can be composted along with the product residue, such as tea bags, should be used.

d. **Redesign**
    - **Innovative Materials**: Use materials such as mushroom-based packaging, seaweed packaging, or bioplastics.
    - **Eliminate Hazardous Materials**: Ensure packaging does not contain toxic components that could contaminate the environment upon disposal.

e. **Rethink Systems**
    - **Bulk Stores**: Create retail environments where customers can buy products in bulk using their containers.
    - **Product-as-a-Service**: When products are leased rather than sold, the manufacturer retains ownership of the packaging, incentivizing them to design for reuse.

Business and Economic Implications

- **Cost Efficiency**: While initial R&D costs might be high, eventually, sustainable packaging can be more cost-effective due to the use of fewer materials or through the economic benefits of brand loyalty from eco-conscious consumers.

- **Regulatory Compliance**: Many authorities are implementing stricter regulations on packaging waste, pushing businesses to adapt.
- **Brand Image and Consumer Demand**: Eco-friendly packaging can enhance a brand's image and cater to a growing demographic of environmentally aware consumers.

Challenges

- **Consumer Behavior**: Changing entrenched behaviors, such as the convenience of single-use packaging, can be challenging.
- **Supply Chain Adaptation**: The entire supply chain, from material suppliers to retailers, must adapt to new packaging practices.
- **Preservation and Safety**: Packaging also serves to protect and preserve products. Sustainable alternatives must meet these criteria, especially for perishables.

Notable Innovations and Initiatives

- **Loop**: A shopping platform where products are sold in reusable containers. Once empty, containers are returned, cleaned, refilled, and resold.
- **Edible** packaging: This is especially true for food items, where packaging such as seaweed-based pouches can be eaten along with the product.

To "rethink packaging" is to challenge the status quo, acknowledging that the way we package goods today is unsustainable in the long run. It is about balancing the need to protect products with the imperative to reduce environmental harm. With innovations in material science, design, and supply chain management, sustainable packaging solutions are not just possible but are already emerging across industries.

B. **Keep Products and Materials in Use**:

- **Extended Producer Responsibility (EPR)**: Companies take responsibility for their product even after it is sold, setting up mechanisms for product return, repair, or recycling.
- **Sharing Economy**: Platforms such as car-sharing, tool libraries, and coworking spaces reduce the need for individual ownership.

C. **Regenerate Natural Systems**

- **Biomimicry**: Design solutions are inspired by nature, ensuring that waste from one process is food for another.
- **Regenerative Agriculture**: Agriculture that does not deplete the soil but instead enriches it.

Economic Implications

A. **Business Models in the Circular Economy**

- **Product-as-a-Service (PaaS)**: Instead of selling products, businesses lease them. This model can be found in industries from fashion to electronics.

- **Reverse Logistics**: Systems that allow products to be returned to manufacturers for refurbishment or recycling.

B. **Economic Opportunities and Challenges**

- **New Market Creation**: New industries and niches emerge, such as recycling technology, remanufacturing hubs, and sustainable product design.
- **Reduced Resource Volatility**: Dependence on finite resources often leads to volatile prices. Circularity can stabilize costs by reusing materials.
- **Skill and Knowledge Gaps**: Transitioning might require new skills and expertise, necessitating investments in education and training.

Technological Innovations

- **Material Science**: Development of materials that are more durable, easily recyclable, or biodegradable.
- **Digital Platforms**: Technologies such as IoT, AI, and blockchain can track and optimize resource use, validate recycled content, and facilitate sharing models.

Policy and Regulation
Governments play a pivotal role in the transition:

- **Incentivization**: Tax breaks or subsidies for sustainable practices and R&D.
- **Legislation**: Mandating EPR, setting minimum recycled content, or banning certain nonrecyclable plastics.
- **Public Awareness**: Educational campaigns about the importance and benefits of a circular economy.

Global Implications

- **Supply Chain Resilience**: A circular approach can localize supply chains, reducing vulnerabilities associated with global disruptions.
- **Biodiversity Conservation**: Reduced extraction and deforestation can alleviate pressure on ecosystems.
- **Climate Change Mitigation**: Resource extraction and waste significantly contribute to greenhouse gas emissions. The circular economy can reduce this footprint.

Barriers and Criticisms

- **Scale**: Some circular practices, while feasible on a small scale, might face challenges when scaled up.
- **Consumer Culture**: Convincing consumers to shift from a culture of constant consumption and disposability can be challenging.
- **Interindustry Collaboration**: Requires cooperation and transparency between industries that might be competitors.

The circular economy is a comprehensive, systems-based approach that offers a roadmap for a sustainable future. However, its adoption requires concerted efforts from businesses, policymakers, and consumers. Through collaboration and commitment, the circular economy can pave the way for a future that is not just sustainable but regenerative and restorative.

2.5 Synergies Between Sustainable Energy and the Circular Economy

Sustainable energy and the circular economy are two progressive paradigms that, while originating from different concerns, intertwine in their objectives and potential outcomes. Both are responses to the unsustainable patterns of the linear "Take, Make, Dispose" model, and both aim for an ecologically sound, economically viable, and socially responsible future. By understanding their synergies, we can better harness their combined potential for a more sustainable future.

Energy as a Key Enabler for CE

The circular economy is an economic model that focuses on leasing, recycling, refurbishing, repairing, reusing, and sharing existing materials and goods for as long as feasible. Extending the lifecycle of products and their components reduces waste and the demand for virgin materials. Global energy demands are transitioning towards electrification and renewables, and the energy and utilities sector is already an active participant in the reduction of extraction-based generation operations. Adopting the circular economy model broadens this perspective to include the assets and materials used to generate, transmit, and distribute energy, as well as the mechanisms used to motivate customers to do the same. As consumer and regulatory expectations rise, participation in the circular economy business model has become a challenge for energy and utility companies (IBM, 2022). To achieve net-zero emissions, clean technologies such as electric vehicles and energy transition apparatus will need to be manufactured from zero-emissions materials and produce no emissions when used. This will be a substantial obstacle. The majority of vehicles will be electric, and the materials used to manufacture them could account for 60% of their total lifecycle emissions, up from 18% in 2020 (World Economic Forum, 2020).

The circular economy can also help to create jobs and economic growth. The circular economy is estimated to create 100 million new jobs by 2030. This is because the circular economy requires a range of new skills and technologies, such as those related to recycling, repair, and remanufacturing. There has been a significant mental transformation regarding the impact of consumption and production on the environment. It is a well-established fact that 75% of natural resource consumption occurs in urbanized metropolitan areas, which also generate 50% of global waste and 60–80% of greenhouse gas emissions (Ellen McArthur Foundation, 2019).

Singapore has been a leader in promoting the circular economy and is examining methods to convert waste materials, such as liquified waste plastics, into jet fuel by 2022. In addition, developments in treatment facilities, transportation infrastructure, and industrial facilities have paved the way for Singapore to achieve its zero waste and low-carbon emission goals.

Achieving net-zero emissions by 2040 will necessitate a six-fold increase in mineral input. Some essential metals, such as lithium, could see growth rates of over forty times, and nickel and cobalt demand could increase by more than twenty times. The price of lithium in February 2021 reached an all-time high of $50,000 per ton, up from $10,000 per ton just 1 year prior (International Energy Agency,

2022). The circular economy, characterized by a closed loop where resources are reused, repurposed, and recycled, has emerged as a paradigm-shifting model for sustainable growth. At its core lies the optimization of resources, reducing wastage and minimizing environmental impact.

Energy is an essential enabler of the circular economy. It is necessary to power the processes that recover and reuse materials and to create new products and services that are designed to be more sustainable. Some of the ways in which energy could be used as an enabler for the circular economy are as follows:

- **Renewable energy:** Renewable energy sources, such as solar and wind, can help to reduce the reliance on fossil fuels, which are a major source of waste and pollution. For example, solar panels can be recycled at the end of their life, and the materials can be used to make new solar panels.
- **Energy efficiency:** Energy efficiency measures can help to reduce the amount of energy needed to power our homes, businesses, and transportation systems. This can lead to a decrease in waste and pollution, as well as a reduction in energy costs. For example, energy-efficient light bulbs use less energy and last longer than traditional light bulbs, which can help to reduce waste.
- **Energy storage:** Energy storage technologies can help to make renewable energy more dependable and affordable. This is important because renewable energy sources are intermittent, meaning that they do not always produce energy when needed. For example, batteries can be used to store energy from solar panels or wind turbines, which can then be used to power homes and businesses during periods of high demand.
- **Smart grids:** Smart grids can help to improve the efficiency of our energy system by better managing the flow of electricity. This can lead to a decrease in waste and pollution, as well as a reduction in energy costs. For example, smart grids can be used to balance the supply and demand of electricity, which can help to reduce the need for peaker plants, which are often powered by fossil fuels.
- **Sustainable transportation:** Sustainable transportation options, such as electric vehicles and public transportation, can help to reduce the amount of energy used for transportation. This can lead to a decrease in waste and pollution, as well as a reduction in greenhouse gas emissions. For example, electric vehicles produce zero emissions, which can help to improve air quality.

By investing in renewable energy, energy efficiency, and other sustainable technologies, we can create a more circular economy that is better for the environment and the economy. In addition to the environmental benefits, the circular economy can create economic opportunities. For example, the renewable energy industry is a growing sector that is creating new jobs and generating economic activity. The circular economy can also help to reduce costs by making more efficient use of resources.

The transition to a circular economy will require a coordinated effort from governments, businesses, and individuals. However, the benefits of the circular economy are clear, and the potential for economic growth and environmental protection is significant.

2.5.1 Challenges and Considerations

- Upfront Costs: While renewable energy costs have declined, initial investments can still be a deterrent for many. Innovative financing models and subsidies can help overcome this hurdle.
- Technological Evolution: The integration of renewable energy into existing grids requires advancements in grid technology, energy storage solutions, and demand-response systems.
- Circular Design and End-of-life Management: To truly embrace circularity, renewable energy infrastructure, such as solar panels and wind turbines, must be designed for recyclability. End-of-life management strategies are essential to ensure that they do not end up as waste.

Energy, with its ubiquitous influence on economic activities, is a cornerstone for the realization of a circular economy. By intertwining energy innovations with circular principles, societies can aspire for a sustainable model that harmoniously balances economic growth, environmental health, and social well-being. The symbiotic relationship between energy and circularity promises a resilient, regenerative, and prosperous future.

2.5.2 Resource Efficiency

The escalating pressures of global demand and the ever-present reality of finite resources have intensified the conversation around resource efficiency. While the general principle – getting more from less – is easy to grasp, the nuances, implications, and challenges require a more detailed exploration.

2.5.3 The Evolving Context of Resource Efficiency

- Historical Perspective
  Historically, human societies were limited by resources within their immediate environment. However, the onset of globalization and technological advances expanded access, leading to a surge in consumption patterns and a false sense of resource abundance.
- The Population and Consumption Conundrum
  The exponential growth in the global population and rising affluence have amplified consumption rates. This surge has highlighted the limits of our planet's resources, making efficiency a pressing concern.
- Environmental Feedback
  The extraction and exploitation of resources have environmental consequences, from habitat destruction to climate change. These feedback loops further emphasize the need for efficiency.

Unpacking the Principles

- Reduce: Beyond Minimalism
  Reduction is not merely about consuming less. It is about innovative approaches to product design, lean manufacturing, and waste minimization, all of which can yield the same output with fewer inputs.
- Reuse: The Second Life of Products
  Reusing goes beyond merely repurposing. In the modern context, it involves creating modular products that can be easily repaired, upgraded, or repurposed, extending their functional life.
- Recycle: Closing the Loop
  Modern recycling is not just about breaking down products but involves creating high-quality secondary materials that can replace virgin resources without compromising product integrity.
- Rethink: A Paradigm Shift
  This principle challenges the foundational ideas of consumption. It promotes the shift from ownership to access, valuing experiences over possessions and prioritizing long-term sustainability over short-term gains.

Socioeconomic Implications of Resource Efficiency

- Economic Resilience
  Resource-efficient industries can better navigate global market volatilities with diversified supply chains and reduced dependency on scarce materials. Economic resilience is about foresight, adaptability, and collective effort. While it is impossible to predict every challenge an economy might face, the ability to adapt and recover is a testament to its foundational strengths. In an era where global challenges are complex and multifaceted, building resilient economies is not just desirable but imperative for sustainable development and prosperity.
- Job Creation
  While certain traditional jobs may become obsolete, resource efficiency can birth new industries, from advanced recycling facilities to remanufacturing hubs.
- Social Equity
  Resource efficiency can promote more equitable distribution as goods become more durable and accessible, and waste reduction leads to a cleaner environment for all.

Challenges in the Path Forward: Navigating the Complexities of Resource Efficiency

Resource efficiency, while recognized as a critical tenet for sustainable development, presents numerous challenges in its widespread implementation. These hurdles range from technological limitations to deep-rooted societal behaviors. By understanding these challenges, stakeholders can develop strategies to navigate them effectively and pave the way for a more resource-efficient future.

A. **Technological Barriers**

- Research and Development Costs
  Developing modern technologies for resource efficiency, such as advanced recycling processes or materials with extended lifespans, requires significant upfront investment. This can deter businesses, particularly small and medium enterprises, from pursuing innovative solutions.
- Adoption Rate
  Even when resource-efficient technologies are available, their adoption can be slow. This can be due to a lack of understanding, infrastructure compatibility issues, or simply resistance to change.

B. Technological Obsolescence
The rapid pace of technological improvements has the potential to render recently implemented resource-efficient solutions obsolete, necessitating stakeholders to continually update and modify their approaches.

C. Economic Structures

- Market Inertia
  Existing market structures, particularly those built around a linear economy, can resist change. This is often due to entrenched interests, established supply chains, and short-term profit motives.
- Externalized Costs
  Many environmental and social costs associated with resource exploitation are not reflected in market prices. Without mechanisms such as carbon pricing or waste disposal fees, there is little financial incentive for businesses to adopt resource-efficient practices.
- Risk Aversion
  The transition to resource efficiency might involve uncharted territories for many businesses, leading to perceived or real risks. Without clear returns on investment or case studies highlighting success, many entities hesitate to embark on the resource efficiency journey.

D. Societal Norms and Behavior

- Consumption Patterns
  Consumer cultures, especially in affluent societies, often prioritize novelty over durability. The allure of the ‘new’ can discourage consumers from choosing long-lasting, repairable, or upgradable products.
- Awareness and Education
  Many consumers and businesses are unaware of the implications of their consumption patterns or the potential benefits of resource efficiency, underscoring the need for comprehensive education campaigns.
- Value Systems
  In some societies, consumption is deeply tied to status, identity, and self-worth. Challenging these values can be a daunting task, necessitating broader cultural shifts.

E. Policy and Regulation

- Inadequate Regulatory Frameworks
  Existing regulations might inadvertently favor wasteful practices, either through subsidies for unsustainable industries or a lack of support for resource-efficient alternatives.
- Enforcement Challenges
  Even when regulations promoting resource efficiency are in place, enforcing them consistently, especially across borders, can be challenging.
- Short-Term Political Cycles
  Political leaders, often working with short electoral cycles, may prioritize immediate concerns over long-term resource efficiency goals, leading to inconsistent policy directions.

F. Global Inequities

- Varying Capacities
  While developed nations might have the means to invest in resource efficiency, developing countries might struggle with basic needs, making it challenging to prioritize resource-efficient strategies.
- Transfer of Burdens
  Sometimes, resource-efficient practices in one country can shift environmental burdens elsewhere, as seen in the export of waste to countries with lax regulations.

The path to resource efficiency is fraught with complexities and challenges that require multidimensional solutions. Collaborative efforts spanning governments, industries, civil society, and consumers are crucial for overcoming these barriers and ushering in a future where resources are used optimally and sustainably.

2.6 Case Studies: Resource Efficiency in Action

The Ellen MacArthur Foundation The Ellen MacArthur Foundation is a non-profit organization that promotes the circular economy. The foundation has a number of case studies on its website that demonstrate the economic benefits of the circular economy. For example, one case study shows how the circular economy could save the European Union €600 billion annually by 2030.

The World Economic Forum The World Economic Forum is an international organization that brings together business, political, and other leaders to discuss global issues. The forum has several reports on the circular economy, including one that estimates that the circular economy could create 100 million new jobs by 2030.

The Circularity Gap Report The Circularity Gap Report is an annual report that assesses the progress of the circular economy. The report estimates that the global economy is only 8.6% circular and that there is a $108 trillion gap between current economic activity and a fully circular economy.

Scandinavian Approaches to Waste

Countries like Sweden have integrated resource efficiency into their waste management, where waste is viewed as a valuable resource, leading to high recycling rates and waste-to-energy initiatives. Resource efficiency, while conceptually simple, unravels layers of complexity when delved into deeply. It challenges not only the way products are designed, produced, and consumed but also societies' underlying values and structures. Adopting it does not just ensure resource sustainability; it paves the path to a more equitable, resilient, and sustainable global society.

Scandinavian countries are known for their progressive environmental policies, and waste management is no exception. Scandinavian countries employ many crucial strategies in managing waste:

- **Waste prevention:** Scandinavian countries prioritize waste prevention as a primary objective. This is achieved through a range of strategies, including product design, reutilization, and restoration.
- **Recycling:** The recycling rates in Scandinavian countries are among the highest globally. This is accomplished by the implementation of governmental programs, dissemination of public education, and development of infrastructure.
- **Waste-to-energy:** Scandinavian countries employ waste-to-energy facilities to transform waste into thermal energy and electrical power. This contributes to waste reduction by diverting waste from landfills and generating renewable energy.
- **Landfill:** Scandinavian countries enforce stringent landfill rules, which effectively mitigate the environmental consequences associated with waste disposal.

The implementation of these strategies has contributed to the remarkable success of waste management in Scandinavian countries. Denmark boasts a recycling rate of 50%, while Sweden has achieved a waste-to-energy rate beyond 50%. Below are concrete instances of how Scandinavian countries are putting these methods into practice:

- In Sweden, every household is obligated to segregate their waste into six specific categories: organic waste, paper, cardboard, metal, plastic, and glass. This aids in guaranteeing that a maximum amount of garbage is recycled or composted.
- In Norway, there is a levy imposed on the disposal of waste in landfills. This serves to deter individuals from depositing waste in landfills and promotes the practice of recycling or composting instead.
- In Finland, waste-to-energy plants receive government subsidies. This enhances the economic feasibility of these plants and promotes their growth.

The waste management strategies employed by Scandinavian countries serve as an exemplary model for other nations. By adhering to these methodologies, we can diminish waste, preserve resources, and safeguard the environment.

Sustainable energy and the circular economy are not just isolated solutions to distinct problems. Their synergies offer a pathway to a comprehensive, integrated approach to sustainable development. By recognizing and acting upon these synergies, stakeholders, from policymakers to industries, can pave the way for a future that harmonizes human aspirations with the planet's well-being. The economics of sustainable energy transition and the circular economy are intertwined concepts that aim to address environmental challenges, promote renewable energy adoption, and create a more sustainable and resilient economy. The circular economy supports the energy transition by conserving resources, reducing waste, and creating low-carbon materials. Collaboration between governments, businesses, and civil society is essential to address these challenges and harness the full potential of sustainable energy transition and the circular economy. Supportive policy frameworks, technological advancements, and investment in research and development can accelerate the transformation towards a more sustainable and circular future.

2.7 Environmental Benefits of Sustainable Energy

- **Reduction in Greenhouse Gas Emissions:** Renewable energy sources produce little to no greenhouse gas emissions during electricity generation, making them a crucial tool in the fight against climate change. By replacing fossil fuels with renewables, sustainable energy transition contributes to global efforts to limit global warming.
- **Air Quality Improvement**: Transitioning to sustainable energy reduces air pollution, benefiting public health by decreasing respiratory illnesses and other health-related issues associated with air pollutants from burning fossil fuels.
- **Conservation of Natural Resources**: Renewable energy harnesses inexhaustible natural resources, such as sunlight, wind, and water, reducing the demand for finite resources such as coal, oil, and natural gas.

2.8 Pathways for Sustainable Energy Transition

Globally, energy transitions are an ongoing process. While sustainable solutions are envisioned for the future, many societies still rely on a fossil fuel-based energy regime high in carbon emissions and pollution. Designing pathways towards a sustainable energy transition has garnered international attention. The integration of new energy technologies, environmental sciences, economics, and management is required to comprehend the possible transition pathways of the energy system (Chen et al., 2019).

The transition to sustainable energy requires a comprehensive and multifaceted approach involving technological advancements, policy support, and collaborative efforts from governments, businesses, and communities. This section explores the critical pathways for achieving a sustainable energy transition, focusing on the necessary strategies and actions to accelerate the shift to renewable and low-carbon energy sources.

A. **Promoting renewable energy deployment**

- **Scaling up Renewable Energy Capacity**: Governments and private sector stakeholders must work together to scale up the deployment of renewable energy technologies, such as solar, wind, hydropower, geothermal, and bioenergy technologies. This involves investing in constructing new renewable energy projects and upgrading existing infrastructure.
- **Supportive Policies and Incentives**: Supportive policies, such as feed-in tariffs, tax credits, and renewable portfolio standards, can incentivize investments in renewable energy and create a conducive environment for sustainable energy growth.
- **Removing Market Barriers**: Addressing regulatory barriers, grid integration challenges, and administrative hurdles can accelerate the adoption of renewable energy technologies. Streamlining permitting processes and ensuring fair market access are crucial to removing barriers.

B. **Energy Efficiency and Demand-Side Management**

- **Energy Efficiency Programs**: Improving energy efficiency across all sectors is vital for sustainable energy transition. Governments and businesses can invest in energy-efficient technologies and implement energy-saving programs to reduce energy consumption.
- **Demand-Side Management**: Encouraging demand-side management practices, such as demand response programmes and time-of-use pricing, can help balance electricity demand and reduce peak load, thus optimizing energy use.

C. **Decarbonization of Industries**

- **Electrification of Industries**: Promoting the electrification of industries, especially those with high energy consumption, can reduce carbon emissions. Transitioning from fossil fuel-based processes to electric-powered technologies can decarbonize sectors such as manufacturing and transportation.
- **Carbon Capture and Storage (CCS)**: Investing in CCS technologies can capture and store carbon dioxide emissions from industrial processes and power plants, allowing industries to reduce their carbon footprint while continuing essential operations.

D. **Innovative Technologies and Research**

- **Advancements in Energy Storage**: Energy storage technologies, such as batteries and pumped hydro storage, are crucial in integrating intermittent renewable energy sources into the grid. Investing in the research and development of energy storage solutions can enhance grid stability and reliability.

- **Green Hydrogen**: Developing and adopting green hydrogen technologies can enable the storage of renewable energy for use in sectors that are difficult to electrify, such as heavy industries and long-haul transportation.

E. **Fostering International Cooperation**

- **Global Collaboration**: The worldwide issue of achieving a sustainable energy transition necessitates international cooperation. Countries can engage in cooperation with the purpose of expediting the worldwide implementation of sustainable energy practices through research collaboration, technological transfer, and knowledge exchange.
- **Climate Agreements and Commitments**: International climate accords, such as the Paris Agreement, establish a structure for nations to establish goals for reducing emissions and collaborate in pursuit of a sustainable energy future.

F. **Public Awareness and Stakeholder Engagement**

- **Educating the Public**: It is crucial to inform the public about the advantages of sustainable energy and the pressing need for climate action in order to gain support and encourage changes in behavior.
- **Engaging Businesses and Communities**: Engaging businesses, communities, and local stakeholders in the transition process helps cultivate grassroots endorsement and instill a feeling of proprietorship in sustainable energy programs.

The achievement of sustainable energy transition necessitates a comprehensive and synchronized endeavor by all stakeholders. By promoting renewable energy deployment, enhancing energy efficiency, decarbonizing industries, fostering innovative technologies, and fostering international cooperation, societies can accelerate the shift towards a sustainable energy future. Public awareness and stakeholder engagement are vital in garnering support and driving collective action. Sustainable energy transition is not only a pathway to combat climate change but also a foundation for building a more resilient, inclusive, and prosperous global economy.

2.9 The Economics of Renewable Energy

- **Declining Costs and Competitiveness**: Advancements in technology, economies of scale, and supportive policies have significantly reduced the costs of renewable energy technologies. As a result, renewable energy sources are increasingly cost-competitive with conventional fossil fuel-based energy.
- **Job Creation and Economic Growth**: The transition to sustainable energy creates job opportunities across the renewable energy value chain, from manufacturing and installation to operations and maintenance. Renewable energy projects can spur economic growth, attract investments and drive innovation.
- **Investment and Financing Opportunities**: Sustainable energy transition requires significant investments in renewable energy infrastructure. Green bonds, project financing, and public-private partnerships are among the financing mechanisms available to support the scaling up of renewable energy projects.

2.10 Policy and Regulatory Support

- **Renewable Energy Targets and Incentives**: Governments worldwide are setting renewable energy targets and offering incentives to encourage the adoption of sustainable energy. Feed-in tariffs, tax credits, and renewable portfolio standards are examples of policy instruments driving the transition.
- **Carbon Pricing and Emissions Reduction**: Carbon pricing mechanisms, such as cap-and-trade systems and carbon taxes, incentivize businesses to reduce emissions and invest in cleaner energy options.
- **Grid Integration and Storage Support**: Supportive policies for grid integration and energy storage technologies are critical in ensuring the reliable and stable incorporation of intermittent renewable energy sources into the energy grid.

Transitioning to sustainable energy is an essential milestone in attaining a future that is both carbon-neutral and resilient to climate change. Through the adoption of renewable energy sources, the mitigation of greenhouse gas emissions, and the promotion of favorable policies and investments, societies may facilitate the shift towards sustainable energy and provide the foundation for a more sustainable and prosperous global community. The field of renewable energy economics showcases the compatibility of environmental conservation and economic advancement, providing a strong motivation to expedite the shift towards sustainable energy.

2.11 The Case for Sustainable Energy Transition

This section emphasizes the crucial need to transition from energy sources that rely on fossil fuels to sustainable alternatives. The analysis centers on the environmental, social, and economic benefits of implementing renewable energy, such as the mitigation of greenhouse gas emissions, enhanced energy resilience, and the creation of job prospects.

- **Economics of Renewable Energy:** An examination is conducted on the economic aspects of renewable energy sources, such as solar, wind, hydropower, and geothermal energy. This section explores the economic efficiency of renewable technologies, the decreasing expenses associated with them, and the possibility of renewable energy surpassing fossil fuels in the future.
- **Investment and Financing in Sustainable Energy**: This chapter explores the contribution of corporate and public sector investments in facilitating the transition to sustainable energy. This text explores different funding methods, such as green bonds, project financing, and public-private partnerships, to expedite the implementation of renewable energy projects.
- **Policy and Regulatory Frameworks**: This section examines the influence of favorable policy and regulatory frameworks on the shift towards sustainable energy. Case studies from several nations illustrate how policy incentives, carbon pricing, and renewable energy targets can effectively facilitate the transition.

2.12 Circular Economy Economics: Rethinking Consumption and Production

The analysis of contemporary economic systems is occurring within a time characterized by rapid technological advancements, global trade integration, and a growing acknowledgment of environmental limitations. Over the course of history, the worldwide economy has followed a linear model known as "take, make, dispose." In light of an increasing global population and depleting planetary resources, this approach, which depends on an ongoing supply of limited resources and results in waste, is becoming unsustainable.

The advent of the circular economy (CE) concept is a paradigm shift that seeks to redefine development by prioritizing the generation of beneficial societal advantages. The concept emphasizes a closed-loop system in which waste is avoided and resources are consistently reused, remanufactured, and recycled, thus separating economic activity from the consumption of limited resources. This concept aims to consistently optimize the usefulness and worth of products and materials, resulting in a self-regenerating system. The adoption of a circular economy model effectively tackles environmental issues while also creating a fresh avenue for economic expansion and fostering innovation. As we go further into this subject, we will examine the economic consequences, difficulties, and significant prospects that the circular economy presents for reconsidering conventional consumption and production patterns.

2.13 The Linear Economic Model: A Brief Overview

The underpinnings of contemporary industry and economic progress predominantly rely on a linear methodology. The linear economic model, commonly referred to as the "Take, Make, Dispose" formula, has long been influential in shaping consumption and manufacturing behaviors over the course of millennia. Although the paradigm has facilitated swift expansion and technical progress, its fundamental constraints are becoming more apparent in a society struggling with environmental issues and limited resources.

- Origins and Historical Context
- In order to understand the evolution of economic models, specifically the Linear Economic Model, it is necessary to examine the historical context and assess the sociopolitical and technological influences that contributed to its emergence and broad acceptance.
- Industrial Revolution
- The Industrial Revolution is responsible for the origin of the linear economic model. The development of machines and technology has led to the emergence of mass manufacturing and global trade, which has resulted in a substantial rise in resource extraction and a rapid expansion of consumer goods.

- Rise of Global Consumerism
- The upsurge in consumer demand can be attributed directly to the economic policies, globalization, and marketing strategies that arose in the aftermath of World War II. Massive amounts of products were produced to meet the worldwide demand for goods, solidifying the supremacy of the linear model.

2.13.1 Core Principles of the Linear Model

- Resource Extraction
  The first phase of the process involves obtaining the essential raw ingredients. Oftentimes, these resources are limited in quantity, such as minerals and fossil fuels, and their extraction can lead to disruptions in the environment.
- Production
  Raw materials undergo several transformative processes, which result in the production of either finished or semi-finished products. This stage is characterized by high energy use, frequently dependent on non-renewable resources, resulting in substantial waste and emissions.
- **Consumption**
  Following the manufacturing process, products are distributed to consumers for their consumption. The linear model thrives because it depends on rapid turnover rates, leading to a society marked by frequent replacement of products and disposability.
- Disposal

Following their utilization, products frequently become garbage. Although a percentage of this garbage can be recycled or repurposed, a significant amount ends up in landfills, incinerators, or the natural environment.

2.13.2 Rise of the Global Consumer Culture

- Post-World War Era
  After the devastation of World War II, many economies shifted their focus to rebuilding and growth. There was a surge in consumer demand, fuelled by increasing disposable incomes and aggressive marketing strategies.
- Globalization
  In the late twentieth century, we witnessed an acceleration in global trade. With improved transportation and communication technologies, products manufactured in one corner of the world found markets thousands of miles away.

- Disposable Culture

With the drop in production costs and the rise of consumerism, a culture characterized by disposability emerged. The manufacturing of products shifted away from durability and longevity and instead embraced a culture of regular replacement, thereby establishing the "take, make, dispose" paradigm.

- Contemporary Reflections

  Currently, the repercussions of the linear model, from environmental degradation to resource depletion, are becoming starkly visible. While it is undeniable that this model catalyzed unparalleled economic growth and technological progress, it also sowed the seeds for many of the sustainability challenges we face today.

As we stand on the cusp of potential new economic shifts, understanding the origins and historical context of our past models becomes crucial. It not only provides insights into the patterns of human development but also offers lessons for charting a sustainable and inclusive future.

Socioeconomic Implications The socioeconomic implications of the circular economy are complex and multifaceted. Some of the potential benefits include the following:

- **Economic Growth**: The linear model has undoubtedly contributed to GDP growth in many countries. Mass production and consumption have led to job creation, infrastructure development, and increased corporate profits.
- **Job creation:** The circular economy could create new jobs in areas such as recycling, repair, and remanufacturing.
- **Built-in Obsolescence**: In order to maintain a consistent level of consumer demand, numerous products are intentionally created with finite lifespans. Although this particular approach yields economic advantages, it concurrently exacerbates issues related to waste generation and resource inefficiencies.
- **Externalities:** The linear model frequently fails to adequately consider the environmental and social costs associated with its operations. Pollution, resource depletion, and labor exploitation are externalities that are not well accounted for, yet they possess substantial ramifications for both society and the environment.
- **Reduced poverty**: The circular economy could help to reduce poverty by providing new opportunities for employment and income generation.
- **Improved environmental quality**: The circular economy could help to improve environmental quality by reducing pollution and resource consumption.
- Greater social equity: The implementation of a circular economy has the potential to foster a fairer society by offering inclusive economic participation for all individuals.

2.13.3 Challenges and Unsustainability

- **Resource Scarcity:** The scarcity of numerous raw materials is a significant barrier. Diminishing resource supply can lead to price increases, which in turn can cause economic volatility.
- **Environmental Degradation:** The linear model often results in environmental harm across the whole life cycle, commencing with extraction and concluding with disposal. These factors include the degradation of natural habitats, the emission of toxic pollutants, and the accumulation of waste items.
- **Economic Vulnerabilities:** Economies reliant on a continuous stream of finite resources are susceptible to supply chain breakdowns, price changes, and geopolitical tensions.
- **Overcoming Transition Costs:** To surmount the expenses associated with switching to sustainable models, a comprehensive strategy encompassing financial incentives, regulatory backing, public education, alliances, and innovation is necessary. Additionally, it entails acknowledging the enduring advantages and financial advantages that these environmentally friendly models provide, thereby presenting a persuasive argument for the shift despite the initial expenses.
- **Social and Employment Implications:** The transition towards sustainable energy and circular economies has significant social and employment ramifications. To tackle these problems, it is necessary to develop well-thought-out plans that effectively manage environmental objectives while also promoting social fairness and employment prospects. This will ensure that the transition is advantageous and just for all sectors of society.
- **Externalities and Metrics**: Efficiently handling external factors and establishing strong measurement systems are essential for directing and evaluating the shift towards sustainable energy and circular economy models. These tools offer the essential information and understanding required to make well-informed decisions that effectively address environmental, social, and economic goals.

The socioeconomic ramifications of the circular economy are favorable. Nevertheless, certain obstacles must be confronted in order to fully unlock the potential of the circular economy. Below are few concrete illustrations of how the circular economy could influence various social demographics:

- **Workers in the waste and recycling sector:** The adoption of a circular economy has the capacity to create employment prospects in the waste and recycling sector, as it would require a larger workforce to collect, classify, and treat materials for recycling objectives.
- **Consumers:** Consumers stand to gain advantages from the circular economy through increased availability of environmentally friendly products and services. In addition, cost savings can be achieved by repairing and repurposing items rather than purchasing new ones.

- **Businesses:** Businesses can reap advantages from adopting the circular economy model through cost reduction and enhanced environmental performance. Additionally, they have the potential to establish fresh markets for recycled materials and products.
- **Governments:** Governments stand to gain advantages from adopting the circular economy model, since it enables them to mitigate environmental expenses and enhance public health. In addition, they have the potential to generate employment opportunities and stimulate economic expansion.

The circular economy holds the capacity to yield advantages for all socioeconomic strata. Nevertheless, it is crucial to guarantee that the shift towards a circular economy is just and impartial. Consequently, it is imperative that all individuals have the chance to engage in the circular economy and reap its advantageous consequences.

2.14 Policy Recommendations and Conclusion

In order to facilitate a smooth and effective transition, it is advisable to implement a series of measures:

(a) The internalization of externalities can be facilitated through the implementation of mechanisms such as carbon pricing, which effectively increases the costs associated with unsustainable practices.
(b) To foster the advancement of sustainable technologies, governments can allocate resources towards research and development (R&D) endeavors. By doing so, they may effectively reduce the costs associated with these technologies.
(c) Education and training are essential for equipping the workforce with the necessary skills to adapt to and succeed in the evolving economic paradigms.
(d) Collaborative approaches involve stakeholders from several sectors, including industry, civic society, and academia, to develop comprehensive solutions.

The economics of sustainable energy transition and the circular economy are intricately linked. The circular economy can help reduce energy transition costs by providing low-carbon materials and resources. For example, recycled aluminium emits up to 95% less carbon dioxide than that from virgin sources. Building energy transition infrastructure from secondary materials will help transition to net zero. Businesses and governments can drive the transition towards a more sustainable and circular future by understanding the economic benefits, financing mechanisms, and policy incentives. The integration of sustainable energy transition and the circular economy is not just a moral imperative but an economic one. By harnessing the synergies between these paradigms, societies can pave the way for resilient, sustainable, and inclusive economies that benefit both the planet and its inhabitants. A brighter, circular future awaits as technology, policy, and global cooperation evolve.

While the linear economic model has been instrumental in shaping the economic landscape of the modern world, its limitations, especially concerning sustainability, are becoming starkly apparent. Amidst the global pursuit for solutions to urgent environmental problems, there is a growing focus on comprehensive and sustainable economic models that can benefit both humans and the planet in the long run.

Nevertheless, the economics of the circular economy also present certain obstacles. Transitioning from a linear economy to a circular one might provide challenges. This is due to the fact that the linear economy is firmly entrenched within our existing economic system. In addition, the implementation of the circular economy may necessitate initial capital outlay, which might be a challenge for certain businesses. Notwithstanding these limitations, the economic prospects of the circular economy are progressively advantageous. The price of renewable energy is decreasing, and the advantages of decreasing pollution and resource use are becoming increasingly evident. In the future, the circular economy will likely have a more prominent part in the transition to sustainable energy. Some of the concrete instances of how the circular economy can be implemented in the energy transition::

- Recycled materials are used to make solar panels and wind turbines.
- Repairing and refurbishing electric vehicles instead of replacing them.
- Developing new business models that encourage product sharing and leasing.
- Designing products that are easy to disassemble and recycle.

The economics of transitioning to renewable energy and implementing a circular economy are intricate and always changing. Nevertheless, the available evidence indicates that these two strategies can complement each other and contribute to the establishment of a sustainable and successful future. Through the allocation of resources towards renewable energy, energy efficiency, and the circular economy, we have the potential to establish a more pristine, salubrious, and fair world that benefits everyone.

References

Bocken, N. M. P., de Pauw, I., Bakker, C., & van der Grinten, B. (2016). Product design and business model strategies for a circular economy. *Journal of Industrial and Production Engineering, 33*(5), 308–320. https://doi.org/10.1080/21681015.2016.1172124

Chen, B., Xiong, R., Li, H., Sun, Q., & Yang, J. (2019). Pathways for sustainable energy transition. *Journal of Cleaner Production, 228*, 1564–1571. https://doi.org/10.1016/j.jclepro.2019.04.372

Ellen McArthur Foundation. (2019). Circular economy in cities. In *Ellen MacArthur Foundation*. https://ellenmacarthurfoundation.org/circular-economy-in-cities

IBM. (2022). *The future of energy business in a circular economy*. https://www.ibm.com/downloads/cas/5WKLAP2D

International Energy Agency. (2022). The role of critical minerals in clean energy transitions. In *World Energy Outlook Special Report*. https://doi.org/10.1787/f262b91c-en.

Kabeyi, M. J. B., & Olanrewaju, O. A. (2022). Sustainable energy transition for renewable and low carbon grid electricity generation and supply. *Frontiers in Energy Research, 9*(March), 1–45. https://doi.org/10.3389/fenrg.2021.743114

Martínez-Alier, J. (2018). Circularity gap. In *Dictionary of ecological economics* (pp. 61–62). https://doi.org/10.4337/9781788974912.c.26.

Marwa, M., Euchi, J., & Jerbi, Y. (2020). A review on optimization modeling of hybrid energy systems. In *Advances in logistics, operations, and management science* (pp. 29–62). https://doi.org/10.4018/978-1-7998-0268-6.ch003.

Mayes-Ramírez, M. M., Gálvez-Sánchez, F. J., Ramos-Ridao, Á. F., & Molina-Moreno, V. (2023). Urban waste: Visualizing the academic literature through bibliometric analysis and systematic review. In *Sustainability* (Vol. 15, issue 3). https://doi.org/10.3390/su15031846.

Nwokolo, S. C., Meyer, E. L., & Ahia, C. C. (2023). Credible pathways to catching up with climate goals in Nigeria. In *Climate* (Vol. 11, Issue 9). https://doi.org/10.3390/cli11090196.

Romero-Castro, N., Ángeles López-Cabarcos, M., Miramontes-Viña, V., & Ribeiro-Soriano, D. (2023). Sustainable energy transition and circular economy: The heterogeneity of potential investors in rural community renewable energy projects. *Environment, Development and Sustainability.* https://doi.org/10.1007/s10668-022-02898-z

Schmid, A. G., & Mendoza, J. M. F. (2022). Circular economy and the clean energy transition. *Futureearth..* https://futureearth.org/2022/12/16/circular-economy-and-the-clean-energy-transition/

Solomon, B. D., & Krishna, K. (2011). The coming sustainable energy transition: History, strategies, and outlook. *Energy Policy, 39*(11), 7422–7431. https://doi.org/10.1016/j.enpol.2011.09.009

Velasco-Muñoz, J. F., Mendoza, J. M. F., Aznar-Sánchez, J. A., & Gallego-Schmid, A. (2021). Circular economy implementation in the agricultural sector: Definition, strategies and indicators. *Resources, Conservation and Recycling, 170*, 105618. https://doi.org/10.1016/j.resconrec.2021.105618

World Economic Forum. (2020). *Forging ahead: A materials roadmap for the zero-carbon car* (Issue December). http://www3.weforum.org/docs/WEF_Forging_Ahead_2020.pdf

Chapter 3
Energy Efficiency and Renewable Energy Technologies

Abstract Implementing energy-efficient techniques and adopting renewable energy technology are essential for facilitating the shift towards a sustainable energy system. This chapter thoroughly examines a range of technologies and tactics that can be employed to improve energy efficiency and encourage the adoption of renewable energy sources. The chapter commences with an examination of the advantages of energy efficiency, including diminished energy expenditures, enhanced energy dependability, and decreased emissions of greenhouse gases. Afterwards, it explores many approaches to improve energy efficiency, including energy audits, building codes and regulations, and energy management systems. The following chapter examines renewable energy technologies, specifically exploring the economic and environmental benefits of solar, wind, hydropower, and geothermal technology. A detailed exposition is presented on the many types of renewable energy technology, along with a thorough evaluation of the advantages and disadvantages linked to each. The chapter concludes by analyzing the challenges and potentialities associated with the widespread adoption of energy efficiency and renewable energy technology. This includes the need for regulatory frameworks that offer assistance, avenues for funding, and collaborations between the public and private sectors to encourage the adoption and monetization of the technology.

Keywords Energy efficiency · Renewable energy · Sustainable development · Climate change · Fossil fuels · Greenhouse gas emissions · Green building practices · Solar Photovoltaics (P.V.) · Solar thermal · Wind turbines · Hydropower · Smart grids · Energy storage · Microgrids · Environmental impact

3.1 Introduction

There is a strong need in today's world to actively pursue sustainable development and effectively tackle the negative consequences of climate change. Therefore, it is crucial to shift towards energy efficiency and the adoption of renewable energy sources. The utilization of energy, primarily obtained from fossil fuels, is a

V. Kandpal et al., *Sustainable Energy Transition*, Circular Economy and Sustainability, https://doi.org/10.1007/978-3-031-52943-6_3

significant contributor to the production of greenhouse gas emissions worldwide. Utilizing sustainable energy sources has the capacity to reduce our carbon emissions and offer a dependable long-term energy supply. Energy efficiency is the practice of achieving equivalent or superior outcomes while consuming less energy. The fundamental concept underlying it is to minimize energy use while maintaining or improving the quality of the output or service. The primary goals of energy efficiency are to minimize energy waste, mitigate energy costs, and alleviate the environmental impacts associated with energy generation and utilization.

Energy has a crucial role in driving economic growth in regions. Indeed, it is a major contributor to climate change. As a result, as the Economy and population grow rapidly, countries significantly contribute to their power generation and consumption (Alvarado et al., 2020). The energy sector is currently confronted with escalating issues pertaining to rising demand, enhanced efficiency, insufficient analytics necessary for optimal management, and shifting supply and demand patterns. Artificial intelligence is integrated into renewable energy systems through essential elements such as energy forecasting, energy efficiency, and energy accessibility (Chen et al., 2021). Heat pumps, encompassing both air source and ground source variations, enable the movement of heat into and out of a system. In contrast, air compression systems and chillers offer energy-saving solutions through comprehensive evaluations and efficient operation. The incorporation of these systems is vital in assessing the thermal comfort and indoor environmental quality of buildings, hence establishing their importance in generating sustainable and comfortable living and working environments. Enhancing energy efficiency is vital for decreasing energy usage (Tsemekidi et al., 2023).

3.2 Energy Efficiency

Energy efficiency is the best way to use energy to provide a service that could have been provided using a more conventional, less efficient method. Energy efficiency is the practice of lowering energy needs while still producing the necessary amount of energy. Another way to describe energy efficiency is "efficient energy in use." This entails reducing energy consumption while achieving the same activity or output. It is imperative to diminish greenhouse gas emissions, diminish energy use, and reduce expenses for enterprises and households.

A comprehensive energy policy encompasses strategies that foster the promotion of energy efficiency. The attainment of energy efficiency can be accomplished through the use of renewable energy sources such as solar, wind, or hydropower, or by modifying the power requirements in a manner that reduces the overall energy consumption without compromising the output. The attainment of effective energy utilization can be ascribed to developments in technology, improved system design, and alterations in human behavior.

For instance, using suitable insulation material at the right places enables a building to maintain a pleasant temperature while consuming less energy for heating and cooling. These features encompass LED lighting, high-efficiency heating and cooling systems, and appliances that are Energy Star-rated.

Following the right energy-efficiency procedures lowers the energy cost for businesses, thus saving money for companies and their clients. Additionally, the environment benefits since greenhouse gas emissions are decreased and environmental corrosion attacks on equipment, buildings, and structures are less likely.

Enhanced insulation, energy-conserving windows, and intelligent architectural designs have the potential to substantially diminish energy usage in residential and commercial structures. Industries have the ability to implement more streamlined manufacturing processes, while the transportation sector can transition to using more fuel-efficient vehicles and public transit systems.

Importance of Energy Efficiency

Energy efficiency benefits the environment by requiring less energy to complete specific tasks. By doing so, it may be possible to lessen the air and water pollution that some forms of energy production generate while preventing harm to essential ecosystems. The strain on the electrical grid may also be reduced. The reduction of energy use is of utmost importance due to its significant implications across various interconnected domains, including the environment, economy, and society. There exist several critical rationales for the significance of reducing energy consumption:

A. **Environmental Impact**:

Environmental protection comprises a diverse array of practices that are designed to maintain and preserve the natural environment from harmful human activity. The notion encompasses the mitigation of degradation, the preservation of resources, and the advancement of sustainable ecosystems. The primary objective is to secure the long-term sustainability of our planet for future generations while preserving its current level of vibrancy, biodiversity, and overall welfare.

- **Reduction in Greenhouse Gas Emissions**:
  Burning fossil fuels for energy generation releases substantial quantities of carbon dioxide and many other greenhouse gases. These gases can retain thermal energy inside the Earth's atmosphere, resulting in the phenomenon of global warming and subsequent alterations in climatic patterns. By implementing energy conservation measures, it is possible to decrease these emissions and contribute to the amelioration of climate change.
- **Decreased Air Pollution**:
  The generation of energy, mainly derived from non-renewable sources such as coal, contributes to the emission of pollutants that have a detrimental impact on air quality. The aforementioned phenomenon has the potential to induce respiratory ailments and inflict damage on the natural surroundings.

The preservation of the environment is a collective obligation. Each and every individual, community, and nation bears a responsibility to safeguard the habitability and vitality of our planet for future generations. Despite several hurdles, the collaborative endeavors of governments, organizations, communities, and individuals on a global scale have the potential to yield significant outcomes.

B. **Cost Savings:**
Implementing energy-efficient and weatherization measures within residential settings, such as the incorporation of insulation, adoption of LED lighting, and installation of heat pumps, can yield financial savings on energy expenditures while concurrently augmenting overall comfort levels. Energy-efficient buildings are characterized by lower costs for heating, cooling, and overall operation, while factories can achieve cost savings in the production of commodities. The adoption of transportation modalities characterized by reduced energy consumption leads to substantial fuel preservation.

C. **Community Benefits:**
Energy-efficiency initiatives play a significant role in bolstering community resilience and addressing energy inequity. These initiatives achieve this by facilitating the deployment of cost-effective and efficient technology and infrastructure in historically underserved communities, with a specific emphasis on communities of color. These specific geographic areas experience a greater energy burden, defined as the proportion of household income dedicated to energy costs, and are disproportionately affected by air pollution.

D. **Resilience and Reliability.**
Energy-efficiency upgrades decrease the load or the total amount of electricity on the grid simultaneously, reducing congestion and stress on the U.S. electric grid. Power outages are avoided with less load.

E. **Health Benefits:**
Reduced use of fossil fuels leads to cleaner air, water, and land, all of which have an immediate impact on human health, especially for those living in underserved areas and those whose conditions are made worse by pollution.

F. **Technologies in energy efficiency.**
Numerous energy-efficient devices are frequently installed and used in effective energy-efficiency programs. Technology can decrease energy consumption and increase energy reliability, which can help various industries.

In order to reduce energy consumption, a variety of technological advancements and techniques focused on enhancing energy efficiency can be employed. Presented below is an illustrative compilation of commonly seen options.

Follow the instructions provided in 4.0 Developing Programs to determine the technologies that are most appropriate for any given program or application. Energy savings, accessibility, supportive legislation, and other crucial considerations all play a role in determining the optimum technological option.

Follow the instructions provided in 4.0 Developing Programs to determine the technologies that are most appropriate for any given program or application. The most effective technology is chosen based on a mix of energy savings, accessibility, and supportive laws.

3.3 Building Envelope

An enclosed building's exterior is known as the building envelope. The envelope is crucial to preserving a dry, heated, or cooled internal environment because it protects a building's interior from the elements. The building's roof, walls, and foundation, as well as the window assemblies, make up the building envelope.

Insulating the walls and roof Insulation for the roof and walls can aid in lowering the energy needed to cool, heat, or change humidity. The rate of heat transfer to the outside environment is slowed by insulation. Insulation is often given an "R-value" rating; higher values correspond to insulation having better qualities. Air Infiltration is lessened Including heating and cooling ducts, sealing gaps, cracks or other leakage spots in the building envelope can reduce energy losses.

Lighting
Energy efficiency is crucial to lighting, which is frequently the first area targeted for improvements. Due to its affordability and price alignment with other cost-effective solutions like compact fluorescent lights (CFLs), light-emitting diode (LED) bulbs are currently leading the way. With a lifespan that is more than three times longer, significant energy savings of up to 90%, and increased light quality, LEDs outperform CFLs and incandescent bulbs. In areas with plenty of natural light or periods of emptiness, LEDs significantly save energy use when combined with daylight and occupancy sensors. LEDs compete with compact fluorescent lamps due to their efficiency, although they use substantially less electricity and last 8 to 15 times longer than incandescent lamps. CFLs do, however, contain mercury, which makes disposal difficult. High-performance T8 (HPT8) systems offer energy savings and enhanced light quality, making them suited for commercial applications. They combine electronic ballasts with long-life T8 bulbs. High-intensity discharge (HID) lamps illuminate large indoor and outdoor spaces because of their long lifespan and great efficiency, while high-output T5 bulbs (T5HOs) release twice as much light as T8 bulbs, making them perfect for structures with lofty ceilings.

Appliances
Energy-saving equipment, like dehumidifiers with efficient parts, high-capacity washers, and clothes dryers with moisture sensors, can lower energy usage without sacrificing functionality. Reduced energy costs and lessening of the environmental impact are made possible by refrigerators with increased insulation and compressors, dishwashers with dirt sensors, advanced filtration, effective jets, and dishwashers. Motion-detecting vending machines, switch-mode power supplies, "smart" chargers, and energy-saving modes are all used in electric and electronic equipment to further improve energy efficiency. These gadgets help users save money while promoting sustainability by reducing greenhouse gas emissions and energy use. Make environmentally responsible decisions while shopping for energy-efficient products and seek the ENERGY STAR label.

3.4 HVAC (Heating, Ventilation and Cooling Systems)

Heating and cooling, or HVAC (Heating, Ventilation, and Air Conditioning), are indispensable for ensuring occupant comfort and productivity in various climates and building types. Effective ventilation systems can also enhance indoor air quality, positively impacting occupants' health. HVAC technologies encompass a wide array of applications, serving residential, commercial, and industrial needs. Air conditioning, akin to refrigeration, regulates indoor air temperature and humidity for improved comfort and air quality. Boilers and furnaces powered by gas or oil, heat water or air are rated by annual fuel utilization efficiency (AFUE). Heat pumps, including both air source and ground source variants, facilitate the transfer of heat in and out of a system. Conversely, air compression systems and chillers provide energy-efficient alternatives by means of thorough assessments and effective functioning. The integration of these systems plays a crucial role in determining the thermal comfort and indoor environmental quality of buildings, hence establishing their significance in creating sustainable and comfortable living and working environments.

Water Heating

Water heaters are divided into groups based on their fuel (electric, natural gas or solar) and how they heat the water (tank or on-demand, tankless systems). Electric storage water heaters, sometimes referred to as heat pump water heaters, use effective heat pumps to transport warmth from the outside air to a storage tank. They also feature energy-saving modes and timers for increased efficiency. On the other hand, gas storage water heaters use a burner to heat a steel tank; high-efficiency models have improved insulation and condensing technology. Whole-home gas tankless water heaters use secondary heat exchangers to increase efficiency while providing hot water on demand, preventing standby energy loss. While this is going on, solar water heaters use a variety of collector designs to heat water using solar thermal energy, which helps make them environmentally friendly and energy-efficient hot water solutions.

Energy Efficiency and Digitization

Digitalization encompasses the widespread adoption of information and communication technology (ICT) throughout all sectors of the economy. This has led to an increase in the volume of data, rapid progress in advanced analytics, and enhanced communication between individuals, devices, and machines (including machine-to-machine interaction) (Taghizadeh-Hesary & Zhang, 2023). Enhancing energy efficiency is crucial to decrease energy usage and carbon emissions. Digital technology is believed to significantly enhance energy efficiency. Examining the correlation between digitization and energy efficiency can provide a useful benchmark for the global attainment of the carbon neutrality objective. Developed nations have already achieved a level of advancement where the implementation of digitalization can enhance energy efficiency. In contrast, poorer nations have not yet attained this level of progress (Lin & Huang, 2023).

The widespread adoption of autonomous vehicles and the incorporation of sensor technology in oil and gas reservoirs illustrate the significant impact of digitalization on worldwide energy production and consumption. The impact of digitalization on demand is intricate. Digital gadgets have the potential to enhance energy efficiency in the building, transportation, and industrial sectors. However, if not effectively controlled, the widespread usage of additional devices—and the corresponding rise in servers required to store the data they generate—may lead to substantial overall growth in energy use.

The process of digitalization is expected to continue without interruption. The primary responsibility of policymakers is to guide the energy system in a manner that optimizes favorable outcomes while minimizing detrimental repercussions. Digitalization has the potential to enhance energy efficiency by utilizing technology that gathers and analyzes data to make improvements in the physical environment, either automatically or with human participation.

Sensors and smart meters, for instance, gather data on energy consumption and other variables (such as the surrounding conditions) that impact energy usage. Data analysis technology, such as artificial intelligence algorithms, is utilized to convert data into meaningful information. The data is subsequently transmitted to machinery capable of implementing tangible alterations to enhance energy efficiency. Certain devices require human involvement to optimize energy utilization. For example, a smartphone application may suggest a route to work that is environmentally beneficial, but the commuter is required to adhere to it. Additional devices have the capability to enhance energy efficiency with greater autonomy: For instance, let us consider the utilization of switches in a building's cooling system or the implementation of robots in a manufacturing line (Fig. 3.1).

The utilization of digital technologies is already widespread across all energy end-use sectors. Residential and corporate buildings commonly have intelligent energy management systems and smart appliances installed. The industrial industry is increasingly adopting advanced robots and 3D printing technologies. The future energy consumption in the transportation sector will be determined by the interplay between automated, connected, electric, and shared (ACES) mobility.

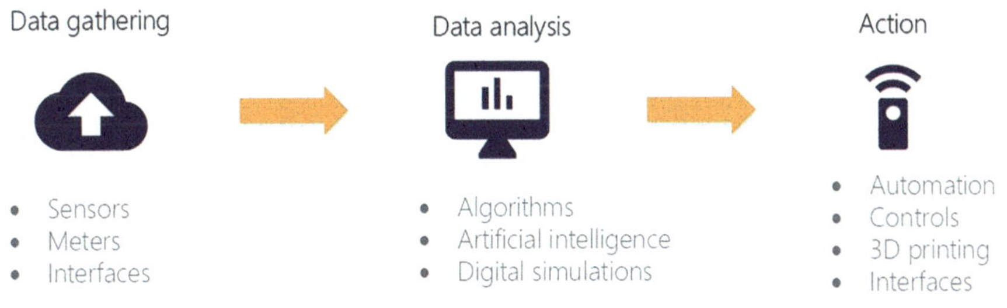

Fig. 3.1 How digitalization can improve efficiency through a combination of technologies

Various energy-intensive operations, such as the manufacturing of industrial items and the cooling of residential residences, have the potential to utilize digital technology to maximize energy efficiency. This represents an enhancement in energy efficiency as generally defined: a reduction in the quantity of energy consumed per unit of activity. The progress of global energy transitions remains highly reliant on the substantial enhancement in end-use efficiency, yielding benefits for both established and emerging nations.

Nevertheless, the benefits of enhanced connectivity brought about by digitalization enable digital technologies to enhance both the efficiency of end-use and the overall efficacy of the energy system.

The integration of both centralized and decentralized variable renewable energy sources into the grid is causing a substantial transformation in the worldwide energy systems. Moreover, there is a noticeable shift towards the adoption of electric power for energy consumption and a rise in the population of "prosumers," those who actively engage in both energy consumption and production. It is essential to optimize the energy system's efficiency by supplying energy when it is required and consuming it when it is accessible. In this perspective, demand-side flexibility is increasingly vital.

Digitalization enables the integration of "smart" infrastructure, such as "smart" homes, "smart" cars, and "smart" factories, which introduce additional sources of fluctuating energy demand into the energy system. This can encourage the utilization of energy generated "behind the meter" by communities and mitigate the curtailment of renewable energy on the supply side. An energy system that incorporates a higher proportion of renewable sources and emphasizes community self-consumption will exhibit enhanced efficiency, owing to reductions in energy production and distribution losses.

Aligned with an escalated demand for diminishing carbon emissions and mitigating the impacts of climate change, scientists and experts have turned to information communication technology (ICT) and digitalization as pivotal instruments for enhancing the effectiveness of energy utilization and generation (Charfeddine & Umlai, 2023). Digital technologies have the capacity to enhance the efficiency of both the system and end-use of energy. This has several advantages for the overall energy system, including a reduction in the requirement for energy infrastructure investments, such as peaking plants. Additionally, it facilitates the integration of renewable energy sources and contributes to the improvement of energy security, among other positive impacts.

3.5 Cost of Energy Efficiency

Energy efficiency enables the reduction of energy prices by decreasing the requirement for costly new power generation or transmission capacity and by alleviating the burden on energy resources. The decline in demand for energy services in many sectors could lead to a decrease in energy prices.

Decreased consumption resulting from improved energy efficiency can contribute to a reduction in prices. This holds especially true if energy efficiency programs, such as fuel economy requirements for automobiles, are widely and significantly adopted. Changes in demand in a single location may not have a substantial impact on energy prices for certain energy sources, such as oil, which are traded globally as commodities. Nevertheless, if endeavors to enhance energy efficiency liberate the availability of energy sources and bolster the security of the energy supply, localized limitations on supply may lead to alterations in local energy expenses. Increased interconnectivity in international energy markets has the potential to reduce the cost of energy commodities globally through the implementation of energy efficiency programs.

Price elasticities are employed in economic models to illustrate the response of consumers and businesses to fluctuations in the prices of products and energy. Policymakers must comprehend the fundamental pricing elasticities and model assumptions, as these factors can exert a substantial influence on the results of macroeconomic analysis. The duration of fluctuations in energy prices has a substantial impact on behavior, particularly when there are fluctuations in revenues and return on capital. This reduces the motivation to purchase new stocks or invest in existing stocks. Research suggests that the long-term impacts of rising energy bills or taxes could be three to four times more significant than the short-term effects. The clients of these services exhibit a reduced propensity to respond to abrupt changes in prices.

Through the utilization of PV/T technology and the implementation of optimization methodologies, it is feasible to attain substantial enhancements in the cost-effectiveness, net output power, and exergy efficiency of a system. These findings enhance the progress of sustainable energy solutions and offer useful insights for constructing efficient renewable energy systems in diverse climates (Assareh et al., 2023).

3.6 Energy Efficiency for Cost-Effective Service

Enhancing the efficiency of the energy supply sector enables energy suppliers to offer improved service to clients, decrease operational expenses, boost profit margins, and mitigate risks. Utility companies that encourage energy efficiency among their customers can achieve significant cost savings by delaying or canceling costly system upgrades and avoiding the necessity of investing in energy generation, transmission, and distribution infrastructure. Additional benefits include enhanced system reliability and decreased price fluctuations in wholesale markets. Providers can see indirect benefits when consumers receive cheaper energy services, as this can lead to a decrease in utility arrears and associated administrative expenses.

To maximize the advantages gained from energy efficiency, utility companies can implement strategies that efficiently decrease energy usage, prioritize reducing and shifting loads, offer consumers valuable guidance on optimal energy-saving methods, and simplify their access to financial incentives. Potential future endeavors include the dissemination of energy-efficient products and their subsequent installation. Promoting the adoption of information and communication technology (ICT) tools can facilitate the shift towards more energy-efficient energy consumption, benefiting both energy providers and their customers. This would enable consumers to effectively regulate their energy consumption while empowering providers to actively monitor, consolidate, and regulate end-user loads (IEA, 2019).

3.7 Energy Efficiency and Green House Gasses

Energy efficiency provides a wide range of environmental advantages. The reduction of greenhouse gas emissions (GHG) is a significant consequence, including both primary emissions from the burning or utilization of fossil fuels, as well as secondary emissions originating from energy generation.

An effective strategy to address the rising energy demand caused by the growing need for materials is to improve material efficiency. This entails providing the same material service while minimizing overall production. To ensure constant customer service, one can improve efficiency across the entire production and utilization value chain by implementing several tactics. These tactics include lowering the weight of the product, minimizing yield losses in manufacturing, finding new uses for scrap materials without melting them down, reusing and recycling components, and creating more durable product components, among other approaches.

To achieve a reduction in energy demand, measures can be used that attempt to decrease the demand for materials and products requiring high energy inputs. The utilization of reclaimed waste in the manufacturing of metals such as steel, aluminium, and copper results in a substantial decrease in energy consumption, ranging from 60% to 90% reduction, in comparison to the energy-intensive procedure of extracting these metals from metal ores.

Enhancing material and energy efficiency concurrently leads to improved efficiency across the entire production value chain. The differentiation between energy efficiency and material efficiency lies in the production input. Energy efficiency and material efficiency are usually synergistic, mutually reinforcing one other. An automobile's structure utilizes a reduced amount of steel, leading to energy conservation and the creation of a lighter, more fuel-efficient vehicle during operation. Conversely, there exist trade-offs between energy and material efficiency. Extending the durability of appliances that contain steel will result in users adopting more energy-efficient equipment in the future.

3.8 Short-Term Costs of Technologies to Reduce GHG Emission

Economists assess the initial expenses and calculate the ratio between these costs and the amount of carbon dioxide (or comparable) emissions prevented to ascertain the immediate expenses of reducing greenhouse gas emissions. Imagine a situation where a government allocates $20 million to incentivize the development of wind farms with the purpose of generating electricity and reducing carbon dioxide emissions by one million tons. The cost of mitigation is projected to be $20 per ton soon. This strategy enables a useful approach to compare the costs of different emission-reduction technologies.

Undoubtedly, exercising prudence is crucial when evaluating conclusions that solely focus on a single technology or strategy. Interactions between regulations may occur, and the price of a technology can vary based on its specific usage and location. Furthermore, the assessments of these expenses are revised on a yearly basis. Undoubtedly, in the past decade, the price of solar and wind energy has experienced a substantial decline, and this pattern is expected to persist.

The most noteworthy finding is that renewable energy solutions are among the most cost-effective. When considering implicit or explicit subsidies, the cost of wind and solar energy may be lower. Nevertheless, these forecasts neglect to consider the sporadic nature of renewable energy generation, as the presence of sunlight and wind is not continuous. To effectively address increased demand, it is crucial to supplement renewable energy sources with storage technologies like batteries or pumped hydroelectric storage. Furthermore, there should be an alternative method of producing energy that is easily accessible to quickly offset any variations in the availability of wind or solar power.

A power station in the United States that combines gas and steam turbines to improve efficiency offers a financially feasible and ecologically sound alternative to coal. The natural gas combined cycle generation approach involves harnessing the plentiful and economical supply of fracked shale gas. There is one important condition to consider regarding the estimated cost of $27 per ton. This estimate is based on the premise that there are no methane leaks happening from wells, pipelines, or storage facilities. The occurrence of methane, a potent greenhouse gas, during the Aliso Canyon incident in California in 2015 emphasizes the possibility of elevated greenhouse gas emissions and subsequent greater expenses per metric ton of all greenhouse gases linked to natural gas production.

3.9 Energy Audit: Types and Methodology

An energy audit serves as the fundamental basis for a methodical approach to decision-making in energy management. It accurately identifies and manages all the energy sources and consumption within a facility, ensuring a balanced and efficient energy utilization. It quantifies energy consumption based on its specific functions.

The industrial energy audit is a very efficient instrument for developing and implementing a comprehensive energy management program. Energy audits involve the verification, monitoring, and analysis of energy utilization, as specified in the Energy Conservation Act of 2001. This entails creating a thorough technical report that provides suggestions for improving energy efficiency, together with a cost-benefit analysis and an action plan to reduce energy consumption.

Need for Energy Audit

Energy (both electrical and thermal), labor, and materials are commonly regarded as the primary operational expenses in all industries. The management of energy is a crucial area for cost reduction, as it consistently represents the highest controllable expense or possible source of savings in all the specified components.

Gaining insight into the utilization of energy and fuel in different industries may be facilitated through the utilization of an energy audit. This assessment will also identify inefficient practices and areas with potential for improvement.

An energy audit would offer valuable guidance for programs that are crucial for production and utility operations, including the reduction of energy expenses, preventive maintenance, and quality assurance. An audit plan of this nature will help to prioritize the analysis of fluctuations in energy expenses, the reliability and accessibility of the energy source, the selection of an appropriate energy combination, the identification of energy-conserving technology, and the retrofitting of energy-efficient equipment, among other factors.

Energy audits often entail proposing technically viable solutions that consider economic and organizational issues, all within a specified timeframe, to implement energy saving concepts. The primary objective of an energy audit is to uncover tactics to reduce operating costs or energy consumption per unit of production. Energy audits function as a reference point for managing energy inside a company and provide the basis for creating strategies to enhance energy efficiency across all areas.

3.10 Renewable Energy Technologies

Renewable energy is generated from natural resources that are replenished at a faster rate than they are used. Two examples of perpetual replenishment are solar energy and wind power. Presently, society could utilize a wide range of renewable energy sources. However, it is important to acknowledge that non-renewable fossil fuels such as coal, oil, and gas take a considerable amount of time, spanning several hundred million years, to form through geological processes. The utilization of fossil fuels for energy production results in the release of hazardous greenhouse gases, particularly carbon dioxide. These technologies utilize renewable energy sources that are continuously replenished. Renewable energy sources are crucial in decreasing reliance on non-renewable fossil fuels and addressing the issue of climate change.

Renewable energy sources have received significant attention in recent decades as a potential solution to the reduction in energy production and the projected acceleration of environmental problems (Wang, Zhou, et al., 2020a). The economy is predominantly intricate. Due to the fact that energy sources are responsible for 80% of global energy usage. When employing renewable energy sources for the purpose of generating electricity on a big scale, the ecosystem would be more effectively preserved and protected (Bali & Kumar, 2016). The instability of renewable energy sources has posed a continuous challenge. However, microgrids are regarded as a solution for effectively managing the upkeep of renewable energy sources. By utilizing energy storage and optimization systems, companies might potentially decrease their pollution emissions and cut down on expenditures (Wang, Lin, et al., 2020b).

3.11 Source of Renewable Energy

3.11.1 Solar Energy

Solar energy is universally recognized as the most plentiful kind of energy, capable of being captured even in situations of cloud cover. The Earth's rate of solar energy absorption surpasses human energy use by a factor of around 10,000.

Solar systems possess the ability to provide numerous advantages, such as heat provision, cooling capabilities, natural lighting, power generation, and fuel supply, for a wide range of applications. Solar technology can harness sunlight and transform it into electrical energy using photovoltaic panels or solar radiation-concentrating mirrors.

Although the availability of solar energy differs between countries, it is crucial to recognize that harnessing solar energy directly has the capacity to make a substantial contribution to the energy resources of any nation.

Solar panels have become more widely available and are frequently the most economical choice for electricity generation, thanks to a substantial decrease in the manufacturing costs linked to solar panel technology in the past decade. Solar panels come in a wide variety of colors, which are determined by the specific materials used in their creation. Furthermore, these panels often have an average duration of around 30 years. Solar panels transform solar energy into electrical energy. This technique can be used at many magnitudes, ranging from modest rooftop installations to huge solar farms.

3.11.2 Wind Energy

Large wind turbines are deliberately positioned either on land (onshore) or in bodies of saltwater or freshwater (offshore) to capture the inherent kinetic energy of moving air. Wind energy, a source that has been utilized for thousands of years, has experienced considerable progress in both onshore and offshore technology in recent years. The main emphasis of these improvements is to enhance power generation by utilizing higher turbines and larger rotor diameters.

Although typical wind speeds vary significantly across different geographical locations, it is important to highlight that the majority of regions worldwide have the potential for significant utilization of wind energy. Wind energy's technical capacity exceeds that of global power generation.

3.11.3 Geothermal Energy

Geothermal energy utilizes the heat energy that is obtainable from the Earth's interior. Geothermal reservoirs can be enhanced by applying thermal stimulation methods such as drilling wells or employing alternative procedures.

Hydrothermal reservoirs are subsurface formations that occur naturally and have both adequate heat and permeability, boosted geothermal systems, in contrast, are defined by naturally existing formations that have sufficient heat but are further boosted through the use of hydraulic stimulation techniques.

Fluids of varying temperatures possess the capability to produce electrical energy when they reach the surface. The technology employed for extracting energy from hydrothermal reservoirs has been in use for more than a century, making it a firmly established, dependable, and sophisticated process. This utilizes the Earth's geothermal energy to generate electricity and provide heating and cooling for buildings.

3.11.4 Hydropower

Hydropower utilizes the kinetic energy of water as it flows downhill from areas of higher altitude to lower altitude. Hydropower is derived from the natural hydrological cycle, which involves the flow of rivers and the controlled storage of water in reservoirs. Run-of-river hydroelectric facilities rely on the natural river flow, whereas reservoir hydropower plants utilize water stored in a reservoir.

Hydroelectric reservoirs serve multiple purposes, including the provision of energy, potable water, irrigation water, flood and drought management, navigation services, and energy delivery.

Hydropower is currently the dominant renewable energy source in the electricity sector. The determination of rainfall patterns depends on the regularity of precipitation, which can be adversely affected by both droughts caused by climate change and changes to ecosystems that affect the distribution of rainfall. Hydroelectric power is produced by utilizing the kinetic energy of moving water, usually by means of constructing dams.

Hydropower infrastructure possesses the capacity to exert adverse impacts on ecosystems. Consequently, numerous individuals perceive small-scale hydro as a more ecologically friendly option that is suitable for remote communities.

3.11.5 Bio Energy

Bioenergy is obtained from a wide variety of organic materials, which are known as biomass. These resources include commodities like wood, charcoal, dung, and other manures that are used to produce heat and power. In addition, agricultural crops are utilized to produce liquid biofuels. Biomass is used in economically disadvantaged communities in developing countries, especially in rural regions. It is primarily used for cooking, lighting, and space heating.

Contemporary biomass systems employ several sources of biomass, such as specifically grown plants or trees, agriculture and forestry byproducts, and diverse organic waste streams.

The use of biomass as an energy source leads to the production of greenhouse gas emissions, although to a lesser extent than the burning of fossil fuels like coal, oil, or gas. However, because of the potential adverse environmental effects such as deforestation and alterations in land use, it is recommended to limit the utilization of bioenergy to specialized applications.

3.12 Characteristics of the Source

Solar radiation can be received from any location on the Earth's surface. The maximal irradiance or optical power density of sunshine is approximately 1000 watts per square meter, regardless of your location on Earth. Insolation, also known as solar irradiance, refers to the quantity of energy available per unit area and time, usually measured in kilowatt-hours per square meter per year. It is a commonly used metric for quantifying solar resources. The annual amount of solar radiation received on the Earth's surface varies significantly, with a threefold difference. In locations like northern Scandinavia and Canada, the insolation is approximately 800 kilowatt-hours per square metre per year, whereas in certain arid desert areas, it can reach as high as 2500 kilowatt-hours per square metre per year.

The range of diffuse to total yearly insolation ratios can vary from 10% in areas with intense sunlight to 60% or higher in regions with mild climates, such as Western Europe. The precise ratio plays a crucial role in determining the feasibility of utilizing either concentrating or non-concentrating solar energy technologies.

A. **Potential of photovoltaic solar energy**

Solar radiation has an average power density of 100 to 300 watts per square metre. Solar electric power systems typically have a net conversion efficiency (from sunlight to electricity) of 10% to 15%. To meet energy needs, large regions must be used to capture and transform enormous amounts of solar energy (particularly in industrialized nations, given current energy use). To create an average of 100 megawatts of power—0.9 terawatt-hours of electricity or 3.2 petajoules of energy annually—using a photovoltaic (or solar thermal electricity) system, for instance, an area of 3 to 10 square kilometres is needed.

The average power available at the surface of the Earth in the form of solar radiation is approximately 1500 times greater than the overall average power used by humans. According to calculations, everyone has access to three megawatts of solar electricity on average, compared to an average usage of two kilowatts, which ranges from one hundred watts in the least developed nations to ten kilowatts in the United States. Although these figures offer a helpful general idea of solar energy's absolute limits, their significance for the technology and economic potential is little. A thorough examination of the technical and financial possibilities of solar energy is best conducted on a regional or national level due to variations in the solar energy supply pattern, energy infrastructure, population density, geographic circumstances, and other factors. The total of these national or regional potentials is then the global potential.

B. **Economics of Solar Energy**

How much power does the sun have in a single hour? Anyone who has ever dozed off while tanning can attest that it is sufficient to leave a mark. A report by Sandia National Laboratories claims that the average solar power received by a square metre on Earth is 174.7 watts. The answer is that the sun gives the Earth about 3.21 x 10 20 joules of energy every hour. To find this amount, multiply it by the Earth's surface area (4r 2, where r is approximately 6378 km), then by 60 min (3600 s), and you will get your answer, which is the same as 76,841 megatons of TNT. In conclusion, solar power-generating technology could have tremendous promise as an energy source with efficiency improvements.

The cost and efficiency of solar technologies in the past were relatively high. Its efficiency and costs have increased dramatically due to technological advancements during the past 20 years. As a result, the capacity of solar energy increased quickly. Costs for solar energy systems have decreased by more than half in the last 20 years. However, for solar energy to be financially competitive with other commercial energy sources, the government still needs to provide incentives and assistance. Additionally, the efficiency of widely used c-Si cells has already reached its theoretical maximum, necessitating the development of modern technology. Based on the IHS Markit global scenarios, solar energy accounted for only 5% of the total generation capacity in the top ten nations listed in the ranking (Fig. 3.2).

By lowering its costs and improving its efficiency, the new and forthcoming solar technologies promise to increase the use of solar energy. It is the most efficient method to distribute solar energy production, such as the electricity generated by

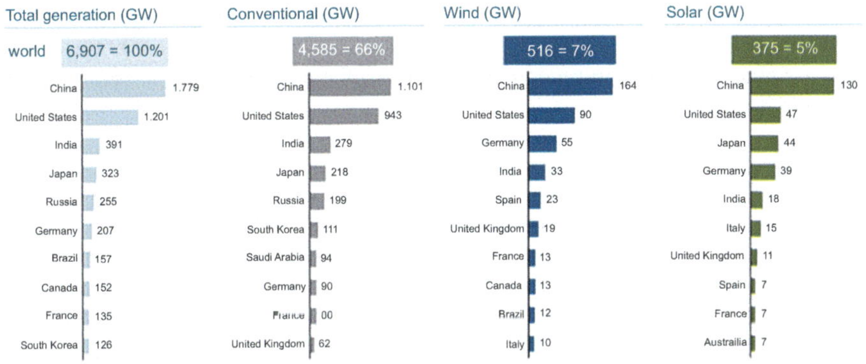

Fig. 3.2 Total generation capacity in the ranking top ten countries

homes with rooftop systems. Distributed solar generating allows individual owners to generate electricity for their needs and sell any excess to a utility company. The advantages of distributed solar power generation include clean energy, cost-effectiveness, reduced grid generation load, and reduced requirements for infrastructure such as transmission and distribution facilities. However, given the way many of today's power grids are configured, many engineers who work in the power generation sector will wince as they read this. Distributed power generation has numerous obstacles.

C. Economic Benefits of Solar Energy

a. Lowers Utility Bills:

An advantageous aspect of implementing solar energy for yourself and your household is a decrease in energy costs. The solar panels installed in your home can offset your electricity consumption, leading to reduced utility expenses. Power companies often offer the opportunity for residents to sell any surplus solar energy back to the firm, a practice referred to as net metering, as a means of providing more environmentally friendly energy to local communities. It is possible that you can decrease your monthly energy expenses consequently.

b. Government Grants for Solar Panel Installation.

Both federal and state governments provide incentives for the installation of solar panels, owing to the evident environmental benefits. These solar tax credits and incentives might assist you in saving money, particularly on the cost of your solar system. Another motivating factor that might assist in financing a solar system or generating financial gains from the energy is the receipt of monetary returns.

c. Source of Independent Energy.

Various natural disasters, such as hurricanes, floods, fires, earthquakes, and others, have a global influence on humanity. Several of these catastrophes possess the capacity to disrupt electricity supply. Utilizing solar energy for electricity in these situations will facilitate the bolstering of your community's economy. Your family and neighborhood will benefit from the ongoing generation of electricity by the sun during emergencies, when alternative energy sources are not accessible. Solar energy can mitigate the impacts of global warming, which contributes to several calamities.

d. **Vast Job Opportunities**

The solar workforce exemplifies the capacity of solar energy to enhance the local economy. The solar sector contributes to the local economy by creating employment opportunities and benefiting local households. The installation of solar panels by a growing number of individuals has resulted in a yearly rise in employment opportunities within the solar industry. Approximately 242,000 individuals are currently working in the solar sector. These roles are non-transferable; hence, they will experience annual growth (VR, 2022).

3.13 Environmental Costs of Solar Energy

Land Use and Ecological Impacts

For solar energy plants to generate power on a large scale, extensive areas for energy collection are required. Consequently, the facilities may not be suitable for existing land uses and can impact the utilization of areas such as wilderness or recreational management zones.

The energy footprints of energy systems may increase when they affect the land through activities such as the exploration, extraction, production, and disposal of materials. Consequently, certain areas may be utilized for energy production in a manner that poses challenges, requires significant time, or both, to fully restore them to their initial condition. Additionally, certain uses are so extreme that the alterations become permanent and cannot be reversed.

Impacts on Soil, Water and Air Resources

The building of solar arrays necessitates extensive land removal and grading, which in turn results in soil compaction, alterations to drainage systems, and heightened erosion. Central tower systems pose a potential hazard due to the need for water cooling, as there is a risk of chemical spills that could pollute groundwater and the surrounding land. Additionally, these systems increase water demand, which can further burden already scarce water resources in arid regions.

Heavy Metals

Some claim that dangerous heavy metals like cadmium telluride are used in the production of some of the most recent commercial innovations, most notably thin-film panels. Although it is true that these dangerous elements are employed in the production of solar panels, coal and oil also contain the same chemicals that are released during combustion.

Additionally, coal power plants pollute up to three hundred times more than solar panel manufacturers by producing significantly more dangerous substances.

Recycling Solar Panels
There are not enough locations to recycle old solar panels right now, and there are not enough of them that are not functioning to make recycling them worthwhile financially. Since the components used to make the solar panels are rare or costly metals and all contain silver, tellurium, or indium, recycling solar panels is especially important. Because there are limitations on recycling the panels, those recoverable metals might be wasted, which could present problems with future resource shortages.

A silicon-based solar cell requires a lot of energy to manufacture, with the amount of carbon the cell emits depending on the energy source, which is frequently coal. Silicon is needed to make the majority of today's photovoltaic cells, and it is currently abundant.

The inability to drive significant change in the recycling of the materials used in solar panel manufacturing, a business that already enjoys great environmental credibility from a power-generation perspective, is due to a lack of awareness about the manufacturing process of solar panels and the issue of recycling them, as well as a lack of significant external pressure.

3.14 Political Implications of Solar Energy

The political economy of India is closely intertwined with electricity. The price of solar energy has significantly declined in recent years, dropping from Rs 12 per unit in 2010 to Rs 2.44 per unit. This reduction has brought solar energy costs on par with the operational expenses of coal-fired power. What implications does this imminent change have for India's enduring nexus between politics and energy? Will the existing pattern of accommodation, characterized by political favoritism and subsidy transfers within the energy sector, encompass renewable energy (RE)? Could the accessibility of renewable energy in small amounts and strategically placed sites trigger a transformation in the realm of electrical governance, and if it does, in what manner?

Energy experts argue that any disturbances in India's energy sector are very probable to cause disruptions in both Indian politics and economics. An examination of energy-induced transformations is crucial for understanding the Indian political economy in the next ten years. The role of electricity evolved from being a facilitator to becoming a source of welfare in a developing nation, because of succumbing to the pressures of democratic politics. Any disturbance in the energy sector is expected to have repercussions on other political and economic domains due to the interconnectedness of the electrical industry with many economic sectors, both preceding and following it in the supply chain.

3.15 Government Regulations' Effect on the Adoption of Solar Energy

Governmental legislation has a substantial impact on the adoption of solar energy. They may strive to foster a favorable corporate atmosphere and incentivize or deter expansion and investment. Government rules on solar energy may manifest as mandates, subsidies, and tax advantages.

The establishment of Renewable Portfolio Standards (RPS) is a government program that has influenced the adoption of solar electricity. Renewable Portfolio Standards (RPS) mandate that utilities produce a designated proportion of their electricity from renewable sources, encompassing solar energy. These policies could create a secure market for solar energy, so promoting investment and facilitating its expansion.

Initial adoption of solar electricity can be made more cost-effective through the utilization of government tax incentives and subsidies. These regulations have the potential to enhance the accessibility of solar power for residential and commercial properties, hence stimulating the demand and use of this renewable energy source.

Nevertheless, rules imposing restrictions on the utilization of solar energy, particularly those accompanied by exorbitant licensing fees, have the potential to deter investment and hinder progress in this field. Utility laws that impose restrictions on net metering, which refers to the practice of selling surplus solar energy back to the power grid, might exacerbate the financial drawbacks associated with the use of solar energy.

Government Support for the Use of Solar Energy

Government endorsement of solar energy could facilitate cost reduction and broaden the availability of solar energy for residential and commercial use. The government could provide many forms of incentives, such as tax credits, rebates, and performance-based rewards.

- **Performance-based incentives**
  Many governments and utility companies provide performance-based incentives to promote the adoption of solar electricity. The energy generated by the solar panels is offset by the utilization of these incentives. The contribution may assist in offsetting the expenses associated with the installation, depending on the geographical area.
- **Rebates**
  Another way to promote the usage of solar energy is through rebates. You might be eligible for incentives if you install solar panels at your home or business. Utility companies as well as state and local governments frequently offer rebates.
- **Net metering**
  Net metering is another incentive that encourages the usage of solar power. Because of net metering, extra solar energy generated by residences and businesses may be resold to the grid, reducing energy costs.

- **Tax credits**
  A common incentive for consumers to use solar energy is a tax credit. A tax credit can decrease the initial expense of installing solar electricity and the tax obligations of a household or business. Solar energy installation qualifies for a tax credit of up to 26% from the U.S. federal government. Furthermore, numerous governments provide tax incentives for the utilization of solar energy (VR, 2023).

3.16 Wind Energy

Wind energy is a noteworthy and sustainable investment for the next period, since it facilitates the conversion of wind energy, an exhaustible resource, into electricity. To harness wind energy, it is necessary to construct wind farms consisting of a multitude of wind turbines, either on land or in offshore locations. The integration of these colossal organisms with the surrounding environment has occurred in recent times, however our comprehension of their operational mechanisms remains limited.

What is the mechanism responsible for the generation of wind? The distribution of solar energy on the Earth's surface is not uniform, resulting in variations in temperature. In regions where the temperature is higher, the air tends to ascend due to its lower density, leading to the formation of low-pressure zones. Conversely, in areas where the temperature is lower, the air wants to descend as it has a higher density, resulting in the formation of high-pressure areas. The movement of air is a consequence of the disparity in pressure, giving rise to the phenomenon of wind, which possesses significant potential for harnessing as a source of energy.

3.16.1 What Is Wind Energy

Wind energy is power generated by the wind's force. How? With the aid of a wind turbine, which converts the kinetic energy of air currents into electrical energy. The two basic mechanisms used to extract the energy are the rotor, which converts kinetic energy into mechanical energy, and the generator, which converts this mechanical energy into electrical energy. We are talking about an energy source that is secure, mature, efficient, and sustainable that is essential to the decarbonization of the economy and the energy transition.

3.16.2 *Workings of Wind Power*

As we just discussed, a wind turbine is required to harness the kinetic energy of the wind and transform it into electrical energy. The best way to use these enormous structures—typically between 80 and 120 metres high—depends on how strong the wind is. As a result, wind farms that pool several wind turbines and enable the production of vast amounts of this energy must be built in areas with a predominance of windy circumstances.

A vane on the nacelle is used to position the wind turbines so that they face the direction of the wind. The three major components of the wind turbine will then begin to rotate due to the force of the air currents:

The rotor: The purpose of the rotor, which consists of three blades and a bushing connecting them, is to collect wind energy and transform it into mechanical rotational energy.

The multiplier: The multiplier's job is to raise the rotational speed from thirty revolutions per minute (rpm) to 1500 rpm. It is connected to the engine by a shaft.

The generator: The generator is the component that transforms rotational mechanical energy into electrical energy.

Underground cables connect each wind turbine that makes up a wind farm, and these cables then transport the electricity to a transformer substation. Through the distribution networks of the numerous energy providers, it is then delivered to residences, businesses, or educational institutions, among other destinations (Fig. 3.3).

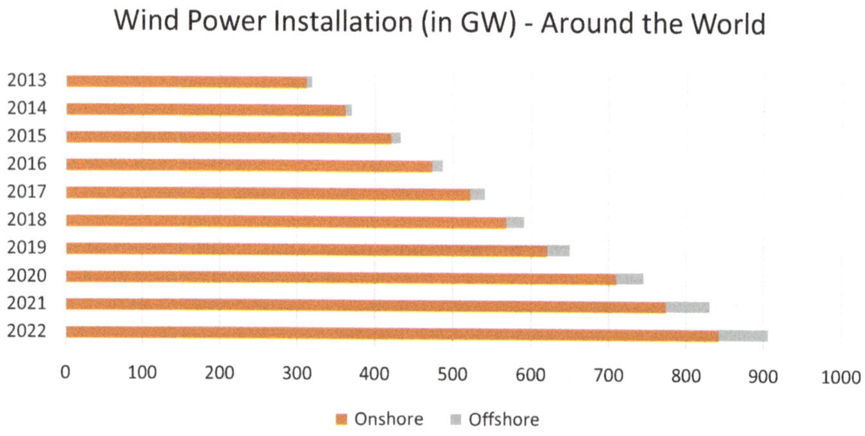

Fig. 3.3 Wind Power Installation. (Source: Global Wind Energy Council, Global Wind Report 2023)

3.16.3 Wind Energy Types

Currently, depending on where the wind turbines are installed, there are two types of wind energy:

- **Onshore wind energy**
  Onshore wind energy uses wind farms on land to capture the wind and turn it into power. To achieve this, we erect wind turbines that can convert wind kinetic energy into usable power and transfer it to the distribution network.
- **Offshore wind energy**
  Offshore wind energy is the energy obtained by harnessing the force of the wind that is produced on the high seas, where it reaches a higher and more constant speed than on land due to the absence of barriers. To make the most of this resource, mega-structures are installed that are seated on the seabed and equipped with the latest technical innovations (Iberdrola, 2022).

3.16.4 Advantages of Wind Energy

Numerous advantages of wind energy also apply to other renewable energy sources. The first is its contribution to the fight against climate change: using wind energy means using fewer fossil fuels, which reduces emissions of carbon dioxide, fine particles, and other substances that affect the climate and cause the greenhouse effect. In addition, since wind power is virtually free once the wind turbine is constructed, it can aid in achieving energy independence and providing countries that adopt it with undeniable economic benefits. This is a step in the direction of sustainable development. In addition, wind is a plentiful energy source that can be found on most of the Earth's surface and is unbounded in terms of supply throughout time.

However, there are some characteristics of wind energy that set it apart from other renewables like solar, geothermal, and hydroelectric power. These are broken down into ten principles to show how, from now on, wind energy can become the second-most significant green energy source globally and the driving force behind a sustainable future for many nations.

1. **Everywhere there is wind**

Not only is there wind practically everywhere on the planet, but we also know precisely where it blows most frequently and intensely. It comes down to availability: as it is generally known where the wind blows the greatest, wind farms may be installed based on our extensive knowledge of the regional conditions. A location may create more energy the more wind there is, which helps pay for the initial investment.

2. **Wind power is excellent in remote areas, wherever they may be**

Wind power has a similar benefit to many other renewable energy sources in that it may be used in locations with no grid connection by adopting micro-grid solutions. This presents an enormous potential in more remote places far from towns or cities because it can result in large savings by avoiding the need to construct pricey infrastructure. Except for the few places where installing wind turbines is not advantageous, wind energy may be used globally, unlike geothermal or hydroelectric electricity.

3. **In the medium and long term, wind is reliable.**

One of the criticisms of wind power is frequently the erratic nature of the wind from one second to the next or from one day to the next. However, there is another side to this coin, just like with solar power: while wind does have a seasonal or annual rhythm, it often does not depend on particular times of day or the alternation of day and night. It is a resource that, unlike solar energy, offers an excellent guarantee of regularity over the medium to long term and distinguishes itself by having variability that does not follow the same pattern.

4. **Excellent conversion efficiency**

Wind energy conversion to electricity is already performing at outstanding levels. Efficiency is between 40% and 50%, which is extremely near to the maximum theoretical level of 59% predicted by Betz's formula.

5. **Wind energy uses minimal land.**

Wind turbines do not require a lot of room despite being erected on land. Wind power can coexist with various land uses because the blades extend vertically and are almost insignificant in size at the base, especially when compared to photovoltaic facilities.

6. **The environmental impact is minimal**

Although there is still room for improvement in terms of the visual aspect and the low level of noise pollution brought on by the movement of the turbine blades, it is essential to note that the current impact is still relatively small. Because all emissions and resource consumption are limited to the manufacture, transportation, and installation of the plant, wind energy is the environmentally friendly source with the least total impact. To guarantee that any collateral impacts on flora and fauna are kept to a minimum, certificates are often required in many nations before the development of a wind power project can begin. While wind turbines are typically constructed on hills, mountains, or in the ocean, they can also be placed in fields or on gently sloping terrain without interfering with the use of the area for livestock pasture or crop cultivation.

7. **A genuinely affordable green energy source**

Wind energy is incredibly inexpensive to install and run compared to other energy sources. Depending on the size of the plant, construction takes only a few months—between two and twenty-four. Furthermore, technical advancements are further reducing expenses, with a cost per kilowatt hour of energy produced that has virtually disappeared. Finally, due in part to national incentives that have decreased costs overall and expenses to the end user, wind power is becoming increasingly affordable. This is because technology is becoming more affordable.

8. **Maintenance is simple and only occasionally necessary**

A wind farm's maintenance is relatively easy. Maintenance is modest and inexpensive unless there are breakages or extraordinary events, which are becoming increasingly rare due to models that are becoming increasingly reliable and digital monitoring systems that are enabling even higher performance levels. In fact, towers and turbines can function for many years without any maintenance; with minor tweaks, they can keep running for more than 20 years.

9. **Excellent circularity in the end-of-life phase**

Wind power facilities can be said to be perfectly reversible. Almost everything can be recovered after they have finished producing energy: the land on which they are located can be totally restored to its original state; the plant itself can be broken down into its constituent parts; and the materials may then be recycled in the future. Keeping with the circular economy model, the same raw materials can be used to build a new plant of the latest generation.

10. **Prospects for domestic wind power**

Wind power can have many benefits, even on a much smaller scale, such as for a single residence. Mini-wind and micro-wind power (up to 200 kilowatts and 20 kilowatts, respectively) enable the home use of wind energy to meet a building's or family's energy needs. To maximize the advantages of each, household wind energy is frequently utilized in conjunction with other environmentally friendly energy sources like solar or geothermal (enel, 2022).

3.16.5 Challenges of Wind Power

1. **Wind power must compete with other low-cost energy sources.**

Wind and solar power projects are currently more economically competitive than gas, geothermal, coal, or nuclear power plants when evaluating the cost of energy connected with new power plants. However, wind projects might not be cost-competitive in some places that are not sufficiently windy. Costs can be reduced even more with innovative technology, better manufacturing practices, and a deeper comprehension of the physics of wind farms.

2. **Ideal wind sites are often in remote locations.**

It is necessary to deliver electricity from wind farms to metropolitan areas to meet demand, but installation difficulties must be overcome. The cost of increasing land-based wind energy might be significantly reduced by upgrading the nation's transmission network to connect regions with plentiful wind resources to major population centers. Additionally, grid hookup and transmission capacities for offshore wind energy are developing.

3. **Turbines produce noise and alter visual aesthetics.**

Although the environmental effects of wind farms are distinct from those of conventional power plants, there are still worries about the noise generated by the turbine blades and the visual impact on the environment.

4. **Wind plants can impact local wildlife.**

Although wind energy projects have less impact on wildlife than other energy initiatives, study is still required to reduce wind-animal interactions. The adverse effects of wind turbines on animals are being lessened thanks to technological advancements, strategic placement of wind farms, and ongoing environmental research (Office of Energy Efficiency and Renewable Energy, 2022).

3.16.6 Wind Energy Policies in India

If India is to end poverty and achieve its bigger human development goals, it must maintain economic growth of at least 9% for the next 25 years. For the expansion to be sustained, the primary energy source must rise at a rate of 5.8% annually (this includes non-commercial energy sources like wood and dung that are collected) (Sholapurkar & Mahajan, 2015). The problem of meeting this need must be dealt with through an integrated energy policy. The integrated energy policy's overarching goal is to provide safe, clean, and affordable energy services to dependably meet the demand for energy services across all industries, including the lifeline energy needs of vulnerable households nationwide.

Generation-Based Incentive

In December 2009, a generation-based incentive (GBI) of 50 paise (equivalent to half an Indian rupee) per unit was introduced. With the help of this subsidy or incentive, wind farms were meant to be compensated based on generation rather than installation. Tax benefits in the form of accelerated depreciation were made accessible to the wind farm developers even before the GBI was created. However, this technique could not promote the expansion of wind energy installations. GBI is a strategy for promoting the construction of more effective wind farms (Sholapurkar & Mahajan, 2015).

While the advancement of renewable energy has been considerable, it cannot be compared to the potential that exists, which is another reason why GBI entered the scene. Only 11,000 MW of the possible 45,000 MW of installed wind capacity has actually been put into service. Additionally, it was believed that the financial incentives already in place were insufficient to achieve the RPO targets outlined in the National Action Plan on Climate Change (NAPCC). The GBI was added to support the capacity augmentation by acting as a booster. Less than 4000 MW had been put into service by December 2009, compared to a target of 10,500 MW capacity additions in the 11th plan. By the end of 2014, the total capacity had climbed to 22,465 MW (Sholapurkar & Mahajan, 2015).

State Wise Tariff for Wind Power

When it comes to wind energy in India, everything went well from about 2002 till the end of 2011, when it benefited from accelerated depreciation (A.D.) until April 2012. After being launched in 2011, the generation-based incentive (GBI) was later abandoned. To stimulate the wind energy industry and put it on par with the well-known solar mission, the government has launched its first wind energy mission. Launched recently is the National Wind Energy Mission (NWEM). This might significantly boost the wind energy industry, which has steadily declined since 2011. India installed over 29,536 MW of renewable energy capacity in 2011, including roughly 19,933 MW from wind, 2079 MW from solar, and 3746 MW.

Renewable Energy Certificate Scheme (REC)

The Ministry of New and Renewable Energy (MNRE), the central authority for all policies, rules, and approvals connected to renewable energy, promotes renewable energy.

The Central and State Electricity Regulatory Commissions (CERC and SERCs) and the Ministry of Power support it. While SERCs oversee local distribution and transmission, CERCs deal with the national grid and interstate power transfer/trading. These are crucial in advancing renewable energy because they are the only ones with the power to decide on feed-in tariffs and other policy issues, such as the Renewable Portfolio Standard (RPS).

Energy Development Agencies (EDAs) are the MNRE's state-level representatives. Their primary duties include advising the MNRE, state governments, and SERCs and evaluating and promoting renewable energy policies for specific states. IREDA encourages financial support for energy efficiency and renewable energy initiatives in India. In power exchange, the cost of REC would be determined. In accordance with the forbearance price and floor price set from time to time by CERC, REC would be traded on the electricity exchange. On August 23, 2011, CERC established the floor price and forbearance price that would be effective from April 2012 through F.Y. 2016–17.

3.17 Geothermal Energy

What is Geothermal Energy?

Renewable energy derived from the Earth's core is known as geothermal energy. It originates from heat produced during the planet's initial creation and the radioactive decay of elements. In the earth's core, rocks and liquids contain this thermal energy.

Thermal energy is continuously transferred from the planet's interior to its surface due to the difference in temperature between the earth's core and its surface.

Some of the rock in the Earth's core melts at temperatures over 4000 °C to generate hot, molten rocks known as magma. Since the mantle is lighter than the underlying rock, this heat also forces some of it to convex upward and act plastically.

The rock and water in the crust of the Earth can get as hot as 370 °C.

From surface-level levels all the way down to several miles below, thermal energy is present in the rocks and fluids.

3.17.1 How Is it Used?

For many centuries, certain countries have employed it as a means of providing warmth and facilitating culinary processes. The utilization of subterranean geothermal reservoirs can be employed for electricity generation as well as heating and cooling purposes by harnessing the heat energy stored in the water and steam.

One example of maintaining a consistent temperature for heating and cooling purposes is the implementation of a geothermal heat pump, which is typically installed approximately ten feet below the ground surface. The pipelines are filled with either water or antifreeze. Water is pumped through a closed loop of pipes. Ground source heat pump systems contribute to the thermal regulation of buildings throughout the summer season. This happens as the water and steam return to the building and absorb heat from the soil.

The utilization of geothermal water has been employed for the purpose of district heating in residential and commercial buildings, as well as for facilitating the cultivation of plants in greenhouse environments. In order to facilitate the melting of snow, one alternative method involves the utilization of piping systems beneath road surfaces.

3.17.2 How Is Geothermal Energy Produced?

To access the geothermal resources, wells up to a mile deep or more are dug into underground reservoirs. These resources can be obtained through enhanced geothermal systems, which increase or develop geothermal resources through a procedure known as hydraulic stimulation or from naturally occurring heat, rock and water permeability. Whether they are upgraded or natural, these geothermal resources power turbines that are connected to electrical generators.

In Larderello, Italy, in 1904, geothermal heat was exploited for the first time to create electricity. However, bathing has been done with geothermal heat ever since the Paleolithic period. It has also been demonstrated that monkeys in Japan use hot spring water that has been heated to stay warm during the colder months in hilly areas (DREW, 2022).

3.17.3 Advantages of Geothermal Energy

Unlike the sources we have encountered thus far, geothermal energy has certain characteristics that set it apart from other renewables such as wind, solar and hydroelectric power. Below is a 10-point summary that demonstrates why the natural heat below the Earth's surface still has enormous potential to become one of the protagonists of the energy sector of the future.

I. Readily accessible
In addition to being almost limitless, like many other renewable energy sources, geothermal energy is constantly accessible. Like solar energy, it is not affected by day or night; unlike wind and solar power, it is not affected by the season, climate, or weather. A geothermal power station typically produces energy for 8600 h a year, compared to solar power plants' average of 2000 h. Therefore, we can say that, at least in the short to medium term, the rate of geothermal energy production is constant. It is easier to predict and plan as a result.

II. Requires no big spaces
In contrast to the magnificence exhibited by large-scale wind turbines and photovoltaic panels, geothermal power plants possess the advantage of necessitating just moderate quantities of space. The majority of components, including the heat exchangers, are located below the surface, resulting in little visible structures above ground, regardless of the system's scale, be it tiny or huge. In residential settings, the heat pump typically has dimensions comparable to those of common household appliances. Conversely, in bigger industrial facilities, the cooling towers and turbines emerge as the predominant structural components. In certain scenarios, the presence of plants can potentially exert an aesthetic influence on the surrounding environment. However, it is worth noting that contemporary building designs have mitigated this issue to a certain extent.

III. Silent energy
The noise levels produced by geothermal power facilities are low and undetectable, at least while operating at full capacity. A certain amount of noise is unavoidable while the plants are being built, notably during excavations, but after that, everything is quiet. This is true for both household appliances and huge power plants outdoors that only have a few spinning turbines.

IV. It creates record numbers of jobs
According to data from the energy services manager, GSE, geothermal energy generates more indirect employment than any other renewable energy source for a given installed power. This amounts to 34 jobs per installed megawatt, which is far more than the 19 jobs produced by wind energy and the 12 jobs created by photovoltaic energy. Four thousand persons and 30,000 additional jobs would be permanently secure with 2000 gigawatts of installed power in Italy alone.

V. Provides more energy for a given nominal power
Geothermal energy may operate continuously at full capacity since delivery is constant (maintenance excluded). As a result, the energy consumption will equal the power times the number of hours of use. In contrast, solar, hydropower, and wind systems rarely operate at full capacity. For the same nominal power, more energy is produced.

VI. Double recycling is possible
Resources are optimized via geothermal power. On the one hand, the plants contain parts that can be recovered and used again once the installation has reached the end of its life. On the other hand, during operation, the flows are set up so that any heat that cannot be used immediately is returned to the circuit via the steam pipes that power the plant, aiding in energy conservation.

VII. Limited maintenance requirements
Geothermal plants do not need any extra upkeep, especially for household use. Since they are closed systems, the fluid pressure in the piping self-regulates, and there are very few electrical and mechanical components that could malfunction.

VIII. The Earth's heat can also cool
When geothermal energy is mentioned, heating and thermal energy come to mind first. However, geothermal plants are made to provide both heating and cooling. This is why geothermal power systems can be built in any sort of building, including residences, shopping malls, public buildings, and sports facilities, aside from big stations. The location's favorable effects on the features of the Earth's crust are, of course, the only restriction.

IX. The plants are long-lasting, safe and reliable
Domestic and industrial plants have extremely long average lives, up to 80–100 years. This longevity is astounding compared to the typical 15-year lifespan of a household boiler. Since there are no fuels involved, there is also no fire risk, and due to our experience with this kind of system, we can guarantee exceptional reliability.

X. Additional benefits for the home
In addition to its capacity to provide heating during winter months and air-cooling during summer months, geothermal energy offers many supplementary advantages when utilized inside residential settings. For example, due to its additional capability of serving as a boiler and providing heated water for various domestic purposes such as in the kitchen and bathroom, it effectively reduces the overall energy consumption by a range of 30% to 70%.

3.18 Hydro Energy

Hydroelectric energy, alternatively referred to as hydroelectric power or hydroelectricity, encompasses the use of the kinetic energy possessed by flowing water to generate electrical power, as exemplified by the downward flow of water in a waterfall. Throughout history, individuals have harnessed this force for an extended period. The ancient Greeks utilized the force of moving water over two millennia ago to facilitate the rotation of their mill's wheel, so enabling the process of grinding wheat into flour.

3.18.1 What Is the Process of Hydroelectric Power?

The predominant characteristics of hydroelectric power plants typically encompass a water reservoir, a gate or valve for the purpose of regulating the outflow of water from the reservoir, and an outlet or designated area where the water is discharged downstream. The object acquires potential energy in the moments preceding the flow of water over the crest of a dam or its descent down a hill. The process of water descending a slope result in the conversion of potential energy to kinetic energy. When water causes the rotation of a turbine's blades, the resultant outcome is the generation of electricity, which is subsequently transmitted to the consumers of the power plant.

3.18.2 Types of Hydroelectric Energy Plants

The most popular type of hydroelectric-producing plants is those equipped with impoundment facilities, constituting one of the three fundamental categories. A dam is employed within an impoundment facility to control the discharge of water stored in a pool or reservoir. The dam releases water in instances where there is a need for more energy. After the water is released, gravitational forces assume dominance, causing the water to flow downwards and pass through a turbine. The generator operates through the rotational motion of the turbine's blades.

A diversion facility can be classified as an alternative type of hydroelectric power plant. The uniqueness of this plant lies in its absence of a dam structure. In contrast, the system utilizes a network of canals to strategically guide the flow of the river towards the turbines responsible for generating electricity.

The term "pumped-storage facility" pertains to the third classification of power plant. This facility is responsible for the collection and storage of energy produced from nuclear, wind, and solar power sources, with the purpose of utilizing it later. The process of transferring water from a lower-level pool to a higher-elevation reservoir is employed by the plant as a means of energy storage. The discharge of water from the upper reservoir occurs in response to a substantial requirement for electrical power. The water is utilized to drive a turbine, so generating further electrical energy upon its descent into the lower reservoir.

3.19 Bio Energy

Bioenergy pertains to the energy obtained from living organisms, namely organic substances like plants, animals, and their byproducts. Renewable energy is essential for decreasing reliance on fossil fuels and addressing climate change. The notion incorporates a range of biofuels, including:

Biomass refers to organic substances such as wood, agricultural wastes, and municipal solid waste. These materials can be either burned or transformed into flammable gas or liquid. Biomass can be utilized for the purposes of heating, generating energy, or serving as a primary substance to produce biofuels.

Biofuels are derived from biological matter and serve as a source of fuel. The two predominant variants include bioethanol, derived from crops such as corn and sugarcane, and biodiesel, derived from vegetable oils, animal fats, or recycled greases. Biofuels are utilized in automobiles and frequently mixed with traditional fuels to diminish carbon emissions.

Biogas is generated by bacteria through the anaerobic digestion of organic materials. It consists mostly of methane and carbon dioxide. It has versatile applications, including heating, energy generation, and serving as an alternative for renewable natural gas.

Derived from or related to algae. Biofuels: Algae can be cultivated for the purpose of biofuel production. These plants exhibit a significant productivity per unit area and may be cultivated in many settings, such as freshwater, saltwater, and wastewater.

The benefits of bioenergy encompass its inherent renewability, the capacity to mitigate greenhouse gas emissions, and the capability to use waste resources. Nevertheless, there are additional obstacles to consider, such as the rivalry for land with food crops, the potential consequences on biodiversity and water resources, and the energy equilibrium involved in the production and processing of biofuels.

3.20 Financing Practices/Models in the Renewable Energy Sector

The financing strategies and approaches in the renewable energy sector are varied and essential for the advancement and growth of sustainable energy projects. These models specifically tackle the distinct obstacles and substantial upfront expenses linked to investments in renewable energy. Notable financial practices and models encompass:

- **Government Grants and Subsidies**: Governments often offer direct monetary assistance to renewable energy projects in the form of grants and subsidies. This assistance can mitigate the substantial upfront expenses and enhance the competitiveness of renewable energy in comparison to conventional energy sources.
- **Feed-in Tariffs (FiTs)**: These are government policies where renewable energy producers are paid a set price for the energy they generate and feed into the grid. FiTs guarantee a fixed, premium price for renewable energy producers, encouraging investment.
- **Tax Incentives and Credits**: Tax incentives, such as investment tax credits or production tax credits, are frequently used to encourage investments in renewable energy. These incentives have the potential to substantially alleviate the tax liability of renewable energy companies or projects.
- **Green Bonds**: Green bonds are financial instruments that are especially issued to raise funds for initiatives that have a good impact on the environment, such as renewable energy projects. They appeal to investors who prioritize environmental, social, and governance (ESG) factors.
- **Public-Private Partnerships (PPPs)**: These initiatives entail the cooperation of governmental bodies and commercial enterprises to fund, construct, and manage renewable energy ventures. Public-Private Partnerships (PPPs) can harness the efficiency of the private sector while receiving assistance from the public sector.
- **Power Purchase Agreements (PPAs)**: A Power Purchase Agreement (PPA) is a contractual agreement in which a purchaser commits to buying power produced by a renewable energy plant for a specified duration at a prearranged price. This concept ensures a consistent and reliable source of income for the project.
- **Venture Capital and Private Equity**: Renewable energy businesses and innovative initiatives frequently depend on venture finance and private equity as their primary sources of funding. Investments are crucial for fledgling enterprises that are engaged in the development of innovative technology.
- **Crowdfunding and Community Financing**: This technique entails soliciting modest sums of money from a substantial number of individuals, usually through online platforms. Community funding enables local communities to make direct investments in renewable energy projects within their vicinity.
- **Yieldcos**: A Yieldco is an entity that is listed on the stock market and established to possess operational assets that generate a reliable stream of cash, mainly through extended contractual agreements. Yieldcos in the renewable energy sector enable investors to acquire involvement in clean energy assets.
- **Energy Service Companies (ESCOs)**: ESCOs are responsible for the design, financing, and execution of energy conservation projects, frequently offering a guarantee of energy savings as part of the agreement. They have a substantial impact on enhancing energy efficiency upgrades.

The choice of the funding strategy is contingent upon several criteria, such as the nature of the renewable energy venture, its magnitude, geographical position, and the regulatory framework in place. Effective funding in the renewable energy industry sometimes entails a combination of these approaches, customized to address the unique requirements of each undertaking. Amidst the sector's ongoing development, novel financial strategies are arising to tackle fresh obstacles and prospects in renewable energy.

3.21 Conclusion

As we wrap up this chapter on Energy Efficiency and Renewable Energy Technologies, we acknowledge that the path towards a sustainable energy future is both difficult and absolutely necessary. The integration of energy efficiency and renewable energy technology is a potent means of tackling worldwide environmental issues, specifically climate change and the depletion of resources.

Energy efficiency, as the first foundation of this endeavor, provides a financially viable and rapid approach to diminish energy usage and greenhouse gas emissions. It encompasses a range of activities, from basic tasks such as swapping out incandescent bulbs for LEDs, to intricate endeavors like optimizing industrial operations. The combined effect of these steps might be significant, not only in terms of the environment but also in terms of economic savings.

The utilization of renewable energy technologies is swiftly reshaping the worldwide energy framework. The utilization of solar, wind, hydro, biomass, and geothermal energy sources plays a crucial role in the shift away from fossil fuels. These technologies are advancing rapidly, becoming more effective, and increasingly cost-competitive. Their capacity to scale, ranging from modest, off-grid applications to big, utility-scale installations, allows them to be flexible and suitable for various contexts and requirements.

The amalgamation of energy efficiency and renewable energy is not solely a technical obstacle; it necessitates a comprehensive strategy that includes legislation, financing, market development, and societal involvement. Governments, organizations, and individuals each have crucial functions. Policymakers must establish favorable regulatory frameworks and provide backing for research and development. Financial innovations and models play a crucial role in gathering the necessary funds for the widespread execution of projects. Increased public awareness and active participation can expedite acceptance and cultivate a societal mindset focused on sustainability.

As we conclude this chapter, it is evident that the way ahead is neither uncomplicated nor direct. Nevertheless, the swift progress in technology, along with the growing worldwide consciousness and determination of governments, provide a sense of optimism. The shift towards a more effective and sustainable energy future is not only feasible but also imperative for the well-being of our planet and future progeny. The path is intricate and diverse, yet it is imperative that we embark upon it with promptness and resolve.

References

Alvarado, R., Ortiz, C., Bravo, D., & Chamba, J. (2020). Urban concentration, non-renewable energy consumption, and output: Do levels of economic development matter? *Environmental Science and Pollution Research, 27*(3), 2760–2772. https://doi.org/10.1007/s11356-019-07158-8

Assareh, E., Mousavi Asl, S. S., Ahmadinejad, M., Parvaz, M., & Ghodrat, M. (2023). Optimization of a solar energy system integrating cooling, hot water, and power units in Australian cities: A climate-based analysis and cost-efficiency investigation. *International Journal of Hydrogen Energy*. https://doi.org/10.1016/j.ijhydene.2023.08.035

Bali, R. S., & Kumar, N. (2016). Secure clustering for efficient data dissemination in vehicular cyber–physical systems. *Future Generation Computer Systems, 56*, 476–492. https://doi.org/10.1016/j.future.2015.09.004

Charfeddine, L., & Umlai, M. (2023). ICT sector, digitization and environmental sustainability: A systematic review of the literature from 2000 to 2022. *Renewable and Sustainable Energy Reviews, 184*, 113482. https://doi.org/10.1016/j.rser.2023.113482

Chen, C., Hu, Y., Karuppiah, M., & Kumar, P. M. (2021). Artificial intelligence on economic evaluation of energy efficiency and renewable energy technologies. *Sustainable Energy Technologies and Assessments, 47*, 101358. https://doi.org/10.1016/j.seta.2021.101358

DREW. (2022). *Geothermal energy: building from sustainability*. https://blog.wearedrew.co/en/geothermal-energy-building-from-sustainability

enel. (2022). *All the benefits of wind power*. https://www.enelgreenpower.com/learning-hub/renewable-energies/wind-energy/advantages-wind-energy

Iberdrola. (2022). *What is wind energy, how is it converted into electricity and what are its advantages?* https://www.iberdrola.com/sustainability/renewables-energy-wind-power

IEA. (2019). *Multiple Benefits of Energy Efficiency*. https://www.iea.org/reports/multiple-benefits-of-energy-efficiency/energy-prices

Lin, B., & Huang, C. (2023). Nonlinear relationship between digitization and energy efficiency: Evidence from transnational panel data. *Energy, 276*, 127601. https://doi.org/10.1016/j.energy.2023.127601

Office of Energy Efficiency and Renewable Energy. (2022). Advantages and challenges of wind energy. Energy.Gov. https://www.energy.gov/eere/wind/advantages-and-challenges-wind-energy

Sholapurkar, R. B., & Mahajan, Y. S. (2015). Review of wind energy development and policy in India. *Energy Technology & Policy, 2*(1), 122–132. https://doi.org/10.1080/23317000.2015.1101627

Taghizadeh-Hesary, F., & Zhang, D. (2023). The handbook of energy policy. *Springer Nature Singapore.*. https://link.springer.com/referencework/10.1007/978-981-19-6778-8

Tsemekidi, S., Bertoldi, P., Economidou, M., Luca, E., & Gonzalez-torres, M. (2023). Determinants of energy consumption in the tertiary sector : Evidence at European level. *Energy Reports, 9*, 5125–5143. https://doi.org/10.1016/j.egyr.2023.03.122

VR, A. (2022). *What Is The Economic Impact Of Solar Power In India?* https://arka360.com/ros/economic-impact-of-solar-power-india/?cv=1

VR, A. (2023). *The Role of Government Policies in Promoting the Adoption of Solar Energy*. https://arka360.com/ros/government-policies-promoting-solar-energy/

Wang, S., Zhou, A., Yang, M., Sun, L., Hsu, C.-H., & Yang, F. (2020a). Service composition in cyber-physical-social systems. *IEEE Transactions on Emerging Topics in Computing, 8*(1), 82–91. https://doi.org/10.1109/TETC.2017.2675479

Wang, Z., Lin, X., Tong, N., Li, Z., Sun, S., & Liu, C. (2020b). Optimal planning of a 100% renewable energy Island supply system based on the integration of a concentrating solar power plant and desalination units. *International Journal of Electrical Power & Energy Systems, 117*, 105707. https://doi.org/10.1016/j.ijepes.2019.105707

Chapter 4
Circular Economy Principles: Shifting Towards Sustainable Prosperity

Abstract The emergence of the circular economy concept is a transformative strategy aimed at tackling the urgent issues of resource depletion, environmental degradation, and unsustainable consumer behaviors. The present chapter, entitled "Circular Economy Principles: Shifting Towards Sustainable Prosperity," delves into the fundamental principles of the circular economy and its capacity to facilitate a shift towards a future characterized by sustainability and prosperity. The chapter commences with a reevaluation of the concept of value creation, placing emphasis on the significance of developing products with attributes such as durability, reparability, and remanufacturability. The subsequent analysis explores the fundamental tenets of resource efficiency and material circularity, emphasizing recycling, upcycling, and waste minimization tactics. The examination of the transition from ownership to access, which is aided by the sharing economy and the concept of "Product-as-a-Service," is also considered a crucial element of the circular economy. This chapter delves into the need to achieve a closed-loop system for materials by means of recycling and reusing, along with an examination of the policy framework known as Extended Producer Responsibility (EPR). This paper examines the significance of digitalization and technology in facilitating the implementation of circular economy practices, focusing on the transformative capabilities of the Internet of Things (IoT), blockchain technology, and data analytics. Achieving sustainable prosperity necessitates the adoption of a comprehensive approach that considers the economic, social, and environmental aspects. This chapter provides an analysis of the economic advantages associated with the circular economy, as well as its social implications and inclusiveness. Additionally, it explores the significance of regulatory frameworks and international collaboration in this context.

Keywords Circular economy · Principles · EPR · IOT · Blockchain

V. Kandpal et al., *Sustainable Energy Transition*, Circular Economy and Sustainability, https://doi.org/10.1007/978-3-031-52943-6_4

4.1 Introduction

In an era marked by more pressing environmental issues, limited availability of resources, and a rising recognition of the finite nature of Earth's resources, the notion of the circular economy has arisen as a promising prospect. The present chapter, entitled "Circular Economy Principles: Shifting Towards Sustainable Prosperity," sets out on an investigative endeavor to examine the fundamental principles that form the basis of the circular economy and its crucial significance in facilitating a shift towards a future that is both sustainable and profitable.

The prevailing economic model, known as the linear "take-make-dispose" paradigm, has demonstrated its lack of sustainability across successive generations. The aforementioned phenomenon has resulted in the exhaustion of limited resources, the widespread generation of refuse, and an increasing ecological impact that poses a significant risk to the intricate equilibrium of our ecosystems. The concept of the circular economy presents a significant alternative approach that entails a comprehensive restructuring of our resource production, consumption, and management methods.

4.1.1 Defining the Circular Economy

The Circular Economy is a system designed to separate economic growth from resource depletion and environmental degradation. It seeks to replace the traditional linear model, characterized by a 'throwaway culture' and excessive consumption, with a more restorative and regenerative approach. Its core principle lies in creating a closed-loop system, where products, materials, and resources are continuously repurposed, reused, and recycled (Qassam, 2023).

4.1.2 Key Principles of the Circular Economy

Design for Longevity and Durability: In a Circular Economy, products are designed to be durable, easily repairable, and upgradeable. By prioritizing quality over planned obsolescence, companies can extend the lifespan of their products and reduce the need for frequent replacements (Martha, 2023). Designing for longevity and durability within the context of a circular economy is essential to create products that can be used for a long time, easily repaired, and eventually recycled or upcycled. Here are some strategies to consider:

(a) **Durable Materials Selection:** Choose high-quality materials that are known for their durability and resistance to wear and tear. This might include materials like stainless steel, solid wood, high-grade plastics, and other materials with a proven longevity track record.

(b) **Modular and Repairable Design**: Design products in a modular fashion so that individual components can be easily replaced or repaired. This extends the lifespan of the product as a whole. Avoid designs that involve irreversible bonding or complex integration of parts, making repairs difficult.
(c) **Standardization of Parts:** Standardize components and parts across different products where possible. This makes it easier to source replacement parts, reducing the chances of products becoming obsolete due to the unavailability of specific components.
(d) **Ease of Disassembly**: Design products with disassembly in mind. Use fasteners and connections that can be easily undone, allowing for efficient repair and material separation during recycling.
(e) **Design for Maintenance**: Consider how maintenance will be performed over the product's lifespan. Make sure access points to critical parts are easily reachable and that maintenance processes are straightforward.
(f) **Prototyping and Testing**: Before mass production, extensively prototype and test products for durability and longevity. Identify potential weak points and areas of wear and tear through rigorous testing.
(g) **User Education**: Educate users about proper usage, maintenance, and repair of products. Clear instructions and resources for troubleshooting and repairs can extend the life of the product.
(h) **Upgradability**: Design products with the potential for upgrades, such as better components or enhanced features. This can incentivize users to extend the life of their products rather than replacing them entirely.
(i) **Warranty and Extended Producer Responsibility (EPR):** Offer extended warranties and establish mechanisms for taking back and responsibly disposing of products at the end of their life. This encourages manufacturers to design for durability.
(j) **Design for Disassembly:** Consider the end of the product's life cycle. Design products in a way that facilitates easy dismantling for efficient recycling or upcycling of components.
(k) **Material Choice for Recycling**: If recycling is the ultimate goal, select materials that can be easily recycled without degrading in quality. Design for material separation and sorting during the recycling process.
(l) **Collaboration and Stakeholder Engagement**: Work closely with suppliers, manufacturers, users, and recycling facilities to ensure that your design aligns with circular economy principles and that the entire lifecycle of the product is considered.
(m) **Lifecycle Assessments:** Conduct lifecycle assessments to understand the environmental impacts of distinctive design choices. This can help identify areas where improvements can be made.
(n) **Aesthetic Timelessness:** Avoid trendy designs that might quickly become outdated. A timeless design can lead to longer product lifecycles.
(o) **Local Manufacturing and Repair Hubs**: Promote local manufacturing and repair facilities to reduce transportation impacts and make repair services more accessible.

4.2 Economic and Environmental Benefits

Reduced Resource Consumption: Designing products for longevity and durability directly aligns with the goal of resource efficiency. By creating products that last longer, the demand for new replacements is reduced. This, in turn, curtails the extraction of raw materials and minimizes the energy-intensive manufacturing processes associated with producing new goods. Consequently, the pressure on finite resources is alleviated, contributing to a more sustainable resource management system.

Waste Reduction: The disposable nature of many products has led to a staggering accumulation of waste. By prioritizing longevity and durability, products remain in use for extended periods, reducing the frequency of disposal and landfilling. This curbs the negative environmental impacts associated with waste disposal, such as pollution and greenhouse gas emissions.

Lifetime Value Maximization: Designing products to be long-lasting increases the value derived from each product throughout its life cycle. Consumers benefit from products that provide reliable performance and require fewer replacements, leading to financial savings over time. Manufacturers also benefit from enhanced brand reputation, customer loyalty, and reduced costs associated with frequent repairs and replacements.

4.3 Embrace Reuse and Refurbishment

Instead of discarding products after use, the Circular Economy encourages the reuse and refurbishment of items. This involves remanufacturing, restoring, and reselling products to extend their lifecycle and keep them in the economic loop. Embracing reuse and refurbishment is a crucial aspect of a circular economy, as it extends the lifespan of products and reduces the need for new production (Niwan & Sharma, 2023). Here is how you can promote reuse and refurbishment within this framework:

(a) **Design for Disassembly and Repair**: Create products with the intention of easy disassembly and repair. Use fasteners and connectors that allow for straightforward access to components and avoid using adhesives or techniques that make disassembly difficult.

(b) **Modular Design**: Design products in a modular way so that individual components can be easily replaced or upgraded without the need to replace the entire product. This also makes it simpler to refurbish products.

(c) **Standardization**: Standardize components and parts across different products or product lines. This makes it easier to source replacement parts, and it encourages repair and refurbishment.

(d) **Refurbishment Centers**: Establish dedicated refurbishment centers where skilled technicians can assess, repair, and upgrade products. These centers can specialize in extending the life of electronics, appliances, furniture, and other items.

(e) **Offer Refurbished Products**: Sell refurbished products alongside new ones. These could be products that have been returned, repaired, and certified for quality. This provides customers with more affordable options while reducing the demand for new manufacturing.
(f) **Educate Users**: Educate consumers about the benefits of refurbished products and how to care for and maintain products to extend their lifespan. Proper usage and maintenance can significantly reduce the need for repairs.
(g) **Extended Warranty and Service Plans**: Offer extended warranty and service plans that cover repairs and refurbishment. This encourages customers to keep their products longer and use repair services instead of discarding items prematurely.
(h) **Collaborate with Manufacturers**: Work with manufacturers to encourage designs that are conducive to refurbishment. Provide feedback based on the challenges and opportunities observed during refurbishment processes.
(i) **Reverse Logistics:** Set up efficient reverse logistics systems to collect, transport, and manage products that need refurbishment or repair. This includes creating channels for customers to return products at the end of their life cycle.
(j) **Component Recovery and Upcycling**: During refurbishment, recover valuable components and materials that can be reused or upcycled. This reduces waste and conserves resources.
(k) **Local Repair and Refurbishment Networks**: Promote local repair and refurbishment networks to minimize transportation impacts and provide accessible services to consumers.
(l) **Partnerships with NGOs and Social Enterprises**: Collaborate with non-governmental organizations (NGOs) and social enterprises that focus on repair and refurbishment. They often have expertise in these areas and can help streamline the process.
(m) **Regulatory Support:** Advocate for supportive regulations and policies that incentivize refurbishment and repair activities. This might include tax incentives or regulations that encourage manufacturers to design for longevity and repairability.
(n) **Consumer Awareness Campaigns**: Launch campaigns to raise awareness about the benefits of refurbished products, including cost savings, reduced environmental impact, and support for local economies.

4.4 Adopt Recycling and Material Recovery

Recycling plays a pivotal role in the Circular Economy, where materials are collected and reprocessed into new products. Effective recycling programs help conserve valuable resources, reduce waste, and minimize the environmental impact of manufacturing processes (Hegab et al., 2023). Adopting recycling and material recovery practices is essential for a circular economy, as they help ensure that valuable materials are kept in circulation and not wasted. Here is how you can promote recycling and material recovery within this framework:

(a) **Design for Recyclability**: Design products with materials that are easily recyclable and separable. Avoid complex combinations of materials that are difficult to disassemble and recycle.
(b) **Material Identification**: Clearly label products and packaging with information about the materials used. This assists recycling facilities in identifying and sorting materials properly.
(c) **Separation at Source:** Encourage consumers to separate recyclable materials from their waste at the source, making it easier for recycling facilities to process them efficiently.
(d) **Recycling Infrastructure**: Invest in and support the development of recycling infrastructure, including collection systems, sorting facilities, and recycling plants. This ensures that materials have a viable path for recovery.
(e) **Collection Programs**: Implement comprehensive recycling collection programs that cover a wide range of materials, including plastics, metals, glass, paper, and electronics.
(f) **Consumer Education**: Educate consumers about proper recycling practices, including what can and cannot be recycled, and how to prepare items for recycling.
(g) **Eco-Design Guidelines**: Develop and promote guidelines for eco-design that emphasize using recycled or recyclable materials, reducing material complexity, and minimizing harmful substances.
(h) **Collaborate with Manufacturers:** Collaborate with manufacturers to create products using recycled materials, which supports the demand for recycled inputs and reduces the need for virgin resources.
(i) **Circular Supply Chains**: Develop circular supply chains that integrate recycled materials back into production processes, reducing the reliance on virgin resources.
(j) **Incentives for Recycling:** Introduce incentives for consumers, businesses, and manufacturers to recycle and use recycled materials. This could include tax breaks, subsidies, or preferential treatment for products with recycled content.
(k) **Waste-to-Energy Conversion**: For materials that cannot be easily recycled, consider waste-to-energy conversion methods, such as incineration with energy recovery, as a way to extract value from waste.
(l) **Recycling Technologies**: Invest in research and development of advanced recycling technologies, such as chemical recycling and upcycling, which can help extract valuable materials from complex products.
(m) **Collaboration with Stakeholders**: Work with local governments, NGOs, industry associations, and other stakeholders to create comprehensive recycling strategies that address regional challenges.
(n) **Promote Circular Business Models:** Encourage businesses to adopt circular business models, such as take-back programs and leasing, which incentivize the return and recycling of products.
(o) **Policy and Regulation**: Advocate for policies and regulations that promote recycling and material recovery, including extended producer responsibility

(EPR) programs that hold manufacturers accountable for managing the end-of-life of their products.
(p) **Investment in Research**: Invest in research to identify new recycling techniques, improve material recovery rates, and develop innovative solutions for recycling challenging materials.

4.5 Foster Collaborative Business Models

The Circular Economy promotes collaboration among businesses, consumers, and governments to create closed-loop systems. Sharing platforms, product-as-a-service models, and collaborative consumption encourage efficient resource use and reduce overall consumption. Fostering collaborative business models in a circular economy involves creating partnerships and networks that facilitate resource sharing, efficient value chains, and the exchange of knowledge. These models encourage businesses to work together towards common goals, such as reducing waste and maximizing resource utilization. Here is how you can promote collaborative business models in a circular economy:

(a) **Multi-Stakeholder Partnerships**: Establish partnerships that involve businesses, government agencies, NGOs, academia, and other relevant stakeholders. These partnerships can leverage diverse expertise and resources to drive circular initiatives.
(b) **Industry Clusters and Networks**: Create industry-specific clusters or networks that bring together businesses from the same sector to collaborate on circular economy projects, share best practices, and jointly address challenges.
(c) **Sharing Platforms**: Develop platforms that facilitate the sharing of resources, assets, and infrastructure. For example, a platform could enable businesses to share equipment, facilities, or transportation services, reducing individual investment and waste.
(d) **Co-Working Spaces**: Provide co-working spaces where businesses can collaborate, share ideas, and potentially work on joint projects related to the circular economy.
(e) **Circular Supply Chains**: Collaborate with suppliers, manufacturers, and distributors to establish circular supply chains that prioritize the use of recycled materials, resource efficiency, and product longevity.
(f) **Take-Back Programs**: Implement take-back programs where businesses take back their products at the end of their useful life for refurbishment, recycling, or proper disposal. Collaboration can make these programs more cost-effective and impactful.
(g) **Product-as-a-Service (PaaS):** Offer products as services, where customers pay for the utility of a product rather than owning it. This encourages manufacturers to design products for durability and easier maintenance, fostering collaboration between manufacturers and users.

(h) **Circular Consortia**: Form consortia or alliances that focus on specific challenges or opportunities within the circular economy, such as material innovation or end-of-life management.
(i) **Educational Initiatives**: Collaborate with educational institutions to promote circular economy principles and practices. This can help develop a skilled workforce capable of implementing circular strategies.
(j) **Resource Mapping and Sharing**: Create platforms that map available resources within a region and enable businesses to share surplus materials, reducing waste and promoting resource efficiency.
(k) **Joint Research and Development**: Collaborate on research and development projects that focus on circular solutions, such as new materials, recycling technologies, or sustainable product designs.
(l) **Joint Marketing and Promotion**: Collaborate with other businesses to promote circular products, services, and initiatives. Joint marketing efforts can amplify the impact and reach of circular economy campaigns.
(m) **Lobbying and Advocacy**: Work together to advocate for supportive policies, regulations, and incentives that encourage circular practices and level the playing field for circular businesses.
(n) **Shared Learning and Training**: Organize workshops, seminars, and training sessions where businesses can learn about circular economy concepts, tools, and implementation strategies.
(o) **Circular Innovation Challenges:** Organize innovation challenges that invite businesses to collaborate on solving specific circular economy challenges, fostering creative problem-solving and collaboration.
(p) **Open Innovation Platforms**: Establish open innovation platforms where businesses can share ideas, challenges, and potential solutions, fostering collaboration across industries.

4.6 Promote Sustainable Consumption

Educating consumers about sustainable choices, encouraging responsible consumption habits, and supporting eco-friendly products are essential elements of the Circular Economy. Consumers play a crucial role in driving demand for sustainable alternatives. Promoting sustainable consumption within a circular economy involves encouraging individuals, businesses, and societies to make mindful choices that prioritize resource efficiency, minimize waste, and reduce environmental impact. Here are strategies to promote sustainable consumption:

(a) **Education and Awareness**: Raise awareness about the concept of a circular economy, its benefits, and how individual actions contribute to it. Provide information about the environmental and social impacts of various consumption choices to empower consumers to make informed decisions.

(b) **Product Information and Labels**: Implement clear and standardized labeling that communicates the environmental footprint of products, including information about materials, energy use, and recyclability. Certifications and eco-labels can guide consumers towards more sustainable choices.
(c) **Consumer Empowerment**: Educate consumers on how to repair, refurbish, and maintain products to extend their lifespan. Provide resources and guides on responsible consumption, such as tips for reducing waste and making conscious purchasing decisions.
(d) **Promote Quality over Quantity:** Encourage consumers to invest in high-quality products that are durable and have longer lifespans, reducing the need for frequent replacements.
(e) **Sharing and Collaborative Consumption**: Support platforms that enable sharing and renting of goods (e.g., car-sharing, tool libraries) to reduce the overall demand for new products.
(f) **Product-as-a-Service Models**: Encourage the adoption of product-as-a-service models where consumers pay for the use of a product rather than ownership. This incentivizes manufacturers to create longer-lasting products.
(g) **Extended Producer Responsibility (EPR):** Advocate for policies that hold manufacturers responsible for the entire lifecycle of their products, including take-back and recycling.
(h) **Incentives for Sustainable Choices**: Provide incentives such as tax breaks or discounts for products with higher eco-friendly attributes, encouraging consumers to make sustainable choices.
(i) **Circular Design Challenges**: Organize design challenges that encourage businesses to create products that are both appealing to consumers and aligned with circular principles.
(j) **Engage Businesses**: Collaborate with businesses to offer repair and refurbishment services, creating a culture of sustainability even after the initial purchase.
(k) **Public Procurement Policies**: Governments and institutions can use their purchasing power to prioritize circular products, sending a signal to the market and encouraging sustainable production.
(l) **Behavioral Nudges**: Use behavioral psychology techniques to nudge consumers towards sustainable choices, such as placing recycling bins in prominent locations.
(m) **Consumer Advocacy and Campaigns**: Launch consumer campaigns that promote sustainable consumption and encourage businesses to adopt circular practices.
(n) **Local Economies and Crafts**: Support local economies and artisans by promoting locally made, handcrafted products that often have lower environmental impact and longer lifespans.
(o) **Collaborative Partnerships:** Collaborate with community groups, NGOs, and industry associations to create collective campaigns and initiatives that promote sustainable consumption.

(p) **Digital Tools and Apps**: Develop apps and tools that help consumers calculate their environmental footprint, find sustainable products, and track their consumption patterns.

4.7 Adopt Renewable Energy Sources

The Circular Economy aligns with the principles of sustainable energy use. Embracing renewable energy sources, such as solar, wind, and hydro power, minimizes reliance on finite fossil fuels, reducing greenhouse gas emissions and mitigating climate change. Adopting renewable energy sources within a circular economy framework is crucial to achieve sustainability and reduce the environmental impact of energy production. Here are steps and strategies to integrate renewable energy sources into a circular economy:

(a) **Energy Efficiency First**: Before adopting renewable sources, prioritize energy efficiency measures to minimize overall energy demand.
(b) **Assess Energy Needs**: Conduct an assessment of energy needs and patterns to determine the appropriate mix of renewable energy sources for different applications.
(c) **Integrate Distributed Energy Generation**: Implement distributed energy generation, such as rooftop solar panels or small-scale wind turbines, to generate clean energy at or near the point of use.
(d) **Solar Power**: Invest in solar power systems for electricity generation, especially in regions with ample sunlight. This can power homes, businesses, and industries.
(e) **Wind Power**: Harness wind energy through onshore and offshore wind farms to produce electricity and power grids.
(f) **Hydropower:** Utilize hydropower where feasible to generate electricity from flowing water. Consider both large-scale hydropower and smaller run-of-river installations.
(g) **Bioenergy**: Use organic waste, agricultural residues, and dedicated energy crops to produce bioenergy through processes like anaerobic digestion, biogas, and biomass combustion.
(h) **Geothermal Energy**: Tap into geothermal energy sources to produce electricity and heating/cooling for buildings and industrial processes.
(i) **Energy Storage Solutions**: Implement energy storage technologies, such as batteries and pumped hydro storage, to store excess energy generated from renewables for use during periods of low production.
(j) **Microgrids and Smart Grids**: Develop microgrids and smart grid systems that integrate renewable energy sources, energy storage, and demand-response mechanisms for efficient energy distribution.
(k) **Circular Energy Systems**: Integrate energy production with other circular practices, such as using waste heat from industrial processes to provide heating or power to nearby facilities.

(l) **Electrification of Transportation**: Promote the adoption of electric vehicles (EVs) and charging infrastructure, allowing EVs to serve as mobile energy storage units and participate in grid balancing.
(m) **Circular Bioeconomy**: Combine circular economy principles with renewable energy production by using organic waste for bioenergy and returning nutrient-rich residues to the soil as fertilizer.
(n) **Green Building Design**: Design buildings and infrastructure with renewable energy integration in mind, using passive solar design and efficient insulation to reduce energy demand.
(o) **Collaboration and Innovation**: Collaborate with energy providers, technology developers, research institutions, and government agencies to drive innovation in renewable energy solutions.
(p) **Regulatory Support**: Advocate for favorable policies, incentives, and regulations that promote the adoption of renewable energy sources and grid integration.
(q) **Community Engagement:** Involve local communities in renewable energy projects to foster support, understanding, and local benefits.

4.8 Benefits of the Circular Economy

Resource Conservation: By maximizing the utility of existing resources, the Circular Economy reduces the demand for raw materials, conserving natural resources for future generations. Resource conservation is a central benefit of a circular economy, as it focuses on reducing the depletion of natural resources, minimizing waste, and optimizing resource use throughout the entire lifecycle of products and materials. Here's how resource conservation is a key advantage within a circular economy:

(a) **Reduced Raw Material Extraction**: In a linear economy, resources are extracted, used, and then discarded. In a circular economy, resources are kept in use for as long as possible, reducing the need for constant extraction of new raw materials. This helps preserve natural ecosystems and habitats.
(b) **Minimized Waste Generation**: Circular economy practices aim to design products and systems that generate less waste. By emphasizing durability, repairability, and recycling, waste is minimized at all stages of the product lifecycle.
(c) **Lower Environmental Impact**: Resource conservation directly leads to a lower environmental impact. It reduces pollution, greenhouse gas emissions, and other negative externalities associated with resource extraction, production, and disposal.
(d) **Extended Product Lifespan**: Circular economy principles encourage designing products that last longer and are easier to maintain and repair. Extending the lifespan of products reduces the frequency of replacements, thereby conserving the resources required for manufacturing new products.

(e) **Closed Material Loops**: Circular economies prioritize closing material loops, where materials are continuously recycled or upcycled. This minimizes the need for virgin materials and reduces the energy and resources needed for new production.
(f) **Energy Savings**: Resource conservation often leads to energy savings, as producing new products from recycled materials requires less energy than processing raw materials.
(g) **Positive Economic Impact**: Resource conservation can lead to cost savings for businesses through reduced material and energy costs. It can also create new business opportunities in recycling, refurbishment, and the development of sustainable technologies.
(h) **Enhanced Resource Security**: By reducing dependence on scarce and non-renewable resources, a circular economy enhances resource security, making industries less vulnerable to supply chain disruptions.
(i) **Preservation of Biodiversity**: The reduced demand for raw materials and the protection of natural habitats contribute to the preservation of biodiversity and ecosystems.
(j) **Innovation and Efficiency:** The need to maximize the use of resources encourages innovation in product design, manufacturing processes, and business models. This drives efficiency improvements and fosters creativity in finding sustainable solutions.
(k) **Climate Change Mitigation**: Resource conservation is intricately linked to reduced greenhouse gas emissions. By using materials more efficiently and adopting sustainable practices, a circular economy contributes to mitigating climate change.
(l) **Resilience to Resource Scarcity**: As the world faces growing resource constraints, a circular economy provides a more resilient framework for continued economic growth by reducing reliance on finite resources.
(m) **Cradle-to-Cradle Thinking**: Circular economy principles encourage "cradle-to-cradle" thinking, where products are designed to be continually recycled or returned to the biosphere, contributing to a regenerative approach to resource use.
(n) **Global Sustainability Goals**: Resource conservation aligns with global sustainability goals, such as the United Nations Sustainable Development Goals (SDGs), by promoting responsible consumption and production patterns.

4.9 Waste Reduction

Implementing circular practices significantly reduces waste generation, preventing pollution and alleviating the strain on landfills and ecosystems. Waste reduction is a fundamental benefit of a circular economy, as it aims to minimize the generation of waste and promote efficient resource use throughout the product lifecycle. Here's how waste reduction is a significant advantage within a circular economy:

(a) **Minimized Environmental Impact**: Waste reduction directly leads to a lower environmental impact. It reduces pollution, greenhouse gas emissions, and the depletion of natural resources associated with waste disposal and production.
(b) **Conservation of Resources**: By keeping products, components, and materials in use for longer periods, a circular economy conserves valuable resources that would otherwise be wasted through premature disposal.
(c) **Less Landfilling and Incineration**: Waste reduction reduces the need for landfilling and incineration, both of which can lead to environmental and health hazards. This improves air and water quality and reduces pressure on waste management infrastructure.
(d) **Energy Savings**: Reducing waste often leads to energy savings, as less energy is required for waste collection, transportation, and disposal. It also avoids the energy-intensive processes of creating new products from virgin materials.
(e) **Circular Material Flows**: A circular economy aims to close material loops, ensuring that materials are continuously recycled or reused. This minimizes the need for untapped resources and reduces waste generation.
(f) **Extended Product Lifespan:** Circular economy practices emphasize designing products that last longer and are easier to repair and refurbish. This extends the lifespan of products, reducing the frequency of replacements and associated waste.
(g) **Promotion of Repair and Reuse:** Waste reduction encourages repair and reuse, which are key components of the circular economy. Repairing and reusing products extend their life and divert them from landfills.
(h) **Positive Economic Impact:** Reducing waste can lead to cost savings for businesses through reduced waste disposal fees and lower material costs. It can also create new business opportunities in waste reduction services and recycling.
(i) **Consumer Awareness:** A circular economy raises consumer awareness about the environmental impact of their consumption choices, encouraging more responsible purchasing decisions and less disposable consumption.
(j) **Job Creation**: The shift towards a circular economy creates job opportunities in recycling, repair, refurbishment, and other waste reduction activities, contributing to economic growth.
(k) **Resilience to Resource Scarcity**: By minimizing waste and maximizing resource utilization, a circular economy enhances resource security, making industries less vulnerable to resource scarcity and supply chain disruptions.
(l) **Alignment with Sustainable Goals**: Waste reduction aligns with global sustainability goals, including the United Nations Sustainable Development Goals (SDGs), by promoting responsible consumption and production patterns.
(m) **Innovation and Creativity**: The need to reduce waste drives innovation in product design, materials, and processes. It encourages creative problem-solving to find new ways to eliminate waste.

(n) **Cradle-to-Cradle Thinking**: Waste reduction is integral to the cradle-to-cradle approach, where products are designed to be continuously recycled or reintegrated into natural systems.
(o) **Waste-to-Value Conversion**: Waste reduction encourages the development of technologies that convert waste into valuable resources, such as turning organic waste into compost or biogas.

4.10 Economic Growth and Job Creation

The Circular Economy stimulates innovation and the development of modern technologies and services. This, in turn, fosters economic growth and generates employment opportunities in green industries. Economic growth and job creation are substantial benefits of a circular economy. Contrary to the misconception that sustainability might hinder economic development, circular economy principles can foster economic prosperity while also addressing environmental challenges. Here's how economic growth and job creation are facilitated within a circular economy:

(a) **New Business Models and Opportunities**: Circular economy principles encourage the development of new business models, such as product-as-a-service, repair services, and remanufacturing. These models create opportunities for entrepreneurs and start-ups.
(b) **Job Diversification**: A circular economy introduces a diverse range of job opportunities, from recycling and refurbishment to design for disassembly and waste management. This diversification reduces dependence on single industries.
(c) **Skills Development**: Circular practices require a skilled workforce for tasks like repair, remanufacturing, and waste management. This encourages the development of specialized skills, enhancing human capital.
(d) **Local Economic Development**: Circular economy practices, such as repair shops and local material recovery facilities, contribute to local economic development by creating jobs and supporting community resilience.
(e) **Innovation and Research**: The transition to a circular economy demands innovation in product design, material science, recycling technologies, and more. This spurs research and development efforts, leading to technological advancements and increased economic activity.
(f) **Resource Efficiency**: Circular practices emphasize maximizing resource use and minimizing waste. This leads to cost savings for businesses and enhances overall economic efficiency.
(g) **Value Retention:** By extending the life of products through repair, remanufacturing, and reuse, value is retained within the economy for a longer duration, benefiting businesses and supporting economic growth.

(h) **Supply Chain Resilience**: Circular practices can make supply chains more resilient by reducing dependence on scarce resources and offering alternatives for raw materials.
(i) **Market Opportunities**: Circular economy strategies create opportunities for companies to differentiate themselves in the market by providing sustainable products and services that appeal to environmentally conscious consumers.
(j) **Green Investment and Financing**: The transition to a circular economy can attract investment from institutions and investors seeking sustainable projects, further contributing to economic growth.
(k) **Circular Industrial Parks**: Circular economy concepts can lead to the development of circular industrial parks, where companies collaborate to share resources, infrastructure, and services, fostering economic synergies.
(l) **Improved Resource Management**: Efficient resource management driven by circular practices reduces the risk of resource scarcity, supporting long-term economic stability.
(m) **Cradle-to-Cradle Design**: Designing products with circularity in mind can open up new markets for design, engineering, and manufacturing services focused on creating durable, recyclable, and upcyclable products.
(n) **Consumer Demand**: As consumer demand for sustainable products and services grows, businesses that adopt circular practices can capture a larger market share and experience revenue growth.
(o) **Global Competitive Advantage**: Companies that embrace circular economy principles early can establish themselves as leaders in sustainability, gaining a competitive advantage in a rapidly changing business landscape.

4.11 Lower Environmental Impact

With a focus on sustainable practices, the Circular Economy mitigates environmental degradation, climate change, and biodiversity loss, safeguarding the planet for future generations. Lowering environmental impact is one of the primary and most compelling benefits of a circular economy. This approach aims to minimize negative effects on the environment by reducing resource consumption, waste generation, and pollution. Here is how a circular economy contributes to lower environmental impact:

(a) **Resource Conservation**: Circular economy practices prioritize the efficient use of resources and raw materials. By extending product lifespans and promoting reuse and recycling, less natural resources are extracted and consumed.
(b) **Reduced Waste Generation**: Circular economy strategies aim to minimize waste at all stages of the product lifecycle. This reduces the need for landfilling, incineration, and other waste disposal methods that contribute to environmental pollution.

(c) **Minimized Pollution**: By emphasizing the reduction of waste and harmful emissions, a circular economy helps minimize pollution in air, water, and soil, leading to improved overall environmental quality.
(d) **Lower Greenhouse Gas Emissions**: Circular practices, such as recycling and using renewable energy sources, result in fewer greenhouse gas emissions compared to linear production and consumption models.
(e) **Energy Efficiency**: A circular economy promotes energy-efficient practices by maximizing the value extracted from products and materials, reducing the energy required for extraction, manufacturing, and disposal.
(f) **Preservation of Ecosystems:** Circular economy principles contribute to the preservation of natural ecosystems by reducing the impact of resource extraction, waste accumulation, and pollution.
(g) **Reduced Water Consumption:** Circular practices, such as water-efficient manufacturing processes and reduced demand for new products, lead to lower water consumption and alleviate stress on water resources.
(h) **Biodiversity Conservation**: Lower resource consumption and waste generation contribute to the protection of biodiversity and habitats, reducing the overall human impact on ecosystems.
(i) **Longer Product Lifespans**: Designing products for durability and repair increases the time products remain in use, reducing the need for frequent replacements and the associated environmental impact of manufacturing new products.
(j) **Closed Material Loops**: Circular economy strategies focus on closing material loops through recycling and upcycling. This minimizes the need for extracting and processing virgin materials, which often comes with significant environmental impacts.
(k) **Sustainable Agriculture**: A circular economy extends to agriculture, encouraging sustainable practices that promote soil health, reduce chemical use, and minimize water pollution.
(l) **Regenerative Practices**: Some circular economy approaches, like regenerative agriculture and restorative forestry, actively restore ecosystems and improve soil health, contributing to overall environmental regeneration.
(m) **Environmental Resilience**: By conserving resources and minimizing waste, a circular economy enhances the overall resilience of ecosystems to human activities and external pressures.
(n) **Climate Change Mitigation**: The reduction in greenhouse gas emissions associated with circular practices contributes to global efforts to mitigate climate change and achieve sustainability goals.
(o) **Community Well-being**: Lower environmental impact improves the quality of life for communities by reducing pollution-related health issues and supporting a healthier environment.

4.12 Enhanced Business Resilience

Circular business models often improve resource efficiency, lower production costs, and reduce supply chain risks, making companies more resilient to economic fluctuations. Enhanced business resilience is a significant benefit of a circular economy, as it equips companies with the tools and strategies to navigate uncertainties, disruptions, and market changes while minimizing risks and maximizing opportunities. Here is how a circular economy contributes to enhanced business resilience:

(a) **Diversified Revenue Streams**: Circular business models, such as product-as-a-service or remanufacturing, diversify revenue sources, reducing dependence on single product lines and enhancing financial stability.
(b) **Resource Security**: Circular practices prioritize resource efficiency and reduce dependence on virgin materials, making businesses less vulnerable to supply chain disruptions and resource scarcity.
(c) **Stable Supply Chains:** The circular economy promotes local sourcing, resource sharing, and collaboration, which can lead to more resilient and less vulnerable supply chains.
(d) **Reduction of Price Volatility**: Circular practices reduce exposure to volatile raw material prices by relying more on recycled materials and durable products, which can help stabilize production costs.
(e) **Eco-Efficiency Savings**: Circular strategies, such as energy and resource efficiency, lower operational costs and improve overall financial resilience.
(f) **Waste Management Cost Reduction**: By reducing waste generation and disposal costs, businesses can allocate resources to more productive areas and build financial buffers.
(g) **Long-Term Relationships**: Circular economy principles encourage longer-lasting relationships with customers, suppliers, and partners, fostering stability and reducing the impact of market fluctuations.
(h) **Adaptability to Regulation**: Circular practices often align with evolving environmental regulations and consumer preferences, reducing the risk of non-compliance and regulatory disruption.
(i) **Innovation Culture**: Circular businesses tend to be more innovative and adaptable, with a culture that embraces change and continuous improvement, making them better equipped to respond to market shifts.
(j) **Futureproofing**: Circular business models are designed to withstand changing market dynamics and consumer behaviour, providing a buffer against future uncertainties.
(k) **Resilient Business Models**: Circular business models inherently focus on durability, repairability, and value retention, which can contribute to the long-term viability of the business.

(l) **Risk Mitigation**: Circular practices, such as reverse logistics and take-back programs, facilitate better management of end-of-life products and reduce the risk of liability and reputation damage.
(m) **Brand Reputation**: Businesses adopting circular practices often enjoy positive brand perception due to their commitment to sustainability and responsible consumption, which can help maintain customer loyalty.
(n) **Market Opportunities**: The circular economy opens up new markets and revenue streams related to sustainable products, services, and technologies, increasing business opportunities.
(o) **Scalability and Adaptability**: Circular business models can be scaled and adapted to various industries and contexts, providing flexibility to diversify or pivot as needed.
(p) **Sustainable Innovation**: Investing in circular economy innovations fosters the development of future-proof technologies that contribute to both environmental and business resilience.

4.13 Challenges and Barriers

Mind-set and Culture: Transitioning from a linear to a Circular Economy requires a fundamental shift in mind-set and a cultural change. It involves challenging the prevailing notions of ownership and consumption patterns deeply ingrained in society. Shifting mindsets and culture is a significant challenge in transitioning to a circular economy. This shift requires changing the way individuals, businesses, and societies perceive and engage with resources, consumption, and waste. Here's how mind-set and culture pose challenges in adopting a circular economy:

(a) **Linear Economy Bias**: People are accustomed to the linear "take-make-dispose" model, which may lead to resistance in adopting circular practices that prioritize reuse, repair, and recycling.
(b) **Short-Term Thinking**: A focus on short-term gains often conflicts with the longer-term benefits of circularity, making it challenging to prioritize sustainable practices.
(c) **Perceived Complexity**: Circular practices, such as disassembly, refurbishment, and material recovery, can be perceived as more complex and time-consuming compared to the traditional linear approach.
(d) **Risk Aversion**: Adopting new practices involves uncertainty, and risk-averse mindsets can hinder the willingness to embrace innovative circular approaches.
(e) **Cultural Norms**: Cultural norms around consumption and ownership can hinder acceptance of sharing, leasing, and repair practices that are central to the circular economy.
(f) **Consumer Behavior**: Encouraging consumers to adopt circular behaviors, such as repairing rather than replacing, can be challenging due to established habits and perceptions.

(g) **Incentive Misalignment**: Current economic structures and incentives may favor linear practices, making it difficult for circular initiatives to compete in the market.
(h) **Educational Gap:** Lack of awareness and education about circular economy principles can prevent individuals from understanding the value and benefits of adopting circular practices.
(i) **Resistance from Industries**: Industries heavily invested in the linear economy may resist change due to established business models, infrastructures, and practices.
(j) **Lack of Role Models**: The absence of prominent examples of successful circular businesses or communities can discourage others from adopting circular practices.
(k) **Mind-set Shift in Organizations**: Within organizations, transitioning to a circular economy requires a cultural shift that may face resistance from employees accustomed to traditional practices.
(l) **Perceived Costs**: Circular practices, such as incorporating recycled materials or designing for disassembly, can be perceived as costlier upfront, impacting adoption.
(m) **Political and Regulatory Challenges**: Political and regulatory frameworks designed for a linear economy can hinder the transition to circular practices, creating obstacles for change.
(n) **Measurement and Reporting**: The circular economy may require different metrics and reporting mechanisms to evaluate success, posing challenges to existing measurement systems.
(o) **Communication and Messaging**: Effectively communicating the benefits of the circular economy and overcoming misconceptions requires tailored messaging and engagement strategies.
(p) **Cultural Shift Timing:** Changing mind-sets and cultures takes time and aligning cultural change with immediate business goals can be challenging.

4.14 Investment and Financing

While the Circular Economy promises long-term benefits, initial investments in sustainable practices may be higher. Lack of access to affordable financing can hinder the adoption of circular business models, particularly for small and medium-sized enterprises (SMEs). Investment and financing are key challenges in transitioning to a circular economy, as the shift requires upfront investments, changes in financial models, and overcoming certain barriers that traditional linear economies do not necessarily face. Here's how investment and financing can pose challenges in adopting a circular economy:

(a) **Upfront Costs:** Transitioning to circular practices, such as redesigning products for durability or setting up recycling infrastructure, often involves initial investments that can be higher than maintaining linear practices.

(b) **Risk Perception**: Investors and financiers may perceive circular projects as riskier due to the unfamiliarity of circular business models, technologies, and potential uncertainties in returns.
(c) **Lack of Short-Term Returns**: Circular projects might not yield immediate financial returns, making it challenging to secure financing from investors focused on short-term gains.
(d) **Valuation Challenges**: Valuing circular assets, such as refurbished or remanufactured products, can be complex and challenging within traditional financial frameworks.
(e) **Complex ROI Calculation**: The return on investment (ROI) for circular projects may be harder to quantify due to multiple benefits, such as reduced waste, energy savings, and longer product lifespans.
(f) **Mismatched Business Models**: Circular business models, like product-as-a-service or leasing, might differ significantly from traditional models, making it harder to fit within existing investment structures.
(g) **Access to Finance**: Small and medium-sized enterprises (SMEs) often struggle to access finance for circular projects due to limited track records, making it challenging for them to invest in circular practices.
(h) **Financial Regulations**: Traditional financial regulations and reporting standards might not align well with the circular economy's emphasis on longer product life, reuse, and resource efficiency.
(i) **Perceived Uncertainties:** Investors may perceive circular economy initiatives as uncertain due to the relatively untested nature of some circular practices and technologies.
(j) **Longer Payback Periods**: Circular projects might have longer payback periods compared to linear projects, which could deter investors looking for quicker returns.
(k) **Lack of Awareness**: Investors might lack awareness about the potential profitability and sustainability benefits of circular economy investments.
(l) **Cultural Shift in Investment Community**: Shifting the investment community's mindset from short-term gains to long-term sustainability can be challenging.
(m) **Financing for Innovation**: Developing and scaling innovative circular technologies often requires patient capital and specialized financing, which might not be readily available.
(n) **Policy Uncertainty**: Lack of clear policies and incentives supporting circular initiatives can create uncertainty for investors and make them hesitant to commit funds.
(o) **Data and Metrics:** Investors may struggle to find consistent and comparable data to assess the environmental and financial performance of circular projects.

4.15 Policy and Regulatory Frameworks

Governments play a vital role in facilitating the transition to a Circular Economy. However, inconsistent policies and inadequate regulatory support can create uncertainty and discourage businesses from adopting circular practices. Policy and regulatory frameworks play a crucial role in shaping the adoption and success of a circular economy. While many policies support linear practices, transitioning to a circular economy requires a re-evaluation and modification of these frameworks. Here's how policy and regulatory challenges can hinder the advancement of a circular economy:

(a) Linear Economy Bias: Many existing policies and regulations are designed to support linear practices, such as single-use products and disposal-based models. Shifting to circular practices can clash with these established norms.
(b) Lack of Alignment: Policy and regulatory frameworks might not align with the principles of a circular economy, making it difficult for businesses to adopt circular practices without encountering barriers.
(c) Incentives for Linear Practices: Some policies unintentionally incentivize linear practices by favoring resource-intensive industries or not adequately accounting for the environmental impacts of disposal.
(d) Risk Aversion: Policymakers might be risk-averse and hesitant to support innovative circular practices due to concerns about economic stability and potential disruptions.
(e) Complexity and Fragmentation: The transition to a circular economy often requires comprehensive changes across sectors, which can result in fragmented policies and regulations that lack coherence.
(f) Lack of Circular Metrics: Policy frameworks may lack standardized metrics and indicators to evaluate circular initiatives, making it challenging to measure and compare progress.
(g) Short-Term Focus: Many policies prioritize short-term economic gains over long-term environmental and social benefits, which can hinder the adoption of circular practices.
(h) Regulatory Uncertainty: Unclear or inconsistent regulations can create uncertainty for businesses investing in circular practices, discouraging innovation and investment.
(i) Legal Barriers: Legal constraints related to property rights, liability, and warranties can hinder circular practices such as repair, reuse, and remanufacturing.
(j) Consumer Protection: Some circular practices, such as using refurbished products or sharing resources, might raise concerns about consumer safety and protection.

(k) Taxation Structure: Tax structures that favor virgin materials over recycled content can discourage the use of recycled materials and inhibit circular practices.
(l) Inertia and Resistance: Policymakers might be resistant to change, especially if there are strong existing industries and stakeholders benefiting from linear practices.
(m) Lack of Education and Awareness: Insufficient awareness about the benefits of a circular economy among policymakers can result in a lack of support for relevant policy changes.
(n) **Sectoral Fragmentation**: Different sectors might have varying needs and challenges in adopting circular practices, making it challenging to create a cohesive policy framework.
(o) **Trade and International Agreements:** Trade agreements might not consider the circular economy's unique supply chain and product lifecycle requirements, potentially creating barriers to circular practices.

4.16 Infrastructure and Technology

Establishing efficient collection, sorting, and recycling infrastructure is crucial for the Circular Economy's success. Upgrading technology and investing in research and development are essential to optimize resource recovery processes. Infrastructure and technology play a crucial role in enabling the transition to a circular economy. However, several challenges can arise when adapting existing infrastructure and developing innovative technologies to support circular practices. Here's how infrastructure and technology can pose challenges in adopting a circular economy:

(a) **Legacy Infrastructure**: Existing infrastructure is often designed for linear processes and may not be suitable for circular practices, requiring significant modifications or replacements.
(b) **Lack of Recycling Facilities**: Adequate recycling and material recovery facilities might be lacking, making it difficult to efficiently collect, sort, and process recyclable materials.
(c) **Inadequate Collection Systems**: Efficient collection systems are essential for retrieving products and materials at the end of their life. Developing and implementing these systems can be challenging.
(d) **Complex Material Flows**: Circular practices require intricate material flows that involve disassembly, separation, and reprocessing. Designing and managing these flows can be complex.
(e) **Technological Gaps**: Developing technologies for advanced recycling, material recovery, and remanufacturing may require significant research and investment.
(f) **Lack of Standardization**: Lack of standardization in materials, parts, and processes can hinder effective disassembly, remanufacturing, and reuse.
(g) **Data and Traceability**: Tracking materials through their lifecycle requires robust data and traceability systems that might not be in place.

(h) **High Initial Costs**: Adopting circular technologies, such as advanced recycling techniques, might involve high upfront costs that deter businesses from investing.
(i) **Innovation and Research**: Developing innovative circular technologies requires research, development, and testing, which might face funding and expertise challenges.
(j) **Consumer Behaviour and Infrastructure**: Circular practices, such as product sharing and repair, require supportive consumer behaviors and accessible infrastructure like repair shops or sharing platforms.
(k) **Education and Training**: Transitioning to circular practices often requires educating workers and technicians about new skills and technologies.
(l) **Scalability**: Scaling up circular technologies and infrastructure to accommodate larger volumes can be challenging and require significant investments.
(m) **Technological Integration**: Integrating new technologies with existing systems can be complex and require careful planning to avoid disruptions.
(n) **Local Contexts**: The suitability of infrastructure and technology solutions varies by location, making it challenging to create standardized solutions for diverse regions.
(o) **Supply Chain Coordination**: Circular practices often involve coordinating various stakeholders across the supply chain, which can be challenging to achieve efficiently.
(p) **Regulatory Compliance:** Circular technologies and practices might need to meet specific regulatory standards, posing challenges in terms of compliance and implementation.

The Circular Economy presents a compelling vision for a more sustainable and prosperous future. By embracing its core principles, we can move towards a regenerative economic system that preserves natural resources, reduces waste, and mitigates the impacts of climate change. However, realizing this vision demands collective efforts from governments, businesses, consumers, and civil society. Through collaboration and innovative solutions, we can build a Circular Economy that not only benefits the environment but also creates a thriving and resilient global community.

4.17 Circular Economy Practices: Implementing Sustainability and Resilience

The Circular Economy is a visionary concept that challenges the conventional linear economic model by promoting sustainable practices that prioritize resource efficiency, waste reduction, and environmental conservation. This essay delves into various Circular Economy practices that can be adopted by businesses, governments, and individuals to create a more sustainable and resilient society. By embracing these practices, we can transition towards a regenerative economic system that harmonizes economic growth with ecological well-being.

4.17.1 Design for Circular Products

At the heart of the Circular Economy is the idea of designing products with circularity in mind. This entails creating goods that are durable, easily repairable, and modular, allowing for components to be replaced or upgraded rather than discarded entirely. By employing eco-design principles, businesses can ensure that their products have a longer lifespan and minimize the need for constant replacements. Designing for circular products is a fundamental practice in a circular economy, as it focuses on creating products that can be easily repaired, reused, remanufactured, and recycled at the end of their life. This approach minimizes waste, extends product lifespans, and conserves resources.

At the heart of the Circular Economy lies the concept of designing for circular products. This concept requires a fundamental reimagining of how products are conceived, developed, and manufactured. Designers and manufacturers must prioritize durability, reparability, and ease of disassembly in their product designs. These considerations extend the lifespan of products, reduce the need for frequent replacements, and contribute to resource conservation.

1. **Extended Product Lifespan**: Circular product design focuses on creating products that are built to last. This involves selecting high-quality materials and employing robust manufacturing techniques that resist wear and tear. By increasing product longevity, consumers are encouraged to use products for a longer period, reducing the overall demand for new products and the resources required to produce them.
2. **Modular Design and Repairability**: Circular products are often designed with modularity in mind. This means that individual components can be easily replaced or repaired when they malfunction or become obsolete. This approach not only reduces the amount of waste generated but also empowers consumers to engage in repair rather than disposal, thereby fostering a culture of maintenance and resource conservation.
3. **Designing for Disassembly and Recycling:** Traditional products are often difficult to disassemble, making the recovery of valuable materials for recycling a challenging task. In contrast, circular products are designed with disassembly in mind, ensuring that materials can be easily separated and recycled at the end of their lifecycle. This practice contributes to a closed-loop system where materials are continuously cycled back into production processes, reducing the need for virgin resources.
4. **Material Selection and Environmental Impact:** Circular product design necessitates careful consideration of material choices. Designers must opt for materials that have a lower environmental impact throughout their lifecycle, from extraction to disposal. This involves evaluating factors such as resource availability, energy consumption, and emissions. Additionally, bio-based and renewable materials can play a crucial role in reducing the ecological footprint of products.

4.17.2 Challenges and Opportunities

While the concept of designing for circular products offers significant benefits, it is not without challenges. Designers may encounter constraints related to existing supply chains, manufacturing processes, and consumer preferences. However, these challenges also present opportunities for innovation and collaboration. Designers can work closely with manufacturers, suppliers, and consumers to overcome these barriers and develop new business models that prioritize circularity. Here is how to implement designing for circular products as a practice in a circular economy:

- **Holistic Design Thinking:**
  Adopt a holistic approach that considers the entire product lifecycle, from raw material extraction to end-of-life treatment.
- **Durability and Longevity:**
  Design products that are built to last, using high-quality materials and robust construction techniques.
- **Modularity and Standardization:**
  Design products with modular components that can be easily replaced or upgraded. Standardize parts to facilitate compatibility and interchangeability.
- **Ease of Disassembly:**
  Ensure products can be disassembled without damaging components, enabling efficient repair, refurbishment, and recycling.
- **Minimal Glue and Adhesives:**
  Avoid excessive use of adhesives, as they can hinder disassembly. Opt for mechanical fastening methods when possible.
- **Design for Repair:**
  Make components that are prone to wear and tear accessible and replaceable. Provide clear repair instructions and support.
- **Material Selection:**
  Choose materials that are easily recyclable, have a low environmental impact, and can be sourced responsibly.
- **Recyclability:**
  Design products with materials that can be easily separated and recycled. Minimize mixed materials and complex composites.
- **Avoid Hazardous Substances:**
  Eliminate or minimize the use of hazardous chemicals or materials that can complicate recycling processes.
- **Design for Upgradability:**
  Allow for easy upgrades to extend the product's lifespan. For instance, in electronics, enable software and hardware updates.
- **Adopt Digital Twin Technology:**
  Use digital twin technology to create virtual models of products, helping manufacturers predict maintenance needs and optimize design.

- **User-Centered Design:**
  Involve end-users in the design process to understand their needs and preferences, ensuring products are functional, appealing, and easy to use.
- **Eco-Labeling and Information:**
  Provide clear information about the product's environmental impact, recycling instructions, and recommended maintenance.
- **Collaboration with Suppliers**:
  Collaborate with suppliers to source sustainable materials and components, aligning their practices with circular principles.
- **Lifecycle Assessment:**
  Conduct a lifecycle assessment to analyze the environmental impact of the product at each stage and identify areas for improvement.
- **Pilot and Iterate:**
  Pilot circular product designs and gather feedback from users and stakeholders to refine and improve designs over time.
- **Educate Designers and Engineers:**
  Train designers and engineers in circular design principles and provide resources to help them implement these practices effectively.
- **Business Model Alignment:**
  Ensure that the circular product design aligns with the chosen business model, such as leasing or selling services instead of products.
- **Regulatory Compliance:**
  Design products in line with relevant regulations, ensuring they can be legally and safely reused, repaired, and recycled.

4.17.3 Implementing Product-as-a-Service Models

Product-as-a-Service models represent a departure from traditional ownership-based consumption patterns. In these models, rather than purchasing a product outright, consumers access the functionality of products as a service, paying for the utility they derive from the product rather than the physical item itself. PaaS models encompass a range of arrangements, including leasing, renting, and subscription-based services. This approach offers several key advantages that align seamlessly with the principles of the Circular Economy.

1. **Extended Product Lifespan**: Under PaaS models, manufacturers have a vested interest in maximizing the longevity of their products. Products are designed, manufactured, and maintained with durability in mind, as their continued use directly impacts the manufacturer's revenue. This contrasts with the traditional model, where planned obsolescence and frequent replacements are encouraged. Longer product lifespans lead to reduced resource consumption and waste generation.

2. **Responsibility for Product Lifecycle**: In PaaS models, the responsibility for the entire product lifecycle, including maintenance, repair, and end-of-life management, often rests with the manufacturer or service provider. This incentivizes manufacturers to create products that are easy to repair, upgrade, and recycle. Materials and components are carefully selected to facilitate disassembly, refurbishment, and material recovery at the end of their useful life.
3. **Resource Efficiency**: PaaS models promote optimal resource utilization. Manufacturers aim to maximize the efficiency of their products to reduce operational costs and enhance customer satisfaction. This frequently involves designing products with modular components that can be easily replaced or upgraded. Consequently, fewer resources are needed to meet consumer needs, reducing the overall demand for raw materials.
4. **Consumer Access over Ownership**: PaaS models emphasize access to products rather than ownership, aligning with the shift towards a more sustainable and less resource-intensive lifestyle. Consumers pay for the utility and experience of using a product without the burden of long-term ownership. This approach encourages a change in consumer behavior, reducing the impulse to constantly acquire new possessions.
5. **Data-Driven Innovation:** PaaS models often leverage digital technologies to monitor product usage, performance, and maintenance needs. This data-driven approach enables manufacturers to make informed decisions about product design improvements, maintenance schedules, and resource allocation. By harnessing data, manufacturers can optimize their operations and enhance product sustainability.

Challenges and Opportunities

While PaaS models offer significant potential for advancing the Circular Economy, their widespread adoption faces challenges. Consumer perceptions and habits may need to shift to embrace the idea of accessing products rather than owning them. Additionally, implementing efficient reverse logistics, refurbishment processes, and end-of-life management systems requires collaboration across industries.

Product-as-a-Service (PaaS) is an innovative business model that shifts the focus from selling products to providing services. Instead of owning a product, customers lease or rent it, while the responsibility for maintenance, repairs, and end-of-life management remains with the manufacturer. This approach incentivizes companies to produce high-quality, long-lasting products since their revenue is tied to the product's continued functionality. Implementing Product-as-a-Service (PaaS) models is a key practice in a circular economy that shifts the focus from selling products to offering them as services. This approach encourages resource efficiency, maximizes product use, and minimizes waste. Here is how to implement PaaS models as a practice in a circular economy:

(a) **Identify Suitable Products**: Begin by identifying products that are suitable for a PaaS model. These could be products with high maintenance or upgrade needs, frequent replacements, or products used intermittently.

(b) **Design for Durability**: Design products that are durable and built to withstand extended use. This reduces the frequency of replacements and maintenance.
(c) **Service Design and Packaging**: Develop service packages that include installation, maintenance, repairs, and upgrades. Ensure the packaging is minimal and designed for easy return and transport.
(d) **Subscription Pricing**: Offer subscription-based pricing models where customers pay for access to the product and the services provided, rather than purchasing the product outright.
(e) **User-Centered Experience**: Focus on delivering a positive user experience by offering seamless access, excellent customer support, and timely maintenance services.
(f) **Product Tracking and Monitoring**: Integrate tracking and monitoring technologies to keep tabs on product usage, performance, and maintenance needs.
(g) **Data Utilization:** Use data collected from product usage to optimize maintenance schedules, improve product design, and enhance customer experiences.
(h) **Reverse Logistics**: Establish efficient reverse logistics systems to retrieve products at the end of their useful life for repair, refurbishment, and recycling.
(i) **Quality Control and Refurbishment**: Develop processes to inspect, refurbish, and upgrade returned products to ensure they can be redeployed effectively.
(j) **Circular Partnerships**: Collaborate with suppliers, manufacturers, and service providers to create a seamless ecosystem for PaaS implementation.
(k) **Consumer Education**: Educate customers about the benefits of PaaS, emphasizing reduced ownership costs, hassle-free maintenance, and environmental impact.
(l) **Risk Management**: Assess and manage risks associated with product maintenance, breakdowns, and customer expectations to maintain service quality.
(m) **Legal and Compliance Considerations**: Address legal aspects such as warranties, liabilities, and intellectual property rights to ensure a smooth PaaS operation.
(n) **Adaptability and Flexibility**: Design PaaS models to be adaptable to changing customer needs and evolving technologies.
(o) **Continuous Improvement**: Continuously gather feedback from customers and monitor performance metrics to refine and improve PaaS offerings.
(p) **Sustainable Financing**: Secure the necessary funding to support the initial investment in product inventory, service infrastructure, and technology.
(q) **Scale Gradually**: Start with a pilot phase and gradually scale up your PaaS offerings as you refine processes and learn from customer interactions.
(r) **Communicate Sustainability**: Highlight the environmental benefits of PaaS models, such as reduced resource consumption and waste, to attract environmentally conscious customers.

4.18 Embracing the Sharing Economy

The Sharing Economy, often referred to as collaborative consumption, refers to the practice of sharing resources, goods, or services among individuals and communities rather than solely relying on ownership. This model taps into the principles of the Circular Economy by extending the lifespan of products, optimizing resource utilization, and reducing overconsumption.

1. **Resource Efficiency**: The Sharing Economy reduces the need for every individual to own items that are infrequently used, leading to a more efficient allocation of resources. Instead of purchasing items that might be used only occasionally, individuals can access them through sharing platforms. This not only reduces the demand for new production but also conserves raw materials and energy.
2. **Product Utilization:** Sharing platforms enable products to be used to their fullest potential, minimizing their underutilization and idle time. Items such as tools, equipment, and vehicles that might otherwise remain dormant in households can be put to use by multiple users, maximizing their functional lifespan and reducing waste.
3. **Reduced Waste Generation**: The Sharing Economy inherently combats the "throwaway culture" by encouraging the reuse of items. Products remain in circulation for longer periods, reducing the rate at which they become wasted. Moreover, when items are shared, they often undergo maintenance and repair to ensure they are in good condition, further extending their useful life.
4. **Behavioral Shifts**: Embracing the Sharing Economy encourages a shift in consumer behavior from ownership to access. Individuals become more conscious of their consumption patterns and the environmental impacts of their choices. This behavioral transformation contributes to a more sustainable lifestyle and supports the Circular Economy's goals.
5. **Innovation and Entrepreneurship**: The Sharing Economy has spurred the development of innovative business models and digital platforms that facilitate resource sharing. This fosters entrepreneurship and supports the creation of new jobs and economic opportunities, contributing to the growth of sustainable industries.

Challenges and Opportunities

Despite its numerous benefits, the Sharing Economy faces challenges that require careful consideration. Issues such as trust, privacy, and regulatory frameworks need to be addressed to ensure the integrity and safety of sharing transactions. Moreover, the Sharing Economy should complement, rather than replace, existing business models, and efforts should be made to ensure inclusivity and accessibility for all members of society.

The sharing economy fosters the shared use of resources, reducing the overall demand for new products. Car-sharing, bike-sharing, and co-working spaces are excellent examples of sharing economy initiatives. By facilitating resource-sharing platforms, businesses can promote sustainable consumption patterns and help maximize resource utilization.

Embracing the sharing economy is a practice that aligns closely with the principles of a circular economy. The sharing economy encourages individuals and businesses to share resources, products, and services, thereby maximizing the utilization of existing assets and reducing the need for excessive production. Here is how to embrace the sharing economy as a practice within the context of a circular economy:

(a) **Identify Sharing Opportunities**: Identify products, assets, or services that are suitable for sharing within your target market. Consider items that are underutilized, infrequently used, or have occasional demand.
(b) **Platform Development:** Create or leverage digital platforms that facilitate peer-to-peer sharing, connecting individuals and businesses willing to share resources.
(c) **Collaborative Consumption Models**: Design sharing models that allow users to access items or services on a temporary basis, paying for usage rather than ownership.
(d) **Quality and Safety Assurance**: Ensure that items or services shared meet quality and safety standards, building trust among users.
(e) **User-Centered Design**: Prioritize user experience by making sharing platforms user-friendly, intuitive, and responsive to user needs.
(f) **Regulatory Compliance**: Navigate local regulations and compliance requirements related to sharing services, such as liability, taxation, and insurance.
(g) **Community Building:** Foster a sense of community among sharers by facilitating interactions, ratings, and reviews among participants.
(h) **Variety of Sharing Models**: Explore various sharing models, including peer-to-peer sharing, business-to-consumer sharing, and business-to-business sharing.
(i) **Educational Campaigns**: Educate users about the benefits of sharing economy practices, such as reduced consumption, cost savings, and environmental impact.
(j) **Circular Business Partnerships**: Collaborate with other businesses to integrate sharing economy practices into existing circular business models.
(k) **Shared Mobility**: Implement shared mobility solutions, such as car-sharing or bike-sharing programs, to reduce the need for individual vehicle ownership.
(l) **Co-Working and Co-Living Spaces**: Create spaces where individuals and businesses can share workspaces, accommodations, and resources, promoting efficiency and collaboration.
(m) **Resource Optimization**: Promote the optimization of resources by ensuring that shared items are consistently in use, reducing waste and idle assets.
(n) **Data Utilization**: Collect and analyse data on sharing patterns to optimize inventory, improve user experiences, and predict demand.

(o) **Circular Education and Awareness**: Communicate how sharing contributes to circular economy goals, fostering a sense of environmental responsibility among participants.
(p) **Scalability and Expansion**: Plan for scalability and expansion of your sharing services based on user demand and market opportunities.
(q) **Feedback and Continuous Improvement**: Regularly collect feedback from users to refine your sharing platform, enhance offerings, and address any concerns.
(r) **Cross-Sector Collaboration**: Collaborate with local governments, NGOs, and other stakeholders to create an enabling environment for sharing economy practices.

4.19 Extending Product Lifecycles Through Refurbishment

In an era of growing environmental concerns and resource depletion, the concept of a circular economy has emerged as a beacon of hope for achieving sustainable development. Rooted in the principles of reducing waste, conserving resources, and promoting responsible consumption, the circular economy advocates for practices that challenge the traditional linear "take-make-dispose" model. Among the pivotal practices that exemplify circular economy ideals, extending product lifecycles through refurbishment stands out as a powerful strategy to minimize waste, conserve resources, and foster a culture of sustainability.

Refurbishment, at its core, involves repairing, restoring, and upgrading products to extend their usable lifespan. This practice resonates deeply with the circular economy philosophy, as it aligns with the principles of reuse, repair, and regeneration. By breathing new life into products that might otherwise be discarded, refurbishment not only reduces the strain on finite resources but also lessens the environmental impact associated with the production of new goods.

One of the central tenets of refurbishment in the circular economy is the concept of design for refurbishment. This approach emphasizes creating products with modular components and easily accessible parts, ensuring that disassembly and repair are straightforward processes. Products are engineered to be "open" rather than sealed shut, enabling technicians to repair or replace specific components without the need for intricate and time-consuming dismantling. By considering the entire lifecycle of a product during the design phase, manufacturers can create items that can be efficiently refurbished, a hallmark of a circular approach.

The benefits of extending product lifecycles through refurbishment are manifold. Foremost among them is the reduction of waste. By diverting products from landfills and incinerators, refurbishment contributes to lowering the volume of waste entering the waste stream, mitigating the environmental impact of waste disposal and reducing the demand for new resource extraction. Moreover, the practice conserves energy and decreases greenhouse gas emissions associated with manufacturing new products, aligning closely with the imperative to combat climate change.

Refurbishment also offers economic advantages, both for businesses and consumers. Consumers benefit from cost savings, as refurbished products often come at a lower price point than brand-new equivalents. For businesses, embracing refurbishment can open new revenue streams through the resale of refurbished items. Furthermore, the practice stimulates job creation by fostering demand for skilled technicians who perform repairs, upgrades, and quality control.

However, implementing an effective refurbishment practice requires a holistic approach. Collaboration among various stakeholders, including manufacturers, suppliers, refurbishment centres, and consumers, is essential to create a circular ecosystem. The development of transparent communication channels is vital, allowing consumers to understand the refurbishment process, the extent of upgrades, and the environmental benefits associated with choosing refurbished products.

The success of extending product lifecycles through refurbishment also hinges on consumer awareness and education. Informing consumers about the environmental and economic advantages of refurbished products can shift consumer behaviour toward more sustainable choices. Building trust is crucial, and providing warranties or guarantees for refurbished items can alleviate concerns about product quality and performance.

4.20 Adopting Sustainable Materials and Recycling

As our world grapples with pressing environmental challenges, the circular economy has emerged as a beacon of hope, offering a transformative approach to production and consumption that is both economically viable and ecologically responsible. One of the pivotal practices within this paradigm is the adoption of sustainable materials and recycling. This practice represents a fundamental shift from the traditional linear model of resource extraction and disposal to a regenerative system that conserves resources, reduces waste, and contributes to a sustainable future.

At the heart of the circular economy's ethos is the idea of breaking free from the "take-make-dispose" mentality that characterizes linear consumption patterns. Adopting sustainable materials is a cornerstone of this paradigm shift. Sustainable materials encompass those that are ethically sourced, have a lower environmental impact, and are designed to be repurposed, recycled, or biodegraded at the end of their lifecycle. These materials, whether they are bioplastics derived from renewable resources or reclaimed metals from discarded products, exemplify the circular economy's commitment to minimizing resource consumption and reducing pollution.

Recycling, a complementary practice, breathes life into the circular economy's ideals. It involves collecting, sorting, and processing materials from discarded products to create new goods, thereby reducing the demand for virgin resources. The practice closes the loop by reintroducing materials into the production cycle, reducing the energy and environmental costs associated with extraction and manufacturing. Recycling embodies the concept of "circularity" as materials are continuously repurposed, thus minimizing waste and conserving valuable resources.

The adoption of sustainable materials and recycling bears numerous benefits, not only for the environment but also for economies and societies. Firstly, it helps mitigate the negative impacts of resource extraction. By opting for materials that are responsibly sourced and have a lower ecological footprint, industries can reduce habitat destruction, water pollution, and carbon emissions. Secondly, it lessens the burden on landfills and incinerators, thereby mitigating environmental degradation and reducing the release of harmful chemicals into the ecosystem.

Furthermore, the practice promotes energy efficiency. Recycling typically requires less energy than extracting and processing virgin materials. This energy savings contributes to a reduction in greenhouse gas emissions, aligning with global efforts to combat climate change. Additionally, recycling supports local economies by creating jobs in collection, sorting, processing, and manufacturing, thus stimulating economic growth and fostering community resilience.

However, embracing sustainable materials and recycling is not without challenges. Achieving a closed-loop system requires overcoming barriers such as inconsistent recycling infrastructure, consumer education, and ensuring the quality of recycled materials. Addressing these challenges demands collaborative efforts among governments, industries, and consumers.

Consumer education is paramount. Raising awareness about the environmental benefits of recycling and encouraging proper waste segregation can enhance the quality and quantity of recyclable materials. Governments must play a role by incentivizing sustainable practices and investing in robust recycling infrastructure. Industries can contribute by designing products for recyclability, using recycled materials in their production processes, and promoting eco-friendly consumption patterns.

4.20.1 Encouraging Local Production and Supply Chains

Localizing production and supply chains can enhance resource efficiency by reducing transportation-related emissions and waste. Supporting local businesses and sourcing materials from nearby suppliers can strengthen regional economies and foster a sense of community.

As the world faces escalating environmental challenges and calls for sustainability grow louder, the circular economy has emerged as a transformative approach that holds the promise of reconciling economic growth with environmental stewardship. Within this paradigm, the practice of encouraging local production and supply chains stands out as a pivotal strategy. By reorienting production processes towards local resources and markets, this practice not only reduces environmental impact but also enhances economic resilience and community well-being.

The essence of encouraging local production and supply chains lies in challenging the traditional model of globalization, which often involves long-distance transportation, resource-intensive manufacturing, and the carbon footprint associated with international trade. The circular economy advocates for a shift towards a regenerative model, emphasizing shorter loops and a stronger connection between production, consumption, and waste management. By sourcing materials locally, producing goods near their intended markets, and minimizing transportation distances, this practice reduces the ecological strain of global supply chains.

At the core of this strategy is the promotion of regional self-sufficiency. By encouraging local production, communities become less reliant on distant sources of materials and goods. This resilience is particularly crucial in times of disruption, such as supply chain interruptions or natural disasters, where local producers are better positioned to meet demand and maintain essential supplies. The COVID-19 pandemic has underscored the importance of local production in ensuring access to critical goods and services during times of crisis.

Moreover, fostering local production aligns with the circular economy's goal of reducing resource consumption. When resources are sourced from nearby regions, the environmental impact associated with transportation decreases significantly, thereby curbing carbon emissions and conserving energy. Local production also promotes the use of sustainable materials and reduces waste, as products are designed to be easily disassembled, repaired, and recycled within the local ecosystem.

Encouraging local production and supply chains also reinvigorates economies at the grassroots level. By supporting local businesses, this practice enhances job creation, stimulates economic growth, and nurtures community cohesion. Small and medium-sized enterprises (SMEs) benefit from reduced transportation costs and can compete more effectively in local markets. This in turn bolsters economic diversity, fostering robust local economies that are less susceptible to the vulnerabilities of global markets.

However, implementing this practice requires overcoming certain challenges. Globalized supply chains are deeply ingrained in existing economic structures and shifting towards local production demands reevaluation and restructuring. Policymakers must craft supportive frameworks that incentivize local production, provide funding for start-ups, and facilitate partnerships between businesses and local communities. Collaborative efforts are essential, as encouraging local production requires the active participation of governments, businesses, and consumers.

Consumer behavior also plays a crucial role. Raising awareness about the benefits of supporting local businesses and choosing locally produced goods is paramount. Consumers can influence markets by prioritizing sustainability and demanding transparency in supply chains. By creating a demand for local products, consumers contribute to a virtuous cycle that promotes local production and supply chains.

4.20.2 Promoting Behavior Change Through Education

Educating consumers about the principles and benefits of the Circular Economy is crucial for driving sustainable behavior. Public awareness campaigns, workshops, and educational programs can empower individuals to make conscious choices, such as opting for eco-friendly products, reducing waste, and embracing sustainable lifestyles.

In an era marked by unprecedented environmental challenges and the urgent need for sustainable solutions, the circular economy has emerged as a transformative framework that addresses both environmental degradation and resource scarcity. Central to this paradigm is the practice of promoting behavior change through education—a powerful strategy that seeks to shift individual and collective mindsets toward more sustainable consumption and production patterns. By harnessing the potential of education, the circular economy endeavors to foster a culture of environmental stewardship, responsible consumption, and long-term sustainability.

At its heart, the circular economy challenges the prevailing linear model of "take-make-dispose" by promoting the principles of reduction, reuse, recycling, and regeneration. However, for these principles to be effective, they must be embraced by individuals at all levels of society. This is where education becomes a vital agent of change. By imparting knowledge, raising awareness, and inspiring action, education equips individuals with the tools they need to make informed choices that align with the circular economy's objectives.

Education's impact on behavior change is profound. It shapes perceptions, influences attitudes, and guides decision-making. In the context of the circular economy, education offers a multifaceted approach to drive change:

Raising Awareness: Education brings attention to the environmental challenges associated with linear consumption and production. It highlights the consequences of overconsumption, waste generation, and resource depletion, fostering a deeper understanding of the urgent need for change.

Informing Choices: Education empowers individuals with the information needed to make conscious decisions. It helps them recognize the benefits of responsible consumption, repair, and recycling, allowing them to actively choose products and behaviors that minimize their environmental footprint.

Fostering Critical Thinking: Education nurtures critical thinking skills that enable individuals to question conventional norms and critically evaluate the impact of their actions on the environment. This critical reflection is essential for breaking free from unsustainable consumption patterns.

Cultivating Sustainable Habits: Education plays a pivotal role in cultivating sustainable habits. By instilling values such as resource conservation, waste reduction, and product longevity from an early age, education sets the foundation for lifelong behaviors that align with circular economy principles.

Creating Advocates for Change: Educated individuals are more likely to become advocates for sustainable practices within their communities and networks. They can influence their peers, family members, and colleagues, creating a ripple effect of behavior change.

However, promoting behavior change through education is not without its challenges. Overcoming deeply ingrained habits and societal norms requires concerted efforts and a multifaceted approach:

Comprehensive Curricula: Education systems must integrate environmental education into curricula at all levels, from primary schools to universities. Subjects such as sustainable development, ecological economics, and circular design should be part of the educational landscape.

Engaging Teaching Methods: Interactive and experiential teaching methods can capture students' attention and create lasting impressions. Hands-on activities, field trips, and real-world case studies enhance students' understanding of circular concepts.

Lifelong Learning: Education is not limited to formal schooling. Continuous learning opportunities for all ages, including workshops, seminars, and online courses, ensure that individuals remain informed and engaged.

Partnerships and Collaboration: Governments, educational institutions, non-governmental organizations, and businesses should collaborate to develop and implement effective education programs. Such partnerships enhance the reach and impact of educational initiatives.

Behavioral Nudges: Employ behavioral economics principles to encourage circular behaviors. Strategies like rewards, incentives, and default choices can nudge individuals towards more sustainable options.

4.20.3 Collaborative Initiatives and Partnerships

Collaboration among various stakeholders is essential for the successful implementation of Circular Economy practices. Governments, businesses, non-governmental organizations, and communities must work together to develop policies, standards, and initiatives that support circular principles. Cross-sector partnerships can unlock new opportunities and accelerate the transition towards a Circular Economy.

In an era defined by resource scarcity and environmental challenges, the circular economy has emerged as a transformative paradigm that offers a promising solution to the complex interplay between economic growth, environmental protection, and social well-being. At the heart of this approach lies the practice of promoting collaborative initiatives and partnerships—a strategy that recognizes the interconnected nature of sustainability challenges and leverages the collective power of diverse stakeholders to drive circular economy principles.

The essence of the circular economy lies in its departure from the linear "take-make-dispose" model and its embrace of reduction, reuse, recycling, and regeneration. Yet, to fully realize the potential of circularity, a paradigm shift in collaboration is imperative. Traditional business models have often focused on individual profit maximization, overlooking the broader impact on ecosystems and society. In contrast, the practice of fostering collaborative initiatives and partnerships cultivates an environment where businesses, governments, non-governmental organizations, and communities work together towards common circular goals.

At its core, this practice emphasizes synergy over silos. Collaborative initiatives bridge the gap between various stakeholders, fostering open dialogue, knowledge sharing, and joint decision-making. For instance, businesses can collaborate with suppliers to ensure the availability of sustainable materials, governments can enact supportive policies, and non-profits can contribute expertise and advocacy. The resulting web of partnerships creates a holistic approach that addresses challenges across the value chain and promotes sustainable practices at multiple levels.

The benefits of collaborative initiatives and partnerships in the circular economy are manifold. They enable the pooling of resources, expertise, and best practices, driving innovation and accelerating the transition to circular business models. Collaborations can lead to the co-creation of new products, services, and processes that align with circular principles. Moreover, partnerships foster resilience, as stakeholders collaborate to overcome challenges and adapt to evolving market conditions, creating a more agile and responsive circular economy ecosystem.

However, forging effective partnerships is not without its challenges. Overcoming differences in objectives, priorities, and organizational cultures requires strong leadership, clear communication, and a shared vision. The diverse nature of stakeholders involved—from governments and corporations to grassroots organizations—demands an inclusive approach that acknowledges varying perspectives while uniting efforts towards a common purpose.

Promoting collaborative initiatives and partnerships also necessitates a shift in mind-set. Businesses must recognize that their success is intertwined with the well-being of ecosystems and communities. Governments must engage in dialogue with industry leaders to enact enabling policies that incentivize circular practices. Civil society organizations must harness their advocacy power to create awareness and hold stakeholders accountable.

Innovative models of collaboration are emerging across industries. Cross-sectoral partnerships, public-private collaborations, and multi-stakeholder platforms are facilitating circularity by aligning interests and creating a space for shared learning. Initiatives like industry consortia, circular innovation hubs, and sustainability alliances bridge gaps, fostering a culture of collaboration that transcends traditional boundaries.

4.20.4 Fostering Innovation and Research

Investing in research and development of new technologies and processes is crucial for advancing Circular Economy practices. Innovations in recycling, material recovery, and sustainable manufacturing can drive resource efficiency and create new business opportunities.

In a world grappling with mounting environmental concerns and depleting resources, the circular economy has emerged as a beacon of hope—an alternative economic model that reimagines production and consumption. At the heart of this transformative paradigm lies the practice of fostering innovation and research. By igniting a relentless pursuit of novel ideas, sustainable technologies, and groundbreaking methodologies, this practice propels the circular economy forward, steering humanity toward a more regenerative and resilient future.

The essence of the circular economy is a departure from the linear "take-make-dispose" model to one that embraces principles of reduction, reuse, recycling, and regeneration. Innovation and research serve as the engines propelling this shift. By leveraging human ingenuity and scientific exploration, the circular economy challenges conventional norms and spurs the creation of new business models, products, and processes that minimize waste and maximize resource efficiency.

One of the pivotal roles of innovation and research in the circular economy is the reimagining of products and systems. Design thinking, biomimicry, and technological advancements have given rise to products that are modular, repairable, and designed for longevity. Innovative materials, such as bioplastics derived from renewable resources or textiles crafted from upcycled materials, reduce the environmental footprint of products from conception to disposal.

Moreover, innovation extends to the realm of supply chains. Collaborative platforms, blockchain technology, and smart tracking systems enhance transparency and traceability, ensuring ethical sourcing and reducing the environmental impact of logistics. Supply chains are optimized to minimize waste, energy consumption, and emissions, aligning with the circular economy's core principles.

Research in the circular economy transcends disciplinary boundaries. Scientists, engineers, economists, sociologists, and policymakers collaborate to tackle complex challenges. Life cycle assessments analyze the environmental impact of products from cradle to grave, informing decision-making. Economic models explore the viability of circular business models and the potential for job creation. Societal studies investigate consumer behaviors and attitudes toward circular products and services.

The benefits of fostering innovation and research in the circular economy are profound. Firstly, it propels economies toward greener growth. Innovations create new markets, drive competitiveness, and stimulate investment in sustainable technologies. Secondly, it enhances resource efficiency. Research uncovers new ways to extract value from waste, turning it into a valuable resource. For instance, organic waste can be transformed into bioenergy or nutrient-rich compost.

However, fostering innovation and research in the circular economy comes with challenges. Funding, especially for early-stage research, can be limited. Short-term economic pressures sometimes overshadow long-term sustainability considerations. Bridging the gap between research and practical implementation requires collaboration between academia, industry, and policymakers.

Education plays a vital role in driving innovation in the circular economy. By imparting knowledge, nurturing curiosity, and fostering a culture of continuous learning, education prepares future generations to tackle the complex challenges of resource scarcity, waste, and climate change. Universities and research institutions are hubs of innovation, where cutting-edge ideas are born and nurtured.

4.20.5 Enabling Policy and Regulation

Governments play a critical role in promoting the Circular Economy through supportive policies and regulations. Providing incentives for circular practices, setting recycling targets, and creating a conducive business environment can accelerate the adoption of circular principles across industries.

In a world confronted by the pressing challenges of resource depletion and environmental degradation, the circular economy has emerged as a beacon of hope—a transformative framework that offers a sustainable alternative to the linear "take-make-dispose" model. Central to the success of this paradigm is the practice of enabling policy and regulation. By creating a conducive environment for circular practices to flourish, policy and regulation play a pivotal role in reshaping economies, industries, and societies toward a more sustainable and regenerative future.

At its core, the circular economy challenges the conventional notion of endless resource extraction and waste generation. It emphasizes principles such as reduction, reuse, recycling, and regeneration. However, the effective implementation of these principles relies heavily on a regulatory framework that encourages and supports circular practices. Enabling policy and regulation provide the guidelines, incentives, and mandates necessary to catalyze the transition toward a circular economy.

One of the central roles of policy and regulation is setting the rules of the game. By incorporating circular principles into legal frameworks, governments signal their commitment to sustainability. Policies can mandate extended producer responsibility, compelling manufacturers to design products that are easily repairable and recyclable. Regulations can also promote eco-labeling, empowering consumers to make informed choices based on a product's environmental impact.

Furthermore, enabling policy and regulation stimulate innovation. By offering financial incentives, grants, and tax breaks to businesses that embrace circularity, governments encourage the development of innovative technologies, processes, and business models. This proactive approach to innovation fosters a culture of creativity that drives economic growth while minimizing environmental harm.

Enabling policy and regulation also creates a level playing field. By internalizing externalities—such as the environmental costs of resource extraction and waste disposal—policies ensure that businesses pay the true cost of their activities. This pricing mechanism encourages resource efficiency and waste reduction, making circular practices economically attractive and competitive.

The benefits of enabling policy and regulation in the circular economy are manifold. Firstly, they provide certainty to businesses. A consistent regulatory environment reduces uncertainty and encourages long-term investments in circular initiatives. Secondly, they bolster consumer confidence. Regulations ensure that products meet specific environmental and quality standards, empowering consumers to make sustainable choices.

However, the path to effective policy and regulation in the circular economy is not without challenges. Striking the right balance between promoting circular practices and allowing flexibility for innovation can be complex. Policymakers must navigate the complexities of industries with differing needs and engage in dialogue with stakeholders to ensure that regulations are practical, achievable, and impactful.

Moreover, policy coherence across different levels of government is vital. Coordination between local, national, and international authorities is essential to prevent conflicting regulations and ensure a harmonious transition to circular practices. Policymakers must also be prepared to adapt and evolve regulations as technologies, market dynamics, and societal expectations change.

4.21 Conclusion

The Circular Economy presents a viable and sustainable path towards a more prosperous and resilient future. By embracing circular practices at all levels of society, from individuals to multinational corporations, we can transition away from the linear "take-make-dispose" model and build a regenerative economic system. The practices outlined in this essay offer a roadmap for integrating sustainability, waste reduction, and resource efficiency into our daily lives. By working together and nurturing a collective commitment to the Circular Economy, we can pave the way for a more sustainable and harmonious relationship between humans and the environment.

References

Hegab, H., Shaban, I., Jamil, M., & Khanna, N. (2023). Toward sustainable future: Strategies, indicators, and challenges for implementing sustainable production systems. *Sustainable Materials and Technologies, 36*, e00617. https://doi.org/10.1016/j.susmat.2023.e00617

Martha. (2023). *Sustainable product design: Creating a better future*. ICSID. https://www.icsid.org/uncategorized/sustainable-product-design-creating-a-better-future/

Niwan, N., & Sharma, B. (2023). Sustainable solutions for a changing world. In *Twist* (Issue 118). Kripa-Drishti Publications.

Qassam, O. (2023). Addressing the electronic waste crisis through circular economy practices. Medium. https://medium.com/@OdaiQassam/addressing-the-electronic-waste-crisis-through-circular-economy-practices-a7d55ea7de0f

Chapter 5
Sustainable Financing for ESG Practices

Abstract The corporate sector has witnessed an increasing incorporation of environmental, social, and governance (ESG) policies as investors, businesses, and consumers become more cognizant of the significance of ethical and sustainable business practices. Achievement of ESG objectives, nevertheless, frequently necessitates substantial financial investments and enduring dedications. This chapter examines the diverse sustainable financing alternatives that organizations can utilize to implement and incorporate ESG standards into their business activities. An examination will be conducted on the manner in which sustainable financing facilitates the execution of environmental, social, and governance (ESG) initiatives and yields positive outcomes. This analysis will encompass both conventional financial instruments and novel funding alternatives.

Keywords ESG practices · Green finance · Sustainable financing · Investor education · ESG reporting · Impact investing · Green bonds

5.1 Introduction

Environmental, social, and governance (ESG) principles have gained growing importance in the financial and corporate sectors in recent years. Integrating environmental, social, and governance (ESG) factors into investment decisions and corporate operations has become a fundamental aspect of sustainable and responsible business practices. Given the pressing global issues such as climate change, social inequality, and ethical governance concerns, it is crucial to prioritize the establishment of sustainable financial structures to support environmental, social, and governance (ESG) activities (PRECEDENCE RESEARCH, 2023). This chapter will explore the core principles of ESG and assess its implications for businesses, investors, and society as a whole.

V. Kandpal et al., *Sustainable Energy Transition*, Circular Economy and Sustainability, https://doi.org/10.1007/978-3-031-52943-6_5

What is ESG?

ESG practices encompass the incorporation of environmental, social, and governance factors into different facets of business operations, investment decisions, and corporate strategies. These approaches acknowledge that a company's influence extends beyond its financial performance and shareholder worth, encompassing its contributions to society, the environment, and overall governance. ESG principles are becoming increasingly important as companies and investors recognize the importance of sustainability, responsible governance, and social impact.

Environmental Considerations (E)

Environmental factors comprise the effect a business has on the natural environment. This consists of its carbon footprint, energy efficiency, waste management, water consumption, and efforts to reduce greenhouse gas emissions. Strong environmental practices are characterized by an effort to minimize negative environmental impacts, adopt renewable energy sources, and develop sustainable products and services.

Social Considerations (S)

A company's interactions with its employees, consumers, suppliers, communities, and other stakeholders are the focus of social factors. Companies with strong social practice prioritize the welfare of their employees, diversity and inclusion, labor standards, health and safety, and community involvement. In addition, they address issues such as human rights, supply chain ethics, and charitable endeavors.

Governance Considerations (G)

Governance factors are an organization's internal structure, policies, and practices that influence its accountability and decision-making. This includes the composition and independence of the board, executive compensation, shareholder rights, financial reporting transparency, and ethical business practices. Strong governance practices ensure management's and stakeholders' transparency, accountability, and alignment of interests.

5.2 Importance of ESG Practices

- **Risk Mitigation**: ESG practices assist businesses in identifying and managing potential environmental and social risks, regulatory changes, and governance deficiencies. By proactively addressing these risks, businesses can increase their long-term resilience and safeguard shareholder value.
- **Enhanced Regulation**: Companies with strong ESG practices typically enjoy a positive reputation with their consumers, investors, and employees. This can increase brand loyalty, attract socially responsible investors, and aid in the recruitment and retention of talent.
- **Access to Capital**: Investors increasingly consider environmental, social, and governance factors when making investment decisions. Strong ESG performance increases a company's likelihood of attracting capital from socially responsible investors and accessing sustainable financing options.

- **Competitive Advantage**: Adopting ESG practices can provide a competitive advantage to businesses in the marketplace. Consumers are increasingly interested in purchasing goods and services from socially responsible businesses, increasing market share and consumer loyalty.
- **Long-Term Value Creation**: ESG practices focus on sustainable and responsible business practices, resulting in long-term value creation for shareholders, employees, and society.

Alignment with Global Goals ESG practices align with the United Nations Sustainable Development Goals (SDGs), a collection of seventeen global goals designed to resolve the world's most pressing environmental, social, and economic challenges.

5.3 Implementing ESG Practices

Implementing ESG principles necessitates a dedicated involvement from top-level executives and a comprehensive, all-encompassing approach. Businesses have the ability to:

- Perform materiality evaluations to identify the most significant environmental, social, and governance (ESG) concerns for their specific industry and stakeholders.
- Establish Environmental, Social, and Governance (ESG) objectives and goals that are in line with the organization's overarching plan.
- Integrate environmental, social, and governance (ESG) considerations into the day-to-day functioning of the organization, the process of making decisions, and the management of potential risks.
- Interact with stakeholders, including investors, consumers, employees, and communities, to understand their ESG (Environmental, Social, and Governance) expectations and concerns.
- Implement stringent ESG reporting and disclosure processes to effectively explain their sustainability initiatives.

ESG approaches enhance environmental and social results while generating enduring value for all stakeholders.

5.4 Interconnectedness of ESG Principles

ESG principles are interconnected, meaning that a robust performance in one area often positively impacts performance in other areas. For instance, a company with robust environmental policies may attract environmentally conscious consumers, leading to improved reputation and customer loyalty. Additionally, good governance practices can result in more sustainable and socially responsible decision-making.

5.5 Importance of ESG Integration

ESG integration is vital for companies seeking long-term success and resilience. By considering environmental and social factors alongside financial ones, businesses can better manage risks, capitalize on opportunities, and strengthen their relationships with stakeholders. Moreover, ESG integration can help companies anticipate and adapt to emerging trends and regulatory changes.

5.6 The Significance of ESG in the Business Landscape

ESG and Corporate Reputation
In today's interconnected world, a company's reputation can be significantly influenced by its ESG performance. Positive ESG practices enhance a company's image, attracting customers, investors, and potential employees. Conversely, negative ESG issues, such as environmental violations or social controversies, can lead to reputational damage, impacting a company's bottom line.

ESG and Risk Management
Integrating ESG considerations into risk management processes allows companies to identify and mitigate potential risks. For example, evaluating environmental risks such as water scarcity or regulatory changes related to carbon emissions can help companies adapt to a changing business landscape and reduce exposure to potential financial losses.

ESG and Financial Performance
Contrary to a common misconception, ESG and financial performance are not mutually exclusive. Many studies have shown a positive correlation between strong ESG performance and financial outperformance in the long run. Companies with robust ESG practices tend to be more resilient, have lower costs of capital, and attract long-term investors.

ESG and Innovation
Embracing ESG principles can drive innovation and lead to the development of new products, services, and processes. For example, companies focusing on sustainability may invest in research and development for renewable energy solutions or eco-friendly technologies, giving them a competitive edge in the market.

5.7 ESG Metrics and Reporting

Measuring ESG Performance
To assess ESG performance, companies use a range of quantitative and qualitative metrics. These metrics vary across industries and sectors, but some common indicators include carbon emissions, water usage, employee turnover rates, diversity metrics, community engagement, and board diversity.

ESG Reporting Frameworks
ESG reporting frameworks provide standardized guidelines for companies to disclose relevant nonfinancial information. Some of the well-known frameworks include the Global Reporting Initiative (GRI), Sustainability Accounting Standards Board (SASB), Task Force on Climate-related Financial Disclosures (TCFD), and United Nations Sustainable Development Goals (SDGs).

5.8 Challenges in ESG Reporting

Despite the progress in ESG reporting, challenges remain. Companies often face difficulties in quantifying certain ESG aspects, as they may lack consistent data or industry benchmarks. Moreover, the lack of standardized reporting requirements can lead to inconsistency in disclosures and hinder comparability.

5.9 ESG Integration in Investment Decisions

Rise of Responsible Investing
The investment landscape has witnessed a significant shift towards responsible investing, where investors consider ESG factors when making investment decisions. Responsible investing strategies include socially responsible investing (SRI), impact investment, and environmental, social, and governance integration (ESG Integration).

ESG Data and Investment Analysis
ESG data play a crucial role in investment analysis. Asset managers and investors use ESG data to assess the sustainability and long-term viability of companies in their portfolios. ESG analysis helps identify risks and opportunities that traditional financial analysis may overlook.

ESG Engagement and Active Ownership.
Responsible investors increasingly engage with companies on ESG issues, seeking to influence positive change and long-term value creation. This engagement involves active ownership practices, such as proxy voting and direct dialogue with company management.

5.10 ESG Challenges and Criticisms

Greenwashing and ESG Misrepresentation
Greenwashing refers to the practice of misleadingly presenting a company's ESG efforts to appear more environmentally or socially responsible than they genuinely are. Companies engaging in greenwashing may face reputational damage and regulatory scrutiny.

Lack of Standardization and Regulation
The lack of standardization in ESG reporting and the absence of mandatory reporting requirements can create challenges for investors and stakeholders seeking comparable and reliable data.

Short-termism and ESG Integration
Some critics argue that the short-term focus of financial markets can hinder effective ESG integration. Investors may priorities short-term financial returns over long-term sustainability considerations.

Traditional Sustainable Financing Instruments
Sustainable financing instruments are financial products and mechanisms designed to fund environmentally and socially responsible projects and initiatives. These instruments align with the principles of Environmental, Social, and Governance (ESG) practices, promoting sustainability and positive impact. In this section, we will explore some of the traditional sustainable financing instruments that have gained prominence in the world of finance.

Green Bonds Green bonds are debt instruments issued by governments, municipalities, and corporations to finance environmentally friendly projects and initiatives. The proceeds from these bonds are exclusively earmarked for projects with positive environmental benefits, such as renewable energy projects, energy efficiency initiatives, sustainable infrastructure development, and climate change mitigation efforts. Green bonds provide investors with an opportunity to support sustainable projects while earning a financial return through interest payments.

Social Bonds Similar to green bonds, social bonds are debt instruments issued to raise funds for projects with positive social impacts. The proceeds from social bonds are used to finance projects addressing social issues, such as affordable housing, access to healthcare and education, employment generation, and poverty alleviation. These bonds attract socially conscious investors seeking to have influence in addressing critical social challenges.

Sustainability-Linked Loans Sustainability-linked loans are credit facilities in which the interest rate is tied to the borrower's performance on predetermined sustainability targets. These targets can be related to ESG metrics, such as reducing greenhouse gas emissions, increasing diversity in the workforce, or improving energy efficiency. If the borrower achieves the set targets, they receive a discount on the loan interest rate, incentivizing companies to improve their sustainability performance.

Green Loans Green loans are similar to sustainability-linked loans but are specific to financing environmentally friendly projects undertaken by businesses. These projects could include investing in renewable energy, adopting eco-friendly technologies, or implementing sustainable supply chain practices. Green loans support businesses in their transition to more sustainable operations and help fund projects that contribute to environmental conservation.

Socially Responsible Investment (SRI) Funds Socially Responsible Investment (SRI) funds are investment vehicles that incorporate ESG criteria into their investment decisions. These funds selectively invest in companies that demonstrate strong ESG practices and align with specific sustainability goals. SRI funds provide individual and institutional investors with opportunities to align their investments with their values and support companies with positive social and environmental impacts.

Microfinance Microfinance is a financial instrument that provides small loans and financial services to low-income individuals and entrepreneurs, particularly in developing countries. Microfinance institutions focus on financial inclusion and poverty alleviation, supporting local communities and fostering economic development (Enwin & Ikiriko, 2023). The repayment rates for microfinance loans tend to be high, making them sustainable and impactful financing instruments for social development.

Development Finance Institutions (DFIs) Development Finance Institutions (DFIs) are specialized financial institutions owned by governments or international organizations. They provide long-term financing to support developing countries economic and social development projects. DFIs are crucial in funding sustainable infrastructure, renewable energy projects, and initiatives promoting inclusive growth and poverty reduction.

Traditional sustainable financing instruments are vital in funding projects that align with ESG principles and contribute to a more sustainable and responsible world. These instruments provide businesses, governments, and organizations with the necessary capital to implement environmentally friendly and socially responsible initiatives. As global awareness of sustainability grows, the demand for these financing instruments is expected to increase, driving positive change and supporting the transition to a more sustainable future.

5.11 Innovative Sustainable Financing Mechanisms

As the importance of sustainability and responsible business practices continues to grow, there is an increasing need for innovative financing mechanisms that support environmental, social, and governance (ESG) initiatives. These mechanisms go beyond traditional sustainable financing instruments and seek to address the evolving challenges of environmental and social issues. These mechanisms leverage modern technologies, investor preferences, and collaborative approaches to fund projects and businesses committed to positive environmental and social impacts.

Impact Investing
Impact investing is an innovative financing approach that pursues the generation of positive social and environmental impacts alongside financial returns. Unlike traditional investments that focus solely on financial profitability, impact investors actively direct their funds towards projects and companies that align with specific ESG goals. This may include investing in renewable energy, healthcare access, education, sustainable agriculture, and other areas with clear social and environmental benefits.

Impact Measurement and Reporting
Impact investors rely on rigorous impact measurement and reporting to assess the effectiveness of their investments. Tools such as the Impact Management Project (IMP) provide a standardized framework for measuring and comparing the impact of different investments, enabling investors to make more informed decisions and track their progress in achieving social and environmental outcomes.

Impact Bonds
Impact bonds, also known as social impact bonds or pay-for-success bonds, are outcome-based financing instruments. In these bonds, private investors provide upfront capital to fund social programs or projects, and if predetermined social outcomes are achieved, the government or other outcome payers repay the investors with a return. Impact bonds incentivize private investors to support social initiatives, shifting the risk from taxpayers to private capital.

5.12 ESG-Focused Venture Capital

(a) **Integrating ESG Considerations in Venture Capital:**
ESG-focused VC firms integrate environmental, social, and governance factors into their investment decisions. They seek out startups and early-stage companies that prioritize sustainability and responsible business practices. These firms support businesses that aim to positively impact society and the environment while seeking financial growth and success.

- **Patience and Long-term Focus:**
  ESG-focused venture capital takes a patient and long-term approach to investments. Unlike traditional venture capital, which often seeks rapid returns and exits, ESG-focused venture capital is willing to support companies through multiple stages of growth and development, fostering sustainable and impactful business models.
- **Social and Environmental Impact Measurement**:
  ESG-focused VC firms actively measure and assess the social and environmental impact of their portfolio companies. They work closely with investees to set and achieve impact-related targets, ensuring that positive outcomes are achieved alongside financial success.

(b) **Crowdfunding and ESG**

- **Crowdfunding for ESG Initiatives:**
  Crowdfunding platforms have become instrumental in supporting ESG-focused initiatives. Entrepreneurs and organizations with sustainability projects can raise funds from a large number of individuals who share their values. This democratization of finance allows projects with strong ESG credentials to gain support directly from like-minded people, regardless of their scale or location.
- **Rewards-based Crowdfunding**:
  In reward-based crowdfunding, backers contribute funds to a project in exchange for nonfinancial rewards, such as timely access to products, exclusive experiences, or recognition. ESG-focused projects can offer rewards that align with their sustainability goals, fostering a sense of community and support among backers who share the project's values.
- **Equity Crowdfunding:**
  Equity crowdfunding allows individuals to invest in early-stage companies in exchange for equity ownership. This mechanism democratizes access to investment opportunities, enabling individuals to support ESG-focused startups and businesses that they believe in.
- **Carbon Markets**
  Carbon markets, also known as emissions trading systems or cap-and-trade programs, are market-based approaches to reducing greenhouse gas emissions. Governments or international organizations set a cap on the total allowable emissions from certain industries or entities. Companies that emit below their allocated limit can sell surplus allowances as carbon credits to those exceeding their cap. This creates a financial incentive for companies to reduce emissions and invest in cleaner technologies and practices.
- **Voluntary Carbon Markets**:
  In addition to mandatory carbon markets, there are voluntary carbon markets where companies and individuals can voluntarily purchase carbon credits to offset their own emissions. These markets enable businesses and individuals to take direct action in addressing their environmental impact and support projects that promote carbon sequestration or emission reductions.

- **Carbon Market Linkages**:
  Innovative carbon market linkages allow different carbon markets to work together, enabling cross-border trading of carbon credits. These linkages expand market liquidity, create more robust carbon pricing, and incentivize emission reduction efforts on a global scale.

(c) **Green Financial Technology (FinTech)**

- **Sustainable Banking and Green Finance**: Green FinTech companies are emerging to provide sustainable banking and financial services that align with ESG principles. These platforms enable individuals and businesses to access green loans, invest in sustainable funds, and track the environmental impact of their financial decisions.
- **Blockchain for Sustainable Supply Chains**: Blockchain technology offers transparent and traceable supply chains, which are crucial for verifying sustainable and ethical sourcing. Blockchain solutions can enhance the visibility of supply chains, helping businesses and consumers make informed decisions and support environmentally and socially responsible products.
- **Renewable Energy Crowdfunding Platforms:** FinTech platforms focused on renewable energy crowdfunding allow individuals to invest directly in renewable energy projects. These platforms democratize access to clean energy investment opportunities, empowering people to become stakeholders in the renewable energy transition.

(d) **Public-private partnerships (PPPs) for ESG projects**

- **Collaborative Financing for Large-scale ESG:** Initiatives Public–Private Partnerships (PPPs) involve collaboration between governments and private sector entities to finance and implement projects of public interest. PPPs are increasingly used to fund large-scale sustainable infrastructure projects, such as renewable energy plants, public transportation systems, and sustainable urban development.
- **Risk Sharing and Expertise:** PPPs allow for risk sharing between public and private partners, making it possible to finance projects that might be too risky or expensive for one party alone. The private sector brings expertise in project management and efficiency, while the public sector provides regulatory support and access to public resources.
- **Sustainable Development Goals (SDGs) and PPPs:** Many PPPs align with the United Nations Sustainable Development Goals (SDGs), a set of seventeen global goals to address social and environmental challenges. By incorporating SDGs into PPPs, governments and private partners work together to create positive impacts on a broad range of issues, from clean energy to affordable housing and healthcare.
- **Green Bonds for Corporations**: Apart from government-issued green bonds, corporations are increasingly issuing their own green bonds to fund their sustainability projects. Corporate green bonds are debt instruments with proceeds allocated exclusively to finance the issuer's environmentally friendly initiatives and projects.

- **Internalizing Sustainability Commitments** Corporate green bonds allow companies to internalize their sustainability commitments by accessing capital markets specifically for green projects. This creates a strong link between a company's environmental objectives and its financial decisions, demonstrating a commitment to stakeholders and investors.
- **Transparency and Reporting**: Corporate green bond issuers must be transparent about allocating and using proceeds to finance green projects. They are often required to report regularly on the progress and outcomes of the funded initiatives, ensuring accountability and building investor trust.

Innovative sustainable financing mechanisms are vital in promoting positive environmental and social impacts while ensuring financial sustainability. Impact investing, ESG-focused venture capital, crowdfunding for ESG initiatives, carbon markets, green FinTech, and public-private partnerships are some of the emerging and rapidly evolving approaches to fund sustainable projects and businesses. These mechanisms cater to the increasing demand from investors, consumers, and businesses for responsible, sustainable, and transparent financial solutions. As the world continues to address pressing global challenges, innovative sustainable financing mechanisms will remain instrumental in driving positive change and creating a more sustainable future for all.

5.13 Public and Private Sector Collaboration

Public and private sector collaboration refers to the partnership and cooperation between government entities and private businesses to address various societal, economic, and environmental challenges. This collaborative approach combines both sectors' strengths, expertise, and resources to find innovative solutions, drive economic growth, and achieve common goals.

Public and private sector collaboration is increasingly essential in a globalized and interconnected world. Many complex challenges, such as climate change, infrastructure development, healthcare access, and education, require coordinated efforts from multiple stakeholders to be effectively addressed.

5.13.1 Complementary Roles of Public and Private Sectors

The public sector, represented by governments and public institutions, plays a crucial role in policymaking, regulation, and providing essential public goods and services. On the other hand, the private sector, represented by businesses and corporations, brings innovation, efficiency, and entrepreneurial spirit to the table. Both sectors can achieve more comprehensive and sustainable solutions by combining their strengths.

5.13.2 Types of Collaboration

Public and private sector collaboration can take various forms, including public-private partnerships (PPPs), joint ventures, knowledge-sharing initiatives, and shared funding arrangements. These collaborations can be at the local, national, or international levels, depending on the nature of the challenge being addressed.

5.13.3 Benefits of Public and Private Sector Collaboration

Leveraging Resources and Expertise Collaboration allows the pooling of resources, knowledge, and expertise from both sectors. Governments can leverage private sector efficiency and innovation, while businesses can benefit from regulatory support and access to the public infrastructure provided by the government.

Enhanced Efficiency and Effectiveness Public-private sector collaboration often results in more efficient and effective project implementation. By combining public sector planning and regulation with private sector execution, projects can be completed more quickly and at a lower cost.

Fostering innovation The private sector is known for its innovative spirit and willingness to take risks. Collaboration with the public sector can create an environment encouraging and supporting innovation, leading to groundbreaking solutions to societal challenges.

Scalability and Sustainability Public-private sector collaboration enables projects to achieve greater scalability and long-term sustainability. Public sector involvement can provide the necessary funding and policy support, while private sector involvement ensures that projects are financially viable and can be scaled up over time.

5.13.4 Challenges and Barriers to Collaboration

Differing Objectives and Priorities The public and private sectors may have different objectives and priorities, leading to potential conflicts or challenges in finding common ground. Government entities may prioritize public interest and social welfare, while businesses focus on profit maximization and shareholder value.

Regulatory and Legal Complexities Collaborating across sectors can be challenging due to regulatory and legal complexities. Different sectors often operate under distinct frameworks, and navigating these legal barriers can be time-consuming and resource intensive.

Risk Allocation and Incentive Alignment Public-private partnerships involve risk-sharing arrangements, which require careful consideration of risk allocation and incentive alignment. Balancing risks and rewards for both sectors is essential to ensure mutual commitment and success in the partnership.

Transparency and Accountability Transparency and accountability are critical to the success of public-private sector collaborations. Ensuring that clear communication, reporting mechanisms, and governance structures are in place is essential to maintain trust and legitimacy.

5.13.5 Successful Examples of Public and Private Sector Collaboration

Infrastructure Development Public-private partnerships have been successful in funding and executing large-scale infrastructure projects, such as toll roads, airports, and water treatment facilities. These partnerships leverage private sector expertise in project management and financing while benefiting from the government's ability to facilitate permits and regulatory approvals.

Healthcare Initiatives Collaboration between pharmaceutical companies and government health agencies has been instrumental in developing and distributing vaccines and medicines during public health crises. The joint efforts of the public and private sectors have contributed to significant healthcare and disease management advancements.

Climate Change Mitigation Public-private partnerships are crucial in addressing climate change. Collaborative initiatives between governments, businesses, and nongovernmental organizations (NGOs) have led to renewable energy projects, sustainable land use practices, and innovative technologies to reduce greenhouse gas emissions.

Education and Skills Development Public-private collaboration has resulted in initiatives to enhance access to quality education and skills development programs. Companies can support education initiatives through funding, mentorship, and vocational training, aligning their workforce needs with societal needs.

5.13.6 Building Effective Public and Private Sector Collaboration

Establishing Clear Objectives and Mutual Goals Successful collaboration requires clear and well-defined objectives that align the interests of both sectors. Identifying mutual goals ensures that the partnership is driven by a shared vision for positive impact.

Engaging Stakeholders and Communities Engaging stakeholders and local communities is crucial in public-private sector collaborations. Understanding the needs and concerns of those affected by the cooperation helps build trust and ensures that the project addresses genuine societal needs.

Robust Governance and Risk Management Effective governance structures and risk management mechanisms are essential to oversee challenges and uncertainties that arise during collaborations. Transparent decision-making processes and risk-sharing arrangements create a solid foundation for the partnership.

Continuous Evaluation and Learning Regular evaluation and learning from both successes and failures are vital to the sustainability of collaborations. Adaptation and improvement based on feedback and data-driven insights enable partnerships to evolve and deliver better outcomes over time.

Public and private sector collaboration is a powerful tool for addressing complex societal challenges and promoting sustainable development. By harnessing the strengths of both sectors, these collaborations can drive innovation, create economic opportunities, and achieve positive environmental and social impacts. However, successful collaborations require careful planning, effective governance, and a commitment to shared goals. As the world faces increasing global challenges, public and private sector collaboration will remain essential in shaping a more inclusive, sustainable, and prosperous future for all.

5.14 Analyzing how PPPs Can Facilitate Sustainable Financing for Large-Scale ESG Projects

Public-private partnerships (PPPs) are critical in facilitating sustainable financing for large-scale environmental, social, and governance (ESG) projects. These partnerships bring together the public and private sectors' resources, expertise, and capabilities to address complex challenges and achieve long-term sustainability goals. In this analysis, we will explore how PPPs enable sustainable financing for large-scale ESG projects and the benefits they offer.

- **Access to Capital**: One of the primary ways PPPs facilitate sustainable financing is by providing access to significant financial resources. Large-scale ESG projects often require substantial upfront investments, which can be challenging for governments or private companies to fund individually. To finance these projects, PPPs allow the public sector to tap into private sector capital, including institutional investors, pension funds, and impact investors. The private sector's involvement not only diversifies the funding sources but also reduces the financial burden on the government, making ambitious ESG projects more feasible.
- **Risk Sharing**: ESG projects, particularly those involving innovative technologies or innovative solutions, can carry significant risks. In a PPP arrangement, risks are shared between the public and private sectors, mitigating the potential financial impact on any single party. Governments often provide regulatory support, risk guarantees, or financial incentives to derisk investments for private partners. This risk-sharing mechanism encourages private sector participation in sustainable projects that they might otherwise consider too risky to pursue alone.
- **Expertise and Efficiency**: The private sector brings efficiency and expertise in project management, design, construction, and operation. Their experience in delivering complex projects within budget and on time can lead to cost savings and better project outcomes. Additionally, the private sector's access to the latest technologies and best practices can enhance the sustainability and effectiveness of ESG projects. The partnership can balance efficiency and social responsibility by collaborating with the public sector, which knows local regulations, community needs, and long-term planning.
- **Scalability and Replicability**: Large-scale ESG projects often require scalability and replicability to have a meaningful impact. PPPs can provide the necessary mechanisms for scaling up successful projects and replicating them in different regions or countries. Leveraging the private sector's ability to expand operations and access markets, PPPs can accelerate the adoption of sustainable practices and technologies beyond a single project's scope.
- **Long-term Commitment and Governance**: Sustainable financing for ESG projects demands a long-term commitment to ensure that the initiatives achieve their intended impacts over time. PPPs typically involve long-term contracts, which can extend for several decades. This commitment encourages the private sector to invest in projects with long payback periods, such as renewable energy plants or sustainable infrastructure and fosters ongoing maintenance and support. Furthermore, PPP governance structures ensure transparency, accountability, and alignment of interests, promoting the sustainability of projects and fostering public trust.
- **Alignment with Sustainable Development Goals (SDGs):** ESG projects often align with the United Nations Sustainable Development Goals (SDGs), which outline global targets for environmental protection, poverty reduction, and social welfare. PPPs can integrate these SDGs into project objectives, ensuring that investments are directed towards projects that have a measurable positive impact on specific social and environmental issues. This alignment attracts public and private sector partners seeking to contribute to broader sustainable development goals while achieving financial returns.

- **Adaptability and Flexibility**: The collaborative nature of PPPs allows for adaptability and flexibility in addressing evolving sustainability challenges. As modern technologies, practices, or regulations emerge, PPPs can adjust project strategies accordingly. The private sector's agility in responding to market changes can enhance the resilience and relevance of ESG projects over time, ensuring that they remain at the forefront of sustainable practices.

Public-private partnerships (PPPs) offer a powerful mechanism for facilitating sustainable financing for large-scale ESG projects. These collaborations leverage the private sector's financial resources, expertise, and innovation, along with the public sector's regulatory support and long-term vision. Through risk sharing, access to capital, scalability, and long-term commitment, PPPs enable the realization of ambitious ESG initiatives that contribute to a more sustainable and inclusive future. The successful implementation of PPPs in sustainable financing has the potential to drive positive environmental and social impacts while fostering economic growth and development.

5.15 Government Incentives

Government incentives are crucial in encouraging environmental, social, and governance (ESG) practices among businesses and organizations. By offering tax incentives, grants, and subsidies, governments can create a favorable environment for sustainable and responsible initiatives, driving positive environmental and social impacts. This investigation will explore how governments can effectively use these incentives to promote ESG practices.

5.15.1 Tax Incentives for ESG Initiatives

Tax Deductions for Green Investments Governments can provide tax deductions or credits to businesses that invest in environmentally friendly projects and technologies. This includes investments in renewable energy, energy-efficient equipment, waste reduction systems, and sustainable infrastructure. Governments encourage companies to allocate more resources to ESG-focused initiatives by reducing the tax burden on such investments.

Carbon Pricing and Cap-and-Trade Implementing carbon pricing mechanisms, such as carbon taxes or cap-and-trade systems, incentivizes businesses to reduce their carbon emissions. Companies that achieve emissions below their allocated limits can sell surplus allowances, while those exceeding their limits face additional costs. This economic incentive encourages businesses to adopt low-carbon practices and invest in clean technologies.

Tax Incentives for Social Impact Initiatives Governments can offer tax incentives to businesses that support social impact projects, such as affordable housing, education programs, healthcare services, and community development. Tax deductions or credits for charitable contributions and philanthropic activities incentivize companies to engage in social responsibility initiatives actively.

5.15.2 Grants and Subsidies for ESG Projects

Funding for Renewable Energy Projects Governments can provide grants or subsidies to support the development and deployment of renewable energy projects. These incentives help reduce the initial capital costs of installing solar panels, wind turbines, or other clean energy infrastructure. By making clean energy more affordable, governments stimulate investment in renewable technologies.

Research and Development (R&D) Grants for Sustainability Governments can offer R&D grants to businesses and organizations working on ESG-focused technologies and solutions to encourage innovation in sustainable practices. These grants support the development of new green technologies, eco-friendly products, and processes that contribute to environmental and social sustainability.

Green Building Subsidies Governments can provide subsidies or reduced permit fees for projects that meet certain sustainability criteria, such as energy efficiency standards or green building certifications. These subsidies incentivize the construction industry to adopt greener building practices and promote sustainable urban development.

5.15.3 Regulatory Incentives for ESG Compliance

Preferential Government Procurement Governments can demonstrate their commitment to ESG practices by favoring suppliers and contractors that adhere to sustainable and responsible standards. Through public procurement policies, governments can encourage businesses to adopt environmentally friendly and socially responsible practices to qualify for government contracts.

ESG Reporting and Transparency Requirements Requiring companies to disclose their ESG performance and impact can promote transparency and accountability. Governments can introduce mandatory ESG reporting frameworks that provide standardized guidelines for businesses to report their sustainability practices. Publicly disclosing ESG metrics allows investors, consumers, and stakeholders to make informed decisions based on a company's sustainability performance.

Fines and Penalties for Noncompliance Governments can impose fines or penalties on companies that fail to meet certain ESG standards or violate environmental and social regulations. This creates a financial disincentive for noncompliance, encouraging businesses to adopt responsible practices and avoid reputational risks.

Government incentives, such as tax benefits, grants, subsidies, and regulatory measures, are powerful tools to promote ESG practices among businesses and organizations. Governments can encourage sustainable investments, innovation, and responsible behaviors by providing financial support and creating a conducive regulatory environment. These incentives align economic interests with environmental and social goals, driving positive impacts on society, the economy, and the environment. As governments worldwide recognize the urgency of addressing global challenges, well-designed ESG incentives will play a vital role in fostering a more sustainable and equitable future.

5.16 ESG Reporting and Transparency

ESG reporting and transparency are integral components of responsible and sustainable business practices. ESG reporting involves disclosing a company's performance in Environmental, Social, and Governance (ESG) factors. It provides stakeholders, including investors, consumers, employees, regulators, and the public, with critical information about a company's sustainability practices, impact on the environment, treatment of its workforce, and adherence to ethical and governance standards. This section will delve into the importance of ESG reporting and transparency, its essential elements, and its benefits to various stakeholders.

5.16.1 Importance of ESG Reporting and Transparency

Informed Decision-Making ESG reporting enables stakeholders to make informed decisions about their engagement with a company. Investors can assess a company's sustainability performance and align their investment strategies with their values. Consumers can support companies prioritizing sustainability and responsible practices, and employees can evaluate a company's commitment to ethical and social issues before joining or remaining with the organization.

Stakeholder Trust and Engagement Transparent ESG reporting builds trust with stakeholders, demonstrating a company's commitment to accountability and responsibility. By being open and honest about their ESG performance, companies can foster positive relationships with stakeholders and engage in meaningful dialogue with investors, customers, employees, and communities.

Risk Management and Resilience Effective ESG reporting helps companies identify and manage ESG-related risks, such as environmental liabilities, labor disputes, and supply chain disruptions. By proactively addressing these risks, companies can enhance their resilience and long-term sustainability.

Regulatory Compliance ESG reporting is increasingly becoming subject to regulatory requirements in many jurisdictions. Companies may be mandated to disclose specific ESG metrics and comply with reporting frameworks set by regulatory bodies. Transparent ESG reporting ensures compliance with these regulations and reduces the risk of legal or reputational consequences.

5.16.2 Vital Elements of ESG Reporting

Environmental Metrics Environmental metrics cover a company's impact on the environment, including greenhouse gas emissions, energy consumption, water usage, waste generation, and biodiversity preservation. Companies may report their efforts to reduce carbon footprints, implement sustainable sourcing practices, and adopt eco-friendly technologies.

Social Indicators Social indicators encompass a company's relationship with its employees, customers, communities, and other stakeholders. ESG reporting in this area may include diversity and inclusion metrics, labor practices, health and safety performance, community engagement, and efforts to support social causes.

Governance Factors Governance factors address a company's internal structures, policies, and practices that guide decision-making and accountability. ESG reporting on governance may include board composition, executive compensation, shareholder rights, anti-corruption measures, and adherence to ethical standards.

Data and Performance Metrics ESG reporting should be data-driven, providing specific and measurable metrics. Companies may use key performance indicators (KPIs) and targets to track their progress on ESG initiatives over time. Consistent data reporting facilitates comparability and benchmarking across companies and industries.

5.16.3 Benefits of ESG Reporting to Various Stakeholders

Investors ESG reporting allows investors to assess a company's sustainability performance and potential risks. It helps investors make decisions aligned with their ESG preferences and can impact investment decisions, portfolio management, and the allocation of capital.

Consumers Transparent ESG reporting empowers consumers to make informed choices and supports companies with responsible practices. It enables consumers to align their purchasing decisions with their values and contribute to positive social and environmental impacts.

Employees ESG reporting helps attract and retain talent. Employees increasingly seek to work for companies that prioritize sustainability and social responsibility. Transparent ESG reporting demonstrates a company's commitment to creating a positive work environment and fostering employee engagement.

Regulators and Policymakers ESG reporting provides regulators and policymakers with valuable data to assess a company's environmental and social regulations compliance. It also helps policymakers develop evidence-based policies to address sustainability challenges and promote responsible business practices.

ESG reporting and transparency are essential for responsible and sustainable business practices. It empowers stakeholders to make informed decisions, builds trust, and drives positive impacts on society and the environment. By reporting on their environmental, social, and governance performance, companies demonstrate their commitment to sustainability, accountability, and ethical behavior. ESG reporting benefits investors, consumers, employees, regulators, and the wider public, making it a powerful tool for driving positive change and creating a more sustainable and responsible business landscape. As the importance of ESG factors continues to grow, transparent reporting will remain a critical aspect of business strategy and stakeholder engagement.

5.16.4 *Overcoming Challenges in Sustainable Financing for ESG Practices*

Overcoming challenges in sustainable financing for environmental, social, and governance (ESG) practices is essential to accelerate the transition towards a more sustainable and responsible business landscape. While sustainable financing mechanisms have gained momentum, several challenges still hinder widespread adoption. Addressing these challenges is crucial to unlocking the full potential of sustainable finance and achieving meaningful impact.

Data and Metrics

Data and metrics are crucial in driving sustainable practices, measuring environmental and social impacts, and assessing progress toward sustainability goals. In the context of environmental, social, and governance (ESG) practices, data and metrics provide quantitative and qualitative insights into a company's sustainability performance. This section will explore the significance of data and metrics in promoting ESG practices and discuss the data and metrics commonly used in ESG reporting.

5.16.5 Significance of Data and Metrics in ESG Practices

- **Informed Decision-Making**: Data and metrics provide stakeholders, including investors, consumers, employees, and regulators, with information to make informed decisions. Investors can assess a company's sustainability performance and align their investments with their values. Consumers can choose to support businesses that prioritize sustainable practices, and employees can evaluate a company's commitment to social and environmental responsibility.
- **Identifying Sustainability Risks and Opportunities**: Data-driven ESG analysis enables companies to identify and manage sustainability risks and opportunities. Companies can proactively address climate change, supply chain risks, labor practices, and governance deficiencies by analyzing relevant metrics.
- **Goal Setting and Monitoring:** Metrics help set clear sustainability goals and track progress over time. With measurable targets, companies can assess their performance, make data-driven decisions, and communicate their progress to stakeholders.
- **Benchmarking and Comparability**: Data and metrics enable benchmarking across companies and industries, facilitating comparisons of sustainability performance. This comparability encourages healthy competition and motivates companies to improve their sustainability practices.

5.16.6 Types of Data and Metrics in ESG Reporting

Environmental Metrics Environmental metrics focus on a company's impact on the environment. Common environmental metrics include:

- **Greenhouse Gas (GHG) Emissions**: Measuring and reporting carbon dioxide (CO2) emissions and other GHGs helps track a company's contribution to climate change.
- **Energy Consumption**: Monitoring energy usage helps identify areas for energy efficiency improvements.
- **Water Usage:** Reporting on water consumption and stewardship practices is essential for sustainable water management.
- **Waste Generation**: Measuring and reducing waste generation supports circular economy initiatives and waste reduction goals.
- **Biodiversity and Land Use**: Assessing the company's impact on ecosystems and biodiversity highlights conservation efforts.

Social Indicators Social indicators assess a company's relationship with its employees, customers, communities, and other stakeholders. Common social metrics include the following:

- **Employee Diversity and Inclusion**: Measuring workforce diversity, equity, and inclusion supports fair and inclusive hiring practices.

- **Labor Practices**: Evaluating labor conditions and worker rights helps ensure the ethical and fair treatment of employees and suppliers.
- **Health and Safety**: Reporting on workplace safety measures safeguards employee well-being.
- **Community Engagement**: Assessing community involvement and impact demonstrates a company's commitment to local communities.
- **Customer Satisfaction**: Monitoring customer satisfaction and feedback helps improve products and services.

Governance Factors Governance metrics focus on a company's internal structures, policies, and practices that guide decision-making and accountability. Common governance metrics include the following:

- **Board Composition**: Assessing the diversity and independence of a company's board of directors promotes effective governance.
- **Executive Compensation**: Evaluating executive pay practices ensures alignment with company performance and sustainability goals.
- **Shareholder Rights**: Reporting on shareholder rights and engagement fosters transparency and shareholder trust.
- **Anti-Corruption Measures**: Evaluating anti-corruption policies and procedures supports ethical business practices.
- **Ethical Standards**: Reporting on adherence to ethical codes and responsible business conduct strengthens corporate integrity.

5.16.7 Challenges in ESG Data and Metrics

- **Data Availability and Reliability**: Obtaining comprehensive and dependable ESG data can be challenging, particularly for smaller companies and emerging markets. Inconsistent data quality and availability can hinder meaningful ESG analysis.
- **Standardization**: The lack of standardized ESG reporting frameworks can lead to inconsistent metrics and hinder comparability across companies and industries.
- **Materiality and Relevance**: Identifying the most relevant ESG metrics for each industry and company can be complex. Metrics must be material, reflecting the most significant sustainability risks and opportunities.
- **Data Interpretation**: Interpreting ESG data and metrics requires expertise in sustainability analysis. Inaccurate interpretations can lead to misinformed decision-making.

Data and metrics are vital in driving ESG practices and promoting sustainable decision-making. By providing quantitative and qualitative insights into a company's environmental, social, and governance performance, data and metrics enable stakeholders to make informed choices and assess sustainability risks and opportunities. Standardization, data availability, materiality, and interpretation are some of the challenges that need to be addressed to enhance the effectiveness of ESG data analysis. As ESG reporting becomes increasingly crucial for businesses and investors, continued efforts to improve data transparency, quality, and comparability will advance sustainability goals and positively impact society and the environment.

5.17 Risk and Return Considerations

Risk and return considerations are fundamental principles in finance and investments. They form the basis of decision-making for investors and businesses alike. In the context of sustainable finance and ESG (Environmental, Social, and Governance) practices, these considerations play a crucial role in driving responsible and sustainable investment decisions. Let us delve deeper into the significance of risk and return considerations in the realm of sustainable finance:

5.17.1 Risk Considerations

Environmental Risks Environmental risks are associated with the impact of climate change, resource scarcity, pollution, and other ecological challenges. Companies operating in industries with high carbon emissions, dependence on non-renewable resources, or vulnerable supply chains may face significant environmental risks. Sustainable finance seeks to address and mitigate these risks by investing in companies with sustainable business practices and environmentally responsible operations.

Social Risks Social risks include labor practices, human rights, employee well-being, and community relations. Companies with poor labor conditions, involvement in controversial activities, or lacking social responsibility commitment may face reputational risks and potential legal liabilities. Sustainable finance considers the social impact of investments and seeks to support companies that uphold strong ethical standards and foster positive social outcomes.

Governance Risks Governance risks relate to a company's internal structures, board effectiveness, executive compensation, and shareholder rights. Weak governance practices can lead to management inefficiencies, conflicts of interest, and lack of transparency. Sustainable finance emphasizes investments in companies with strong governance structures, transparent decision-making, and shareholder-friendly policies.

Transitional Risks Transitional risks arise from shifting towards a more sustainable economy and implementing environmental regulations and policies. Companies heavily reliant on fossil fuels or nonsustainable practices may face financial risks as they transition to more sustainable business models. Sustainable finance acknowledges these transitional risks and encourages investments in proactive companies adapting to a low-carbon and sustainable future.

5.17.2 Return Considerations

Financial Performance Historically, there has been a perception that sustainable investments may lead to lower financial returns. However, this view is changing, as evidence suggests that companies with strong ESG practices can outperform their peers in the long term. Sustainable finance recognizes that sustainable business practices can enhance financial performance and improve the bottom line.

Market Opportunities Sustainable finance identifies emerging market opportunities in industries that address environmental and social challenges. Companies offering innovative solutions to climate change, resource efficiency, and social inclusion will likely tap into growing markets and benefit from increased demand for sustainable products and services.

Brand Value and Reputation Investing in sustainability can boost a company's brand value and reputation. Consumers and stakeholders increasingly value businesses that prioritize ESG practices. Positive brand perception can increase customer loyalty and enhance competitiveness, resulting in improved financial returns.

Long-term Value Creation Sustainable finance emphasizes long-term value creation. Companies prioritizing sustainability are better equipped to manage risks, adapt to changing market conditions, and create enduring shareholder value. Investing from a long-term perspective aligns with the principles of sustainable finance.

Access to Capital As sustainable finance gains traction, companies with strong ESG performance may gain access to a broader pool of capital. Sustainable investment funds, green bonds, and ESG-focused investors actively seek opportunities in companies committed to sustainability. Enhanced access to capital can facilitate growth and expansion for sustainable businesses.

Incorporating risk and return considerations in sustainable finance is pivotal for driving responsible investment decisions and fostering a more sustainable and equitable global economy. The convergence of financial performance with environmental, social, and governance impacts is a hallmark of sustainable finance.

By identifying and managing risks while pursuing attractive financial returns, sustainable finance empowers investors to contribute to positive societal and environmental outcomes. As the world continues to address pressing global challenges, the integration of risk and return considerations in sustainable finance will remain instrumental in driving positive change and creating a more sustainable future for all.

Investor Education

Investor education is crucial in promoting sustainable finance and driving responsible investment practices. By increasing awareness and understanding of environmental, social, and governance (ESG) considerations, investor education empowers individuals and institutions to make informed decisions that align with their values and contribute to positive societal and environmental outcomes. Let us explore the significance of investor education in the context of sustainable finance:

(i) **Raising Awareness of ESG Factors**: Investor education raises awareness about the importance of ESG factors in investment decisions. It highlights businesses' environmental and social impact and the significance of responsible governance. Through educational initiatives, investors gain insights into how ESG considerations can influence financial performance and risk management.

(ii) **Encouraging Responsible Investment Practices**: Investor education fosters a culture of responsible investing. By understanding the broader implications of their investment choices, investors are encouraged to prioritize companies with strong ESG practices, thereby promoting sustainable business models and ethical behavior.

(iii) **Empowering Informed Decision-Making**: Knowledgeable investors can make informed decisions aligned with their values and long-term financial goals. Investor education provides tools and resources to assess ESG risks and opportunities, enabling individuals to invest with a more comprehensive understanding of sustainability factors.

(iv) **Clarifying ESG Metrics and Reporting**: ESG metrics and reporting can be complex and varied. Investor education simplifies these concepts, helping investors interpret ESG data and disclosures effectively. Investors can compare and benchmark companies' sustainability performance by understanding standardized reporting frameworks.

(v) **Addressing Misconceptions and Myths**: There are misconceptions surrounding sustainable finance, such as the belief that ESG investing leads to lower financial returns. Investor education dispels such myths and highlights the growing evidence that responsible investing can align with financial performance.

(vi) Supporting ESG Integration: Investor education encourages financial professionals to integrate ESG considerations into their investment strategies and advisory services. This integration enhances the quality of investment advice and facilitates the mainstream adoption of sustainable finance.

(vii) **Promoting Long-Term Thinking:** Sustainable finance emphasizes the importance of long-term thinking in investment decisions. Investor education reinforces this principle, encouraging investors to look beyond short-term gains and consider the lasting impact of their investments on the environment and society.

(viii) **Engaging Stakeholders**: Education initiatives engage various stakeholders, including asset managers, financial advisors, regulators, and the public. Collaborative efforts to enhance investor education create a more comprehensive understanding of sustainable finance throughout the financial ecosystem.

(ix) **Fostering Investor Activism**: Informed investors may become advocates for positive change by using their influence to engage with companies and promote better ESG practices. Investor education empowers individuals to exercise their shareholder rights effectively.

(x) **Driving Market Transformation**: By increasing demand for sustainable investments, investor education drives market transformation toward a more sustainable and responsible financial system. As more investors embrace sustainable finance, companies are encouraged to improve their ESG performance to attract capital.

Investor education is critical for advancing sustainable finance and responsible investment practices. Education initiatives foster a culture of responsible investing by raising awareness, empowering informed decision-making, and dispelling misconceptions. Investors with a deeper understanding of ESG factors can align their financial goals with their values, contributing to positive environmental, social, and governance outcomes. As sustainable finance continues to gain momentum, ongoing efforts to educate investors and financial professionals will be instrumental in creating a more sustainable and equitable global economy.

5.18 Case Studies

Case Study 1 Tesla Inc. – Leading the Electric Vehicle Revolution

Company Overview
Tesla Inc. is a global electric vehicle (EV) manufacturer founded by Elon Musk in 2003. The company aims to accelerate the world's transition to sustainable energy. Tesla manufactures and sells electric cars, battery storage systems, and solar energy products.

ESG Highlights

1. **Environmental**: Tesla's primary goal is to reduce the world's dependence on fossil fuels and combat climate change. Tesla helps reduce greenhouse gas emissions and air pollution by producing electric vehicles. The company also operates gigafactories, which produce lithium-ion batteries for EVs and energy storage solutions, promoting the adoption of renewable energy.

2. **Social:** Tesla is committed to ensuring the well-being of its employees and maintaining safe working conditions. The company focuses on fostering a diverse and inclusive workforce, with efforts to increase the representation of women and underrepresented minorities in leadership positions.
3. **Governance**: Tesla prioritizes transparency and good governance practices. The company's board of directors includes independent directors and has implemented compensation policies tied to company performance, aligning executive and shareholder interests.

Sustainable Financing Initiatives Tesla has actively engaged in various sustainable financing initiatives to support its mission and growth:

1. **Green Bond Issuance**: Tesla issued green bonds to fund projects with positive environmental impacts, such as gigafactories and renewable energy projects. These bonds are aligned with the company's commitment to sustainability and allow it to raise capital from investors interested in environmentally responsible investments.
2. **Access to Capital Markets**: Tesla's focus on ESG and sustainable practices has attracted a growing base of socially responsible investors. As a result, the company has accessed a broader pool of capital, including ESG-focused funds and institutional investors with sustainability mandates.

Case Study 2 Unilever – Sustainable Sourcing and Social Impact

Company Overview

Unilever is a multinational consumer goods company producing a wide range of products, including food, beverages, cleaning agents, and personal care products. The company has a strong commitment to sustainability and social impact.

ESG Highlights

1. **Environmental:** Unilever is dedicated to reducing its environmental impact across its supply chain. The company has set ambitious targets to achieve net-zero emissions and ensure that 100% of its plastic packaging is reusable, recyclable, or compostable by 2025.
2. **Social:** Unilever places a strong emphasis on social responsibility, focusing on improving livelihoods, promoting gender equality, and enhancing access to education and clean water. Through its "Sustainable Living Plan," Unilever aims to improve the lives of millions of people while reducing its environmental footprint.
3. **Governance:** Unilever's governance structure emphasizes transparency, accountability, and ethical behavior. The company actively engages with stakeholders and regularly reports on its progress toward sustainability goals.

Sustainable Financing Initiatives Unilever has implemented several sustainable financing initiatives to support its social and environmental objectives:

1. **Sustainable Sourcing**: Unilever sources raw materials from sustainable and responsibly managed sources, including certified palm oil, tea, and cocoa. By supporting sustainable agricultural practices, Unilever contributes to biodiversity conservation and helps improve the livelihoods of small-scale farmers.
2. **Social Impact Investments**: Unilever has invested in social impact initiatives, including providing access to clean water, sanitation, and hygiene (WASH) in underserved communities. These investments help address social challenges while creating value for the company and its stakeholders.
3. **Green Bonds and Sustainability-Linked Financing**: Unilever has issued green bonds to finance projects with positive environmental impacts, such as renewable energy and waste reduction. The company has also explored sustainability-linked financing, where the financing terms are linked to achieving specific sustainability targets.

Case Study 3 Danone – Leveraging Sustainable Finance for Positive Impact

Company Overview
Danone is a multinational food and beverage company committed to sustainability and social responsibility. The company's products include dairy, plant-based foods, water, early life, and medical nutrition.

Sustainable Financing Initiatives Danone has successfully implemented sustainable financing initiatives to support its ESG goals:

1. **Green Bonds**: In 2018, Danone issued its inaugural €300 million green bond. The proceeds from this bond were allocated to finance and refinance projects that contribute to sustainability objectives. These projects included investments in sustainable agriculture, renewable energy, water conservation, and packaging waste reduction.
2. **Social Impact Bonds**: Danone has explored social impact bonds to support its social initiatives. These bonds raise capital from private investors to fund projects with measurable social outcomes. For example, through nutrition programs, Danone piloted a social impact bond in France to improve young children's health and well-being.
3. **Sustainable Supply Chain Financing**: Danone collaborates with its suppliers to promote sustainability across its value chain. The company has collaborated with financial institutions to offer sustainable supply chain financing, incentivizing suppliers who adopt sustainable practices, such as reducing carbon emissions or implementing responsible sourcing.

Results and Impact Danone's sustainable financing initiatives have had a significant impact on the company's sustainability performance:

- The green bond issuance has allowed Danone to fund various sustainability projects, supporting its ambition to become a carbon-neutral company by 2050.

- Social impact bonds have facilitated investments in programs to improve child nutrition, positively impacting the health and well-being of vulnerable populations.
- Sustainable supply chain financing has incentivized suppliers to adopt more sustainable practices, contributing to the company's efforts to promote responsible sourcing and reduce environmental impacts.

Case Study 4 Ørsted – Transformation to a Renewable Energy Leader

Company Overview
Ørsted, formerly known as DONG Energy, is a Danish renewable energy company focused on offshore wind power and other renewable energy sources. The company is committed to leading the transition to green energy and sustainability.

Sustainable Financing Initiatives Ørsted successfully used sustainable financing mechanisms to support its renewable energy projects:

1. **Green Finance Framework**: Ørsted established a green finance framework that guides the issuance of green bonds and other sustainable finance instruments. This framework outlines the eligible projects and defines how the proceeds will be allocated to finance renewable energy projects with clear environmental benefits.
2. **Green Bonds:** Ørsted has issued multiple green bonds to finance its offshore wind farm projects and other renewable energy initiatives. The proceeds from these bonds have been used to develop and operate wind farms, contributing to the global transition to renewable energy.
3. **Transition Bonds:** Ørsted also issued transition bonds, a type of sustainable debt instrument that helps companies finance their transition to a low-carbon economy. The proceeds from these bonds are used to fund green and nongreen projects, promoting Ørsted's transformation into a more sustainable energy company.

Results and Impact Ørsted's sustainable financing efforts have been instrumental in its transformation into a global renewable energy leader:

- The issuance of green bonds has allowed Ørsted to secure financing for its offshore wind projects, increasing its capacity to produce renewable energy and reduce greenhouse gas emissions.
- Transition bonds have enabled Ørsted to fund renewable energy projects and gradually phase out fossil fuel-related activities, accelerating the transition to clean energy sources.
- The company's commitment to sustainable financing has enhanced its reputation as a leader in the renewable energy sector, attracting socially responsible investors and driving further sustainable growth.

5.19 Emerging Trends

As the world grapples with pressing environmental and social challenges, several emerging trends shape the landscape of sustainable financing for environmental, social, and governance (ESG) practices. These trends reflect the growing importance of sustainability in the global financial system and the increasing demand for responsible investments. Some of the emerging trends in sustainable financing for ESG practices include:

Green and Sustainable Bonds Market Expansion The green and sustainable bonds market is experiencing rapid growth. These debt instruments are used to finance projects with positive environmental and social impacts. As more companies and governments commit to sustainability goals, the issuance of green and sustainable bonds is likely to increase, mobilizing capital for climate-friendly projects and initiatives.

ESG Integration in Mainstream Investing ESG integration is becoming a mainstream practice among investors and financial institutions. Traditional asset managers are increasingly considering ESG factors in their investment decisions, recognizing the materiality of sustainability risks and opportunities. As ESG integration becomes more widespread, it is likely to influence capital allocation and drive more sustainable business practices.

Impact Investing and Socially Responsible Investing (SRI) Growth Impact investing and SRI are gaining traction as investors seek to align their investments with their values and support positive social and environmental outcomes. These investment approaches prioritize measurable positive impact alongside financial returns, fostering the growth of businesses with a strong commitment to ESG principles.

Sustainable Lending and Financing Products Financial institutions are developing various sustainable lending and financing products. These include sustainability-linked loans, which offer preferential terms based on borrowers' ESG performance, as well as green mortgages and green consumer loans, incentivizing sustainable behavior.

Climate Risk Assessment and Disclosures The focus on climate risk assessment and disclosure is intensifying. Investors and regulators are demanding more transparent reporting on climate-related risks and opportunities. Initiatives such as the Task Force on Climate-related Financial Disclosures (TCFD) drive the standardization and integration of climate-related considerations into financial decision-making.

Biodiversity and Nature-Related Financial Disclosures Like climate disclosures, there is a growing recognition of the need for financial institutions and companies to disclose their impacts and dependencies on nature and biodiversity. The Task Force on Nature-related Financial Disclosures (TNFD) is working to develop a framework for these disclosures, enhancing the understanding of nature-related risks and opportunities.

Thematic ESG Funds and Solutions Thematic ESG funds are gaining popularity, focusing on sustainability themes such as clean energy, water conservation, gender equality, and healthcare. Investors increasingly seek targeted exposure to companies making a positive impact in these areas.

Social Bonds and Social Impact Investing The COVID-19 pandemic has accelerated the demand for social bonds and social impact investing. Social bonds are used to raise funds to address social challenges, such as healthcare, education, and affordable housing. The pandemic underscored the importance of investing in projects with social benefits.

1. **Green Technology and Innovation Financing**: The transition to a sustainable economy relies on innovative green technologies. Financing green technology startups and initiatives is becoming more prevalent as investors seek opportunities to support innovative solutions that can drive environmental progress.
2. **Regulation and Policy Drivers**: Regulators and policymakers increasingly recognize the importance of sustainable finance. New regulations, tax incentives, and disclosure requirements are being introduced to encourage businesses and investors to prioritize ESG considerations.

The emerging trends in sustainable financing for ESG practices demonstrate the growing momentum and integration of sustainability principles into the financial ecosystem. These trends highlight the transformative potential of sustainable finance in addressing global challenges and driving positive environmental, social, and governance outcomes. As sustainable finance continues to evolve, it is poised to play an increasingly crucial role in shaping a more sustainable and responsible global economy.

5.20 Prospects and Challenges

Prospects and challenges in sustainable financing for environmental, social, and governance (ESG) practices are interconnected and reflect the evolving landscape of sustainable finance. As the world grapples with urgent environmental and social issues, the prospects for sustainable financing are promising, but several challenges must be addressed to realize their full potential.

Prospects

I. **Increased Demand for Responsible Investments**: The growing awareness of climate change, social inequalities, and corporate governance issues has increased the demand for responsible investments. Investors seek opportunities to align their portfolios with their values, driving the growth of sustainable financing.

II. **Market Expansion of Sustainable Financial Products**: Green bonds, social bonds, sustainability-linked loans, and other sustainable financial products are gaining traction. As more companies and governments commit to sustainability goals, the market for such products is expected to expand, mobilizing capital for ESG initiatives.

III. **Innovation and Green Technology Financing**: Sustainable financing drives investment in green technologies and innovations. The transition to a low-carbon economy relies on advancements in renewable energy, clean transportation, and circular economy solutions, offering significant prospects for sustainable financing.

IV. **ESG Integration Becoming Mainstream**: ESG integration in investment decision-making is becoming a mainstream practice. Institutional investors and asset managers increasingly consider ESG factors to assess long-term risks and opportunities, influencing capital allocation.

V. **Regulatory Support and Standardization:** Policymakers and regulators recognize the importance of sustainable finance. Governments are implementing regulations, tax incentives, and disclosure requirements to promote ESG practices, while standardization efforts enhance transparency and comparability in ESG reporting.

Challenges

I. **Data Availability and Quality**: Access to dependable ESG data remains a challenge, especially for smaller companies and those in emerging markets. Inconsistent reporting standards and data quality hinder effective ESG analysis and decision-making.

II. **Short-Term Mindset and Lack of Long-Term Metrics**: Traditional financial markets prioritize short-term gains over long-term sustainability. A lack of standardized long-term metrics and performance indicators makes assessing companies' true sustainability performance challenging.

III. **Greenwashing and Lack of Transparency:** Greenwashing, where companies exaggerate their environmental commitments without meaningful action, is a concern in sustainable finance. The lack of transparency and the absence of standardized reporting might pose challenges for investors in discerning authentic sustainable initiatives from cosmetic ones.

IV. **Pricing of ESG Risks and Returns**: Financial markets are currently in the process of creating methods to accurately assess and value environmental, social, and governance (ESG) risks and returns. Measuring the monetary consequences of environmental, social, and governance (ESG) variables can be intricate, resulting in difficulties in precisely determining the worth of sustainable investments.

V. **Policy and Regulatory Uncertainty**: Fluctuations in rules and regulations can generate ambiguity for investors and companies. Establishing and maintaining reliable and unchanging policy frameworks is crucial for instilling trust in sustainable investments.

VI. **Impact Measurement and Reporting:** Quantifying and documenting the concrete effects of sustainable investments might pose difficulties. Additional standardized approaches are required to evaluate the social and environmental impacts of ESG activities.

The outlook for sustainable finance for ESG practices is optimistic, fuelled by growing demand for ethical investments, the proliferation of sustainable financial products, and the widespread incorporation of ESG principles. Nevertheless, there are other obstacles that need to be tackled in order for sustainable finance to achieve its maximum effectiveness, such as the availability of data, the issue of greenwashing, and uncertainty surrounding policies. The cooperation of governments, financial institutions, businesses, and regulators is crucial in order to address these difficulties and establish a global financial system that is both sustainable and fair. The growth of sustainable finance is anticipated to have a substantial impact on tackling environmental and social concerns and fostering good change.

5.21 Conclusion

ESG principles have evolved from a niche concept to a mainstream business imperative. The incorporation of ESG principles into financial decisions has emerged as a key enabler of responsible and sustainable business practices, influencing investment decisions, and reshaping the future global financial landscape. As a result of the inadequacies of traditional financial practices in capturing the entire spectrum of risks and opportunities associated with sustainability, the need for sustainable financing solutions to support ESG initiatives was emphasized.

Traditional instruments such as green bonds, social bonds, and sustainability-linked loans, which have acquired traction recently, were examined as viable means of sustainable financing. In addition, the chapter explored innovative sustainable financing mechanisms, such as green technology financing, sustainable supply chain financing, and impact investing, demonstrating the potential for these emerging trends to generate transformational change in various sectors.

Public-Private Partnerships (PPPs) are essential for funding extensive ESG projects, and the cooperation between the public and private sectors is recognized as a significant catalyst for sustainable finance initiatives. The efficacy of government incentives, including as tax exemptions, grants, and subsidies, in encouraging ESG practices and influencing the overall sustainable finance environment was examined. Sustainable funding for environmental, social, and governance (ESG) activities is not merely a passing fad, but rather a need for achieving a more sustainable and resilient future. Integrating environmental, social, and governance (ESG) factors into financial choices is crucial for effectively tackling the intricate global challenges of our day. Businesses, investors, and governments have the ability to collaboratively promote a fair, enduring, and successful society by ensuring that financial decisions are in line with environmental, social, and governance principles. The development of sustainable finance offers a powerful tool for creating good impact and generating long-lasting value for the environment and future generations.

References

Enwin, A. D., & Ikiriko, T. D. (2023). Eradicating homelessness and improving the living standards of the poorest of the poor in greater Port Harcourt City, Nigeria. *European Journal of Theoretical and Applied Sciences, 1*(3), 85–104. https://doi.org/10.59324/ejtas.2023.1(3).09

PRECEDENCE RESEARCH. (2023). *Sustainable Finance Market*. https://www.precedenceresearch.com/sustainable-finance-market?cv=1

Chapter 6
Sustainable Investment Strategies for Renewable Energy and Circular Economy Projects

Abstract Sustainable investment strategies are investment strategies that seek to generate financial returns while also having a favorable implication on society or the environment. These strategies can be used to finance renewable energy and circular economy projects, which are critical for addressing climate change and other environmental challenges. Sustainable investment, fundamentally, seeks a congruence between positive environmental, social, and governance (ESG) impacts and financial profitability. The chapter delineates the various renewable energy sources, including solar, wind, and hydropower, underscoring associated investment avenues such as direct investments, green bonds, and venture capital. Similarly, the nuances of the circular economy, characterized by waste minimization and resource efficiency, are expounded upon, accompanied by relevant investment strategies like private equity and innovation funds. Influencing factors like regulatory frameworks, technological advances, stakeholder pressures, and financial incentives are analyzed. While acknowledging challenges like initial costs and market unpredictabilities, the chapter postulates solutions through diversification, partnerships, and continuous learning. Illustrated with case studies, this comprehensive exploration positions sustainable investment as a moral choice and a pivotal financial strategy in today's globalized world.

Keywords Sustainable investment strategies · Renewable energy · Circular economy · Impact investing · Green bonds · Circular economy projects · Recycling · Remanufacturing

6.1 Introduction

In an age where global challenges related to climate change, resource depletion, and waste accumulation are intensifying, sustainable investment strategies that focus on renewable energy and circular economy projects have never been more critical. These investment avenues not only promise financial returns but also work towards building a sustainable and resilient future. Sustainable investment refers to the

V. Kandpal et al., *Sustainable Energy Transition*, Circular Economy and Sustainability, https://doi.org/10.1007/978-3-031-52943-6_6

practice of investing funds in projects and businesses that have positive environmental, social, and governance (ESG) impacts while also delivering financial returns to the investors.

Sustainable investment strategies are investment strategies that seek to generate financial returns while also having a positive impact on society or the environment. These strategies can be used to finance renewable energy and circular economy projects, which are important for addressing climate change and other environmental challenges (Yazar et al., 2020). Sustainable investment strategies can be used to finance a wide range of renewable energy projects, such as solar, wind, and hydropower projects. They can also finance circular economy projects, such as recycling, repair, and remanufacturing projects. For example, an impact investor could invest in a company developing a new solar panel technology. The investment would give the company the capital to develop the technology and help reduce greenhouse gas emissions.

Another example would be a green bond issued by a government to finance a new wind power project. The bond would raise money to build the project and help reduce the country's reliance on fossil fuels. Sustainable investment strategies are becoming increasingly popular as investors seek to make their money work for good. These strategies can help finance the projects needed to address climate change and other environmental challenges while generating financial returns for investors. Here are some of the benefits of sustainable investment strategies:

- They can help to address climate change and other environmental challenges.
- They can create jobs and boost economic growth.
- They can improve the lives of people and communities.
- They can help to make the world a more sustainable place.

As we stand at the nexus of environmental, economic, and social revolutions, the world is rapidly transforming. A burgeoning awareness of our planet's finite resources and a desire to secure future generations' well-being is reshaping our perspectives on growth, value, and prosperity. Central to this transformation is the concept of sustainable investment, a comprehensive approach that intertwines profitability with the broader well-being of society and the environment. Gone are the days when monetary return was the sole metric for investment success. Today, forward-thinking investors recognize that true success incorporates environmental, social, and governance (ESG) factors. Sustainable investment is not just about creating wealth in the present but ensuring a viable and thriving planet for the future.

Renewable energy production is a promising social enterprise field in Scotland's desolate northern communities. Community wind energy generates funds for reinvestment in local development initiatives, such as community enterprises, social services, infrastructure, and communications. Strategic reinvestment of revenues in local social services generates approximately ten times more employment and income than wind power production (Okkonen & Lehtonen, 2016).

This chapter unravels the intricacies of sustainable investment strategies, particularly emphasizing renewable energy and circular economy projects. As we navigate the pages, readers will be equipped with an understanding of the varied renewable energy sources, innovative models promoting circularity in economic activities, and the confluence of opportunities and challenges these sectors present. In an information age, knowledge, perspective, and strategic foresight set one apart. This chapter aims to provide these tools, enabling investors, policymakers, and enthusiasts to make informed decisions that shape a sustainable and prosperous future.

6.2 Renewable Energy Projects

The epochal shift from fossil fuels to renewable sources of energy is not just a response to the global energy crisis but an acknowledgment of the looming environmental threats posed by non-renewable energy sources. Renewable energy projects harness naturally replenishing resources, offering a sustainable, clean, and often economically competitive alternative to traditional fuels. Here is a deeper dive into the realm of renewable energy:

6.2.1 Types of Renewable Energy

- **Solar Energy:** Utilizing the sun's radiation, solar panels convert sunlight into electricity. There are various technologies, such as photovoltaic (PV) for direct electricity production and concentrated solar power (CSP) for heat-driven power generation.
- **Wind Energy:** Wind turbines harness the wind to generate electricity. The kinetic energy of the wind turns the blades, which spin a generator, producing electricity.
- **Hydropower:** The oldest form of renewable energy, hydropower uses flowing water to turn turbines that generate electricity. It can be sourced from large-scale dams or smaller run-of-the-river installations.
- **Biomass:** Organic materials, such as wood, agricultural residues, and animal waste, can be burned or biologically processed to produce electricity or biofuels.
- **Geothermal Energy:** This involves harnessing the Earth's internal heat. By tapping into geothermal reservoirs, steam or hot water can be used to generate electricity or provide direct heating.
- **Ocean and Tidal Energy:** Oceans offer energy through tides, waves, and temperature differences. Technologies like tidal turbines, wave energy converters, and ocean thermal energy conversion systems are being developed to harness this potential.

6.2.2 Benefits of Investing in Renewable Energy Projects

- **Environmental Impact:** Reduced greenhouse gas emissions and a smaller carbon footprint.
- **Economic Growth:** Job creation in manufacturing, installation, and maintenance sectors.
- **Energy Independence:** Reduces reliance on imported fossil fuels, leading to greater energy security.
- **Stable Returns:** Once set up, many renewable energy projects, like solar or wind farms, have predictable and stable returns, given the low operational costs.

6.2.3 Challenges

- **Intermittency:** Some renewable sources, like solar and wind, are intermittent, necessitating energy storage solutions or backup systems.
- **Initial Capital Costs:** Some renewable technologies have high upfront costs, though these are often offset by low operational costs over time.
- **Land Use Concerns:** Large-scale renewable installations, especially solar farms, or hydropower dams, might require significant land or disrupt local ecosystems.
- **Technological Evolution:** As technology progresses, earlier investments might become outdated quickly.

Renewable energy projects represent a vibrant and crucial sector in the transition to a sustainable global energy landscape. The growth potential, coupled with the urgent need to mitigate environmental challenges, makes these projects an ethical imperative and attractive investment proposition.

6.3 Circular Economy Projects

The circular economy has emerged as a transformative paradigm in a world grappling with limited resources and escalating waste issues. It eschews the traditional linear "take, make, dispose" model, opting instead for a cyclical approach where products and materials are continually reused, repurposed, and recycled. Circular economy projects aim to extend product lifetimes, reduce waste, and foster sustainable consumption. Here is an exploration of these initiatives:

6.3.1 Core Principles of the Circular Economy

- **Design Out Waste and Pollution:** Products are designed from the outset to minimize waste, use fewer resources, and reduce environmental harm.

- **Keep Products and Materials in Use:** Rather than discarding items, the goal is to find ways to reuse, refurbish, and recycle, thus prolonging their useful life.
- **Regenerate Natural Systems:** Return valuable nutrients to the environment, ensuring a replenished and balanced ecosystem.

6.3.2 Elements of Circular Economy Projects

- **Product-as-a-Service:** Traditional ownership models are replaced by leasing or sharing, ensuring products are returned, refurbished, and redeployed.
- **Sustainable Product Design:** Designing products for longevity, modularity, and recyclability.
- **Recycling and Upcycling:** Transforming waste materials into new products or higher-value materials.
- **Reverse Logistics:** Systems that enable the return of products to the manufacturer for refurbishment or recycling.

6.3.3 Investment Avenues in Circular Economy Projects

- **Direct Investment:** Invest directly in businesses that operate based on circular economy principles or in infrastructures that support the circular model.
- **Private Equity:** Funding mature circular businesses or startups that are championing innovative circular solutions.
- **Circular Bonds:** Financial instruments that are dedicated to supporting circular economy projects.
- **Innovation Funds:** These are tailored for innovative technologies and solutions promoting circularity, such as advanced recycling technologies.

6.3.4 Benefits of Investing in Circular Economy Projects

- **Resource Efficiency:** Reduced dependency on scarce resources, ensuring business resilience against supply shocks.
- **Waste Reduction:** Less waste translates to decreased disposal costs and lesser environmental impact.
- **Innovation and Competitive Advantage:** Circular strategies often involve innovative processes and business models, providing a competitive edge in the market.
- **Regulatory Compliance:** As governments globally push for more sustainable practices, circular businesses are well-positioned to meet regulatory requirements.

6.3.5 *Challenges*

- **Market Maturity:** The circular economy, though gaining traction, is still maturing. This can present uncertainties and risks to investors.
- **Consumer Behavior:** Shifting consumers from ownership models to sharing or leasing requires cultural change and can be challenging.
- **Complex Supply Chains:** Circular practices sometimes require more intricate logistics and supply chains.
- **Technological Barriers:** Some circular processes, especially in recycling, require advanced technologies that may still be in developmental stages.

In summary, circular economy projects signify a profound shift in how businesses and societies view resources, consumption, and waste. They represent a harmonization of economic growth with environmental stewardship. For investors, these projects offer a chance to be at the forefront of a sustainable revolution, driving positive change and realizing attractive returns.

6.4 Factors Influencing Sustainable Investment Decisions

Sustainable investment decisions are a confluence of numerous factors, reflecting the interconnectedness of our modern world. Whether one is considering renewable energy, circular economy projects, or other sustainable ventures, several driving forces shape these decisions. A thorough understanding of these factors is crucial for investors, policymakers, and business leaders. Here is an exploration of the critical determinants:

6.4.1 *Regulatory and Policy Framework*

- **Governmental Incentives:** Tax breaks, subsidies, and grants can make sustainable projects more financially appealing.
- **Regulations:** Stricter environmental regulations or sustainability reporting mandates can push businesses towards sustainable investments.
- **International Agreements:** Commitments such as the Paris Agreement might influence national and regional policies, affecting investment priorities.

6.4.2 *Economic Considerations*

- **Return on Investment (ROI):** The potential for competitive financial returns remains a critical factor.

- **Cost Efficiency:** Over time, certain sustainable technologies (like solar panels) have become more affordable, influencing their adoption rate.
- **Market Demand:** As consumers and businesses increasingly demand sustainable goods and services, it creates more significant investment opportunities in those areas.

6.4.3 Technological Advancements

- **Innovation:** Breakthroughs in technologies can suddenly make certain sustainable investments more viable or attractive.
- **Access to Technology:** The availability and transfer of sustainable technologies across regions can influence investment decisions.

6.4.4 Stakeholder Pressures and Preferences

- **Consumer Behavior:** A shift in consumer preferences towards sustainable products/services can drive businesses to align with those values.
- **Shareholder Activism:** Shareholders increasingly push for more transparent ESG (Environmental, Social, Governance) practices.
- **Supply Chain Standards:** As businesses set sustainability standards, their suppliers are influenced to invest sustainably to meet those requirements.

6.4.5 Risks Management

- **Environmental Risks:** The potential fallout from environmental degradation, like flooding or resource scarcity, might push investments in mitigation or adaptation solutions.
- **Reputational Risks:** Businesses increasingly recognize the reputational damage linked with unsustainable practices.
- **Long-term Viability:** Companies look at the long-term availability of resources and the sustainability of their business models.

6.4.6 Socio-cultural Dynamics

- **Cultural Values:** Societies with deep-rooted environmental or communal values might naturally lean towards sustainable investments.

- **Public Awareness:** Greater public awareness and education about environmental issues can influence both individual and institutional investment behaviors.

6.4.7 Geopolitical Factors

- **Trade Dynamics:** Trade relationships and dynamics can influence the availability and cost of sustainable technologies or solutions.
- **Political Stability:** Stable political environments are often more conducive to long-term sustainable investments, while volatile regions might see more short-term, risk-averse investments.

6.4.8 Access to Information and Transparency

- **Research and Data:** The availability of accurate and comprehensive data on sustainable technologies, market trends, and impacts can guide informed decisions.
- **Transparency and Reporting:** Companies that maintain transparent reporting on their ESG metrics can attract more sustainable investment.

6.4.9 Competitive Advantage

- **Branding and Positioning:** Being recognized as a sustainable brand can offer a competitive edge in the market.
- **Innovation Leadership:** Companies might invest in sustainable projects to be seen as pioneers or leaders in their industry.

6.4.10 Ethical and Moral Commitment

- **Personal or Organizational Values:** Some investors or businesses might prioritize sustainability based on personal, ethical, or organizational values beyond economic considerations.

Sustainable investment decisions are multifaceted and influenced by a myriad of intertwined factors. While each element might vary in weightage depending on the investor's perspective, understanding this complex tapestry is essential for effective decision-making in sustainable investments.

6.4.11 Types of Sustainable Investment Strategies

Sustainable investment strategies are becoming increasingly popular as investors seek to make their money work for good. These strategies can help finance the projects needed to address climate change and other environmental challenges while generating financial returns for investors. There are many different types of sustainable investment strategies, but some of the most common include:

Impact investing: Impact investing is an investment strategy that seeks to generate financial returns while also having a positive impact on society or the environment. Impact investors typically invest in companies or projects that are working to address specific social or environmental problems, such as poverty, climate change, or water scarcity.

ESG investing: ESG investing is an investment strategy that considers environmental, social, and governance factors when making investment decisions. ESG investors typically invest in companies with sound environmental practices, treat their employees well, and are transparent and accountable to their shareholders.

Green bonds: Green bonds are issued to finance projects with environmental benefits, such as renewable or energy efficiency projects. Green bonds are a popular way for governments and businesses to raise money for sustainable projects.

Social impact bonds: Social impact bonds are a financial instrument that combines the public sector's need for social services with the private sector's ability to innovate and deliver those services. Social impact bonds typically involve a government or other public entity issuing bonds to finance a social service program. The bond is repaid with interest if the program achieves its desired outcomes.

Voluntary carbon markets: Voluntary carbon markets are a way for businesses and individuals to offset their carbon emissions by investing in projects that reduce emissions, such as renewable energy or forest conservation projects.

It is important to note that sustainable investing is a relatively new field, and there is still some debate about the best way to measure the impact of sustainable investments. However, the demand for sustainable investments is growing, and many resources are available to help you learn more about sustainable investing and find the right strategy for you.

6.4.12 The Importance of Sustainable Investment

Amidst the rapid economic and technological changes of the twenty-first century, sustainable investment has emerged as a beacon guiding us towards a future that harmonizes financial growth with environmental and social well-being. Its importance cannot be understated, and here is why:

1. **Environmental Protection:**
   - **Mitigating Climate Change:** Sustainable investments, especially in renewable energy, reduce the dependency on fossil fuels, thus curbing greenhouse gas emissions. These investments actively combat global warming and the adverse effects of climate change.
   - **Preservation of Biodiversity:** By directing funds towards sustainable agriculture, forestry, and fisheries, we ensure the conservation of essential ecosystems and species.
2. **Socio-economic Benefits:**
   - **Job Creation:** Renewable energy projects and circular economy initiatives often create new jobs, stimulating local economies and reducing unemployment.
   - **Community Development:** Sustainable investments frequently target and uplift marginalized communities, ensuring they access essential services like clean energy, clean water, and waste management.
   - **Health and Well-being:** Reducing pollution and promoting cleaner industries have direct health benefits, decreasing medical expenses and increasing the overall quality of life.
3. **Risk Management:**
   - **Long-term Stability:** Companies with sustainable practices are less susceptible to environmental liabilities or costly litigations, ensuring more predictable and stable returns.
   - **Reputation and Brand Value:** Consumers gravitate towards brands that align with their values as they become more environmentally conscious. Sustainable investment boosts brand image, leading to increased customer loyalty and sales.
4. **Financial Returns:**
   - **Emerging Opportunities:** Many renewable energy and circular economy projects are at the forefront of innovation. Investing in these sectors can yield significant returns as demand for sustainable solutions grows.
   - **Government Incentives:** Several governments worldwide offer tax breaks, grants, or subsidies for sustainable initiatives, making such investments financially attractive.
5. **Ethical Responsibility:**
   - **Intergenerational Equity:** Sustainable investments ensure that future generations inherit a world with ample resources and a stable environment.
   - **Moral Obligation:** Beyond financial considerations, there is an intrinsic value in doing good and acting as stewards of the planet.

6. **Foresight and Future-Readiness:**
   - **Adaptation to Market Dynamics:** As global markets pivot towards sustainability, businesses that fail to adapt to risk obsolescence. Sustainable investment prepares and positions organizations for future market trends.
   - **Resilience against Shocks:** Sustainable businesses often have diversified supply chains and resource pools, making them more resilient against market shocks or resource scarcities.

The importance of sustainable investment lies in its dual nature: it is both a moral imperative and a strategic financial decision. In an interconnected world, it bridges the gap between immediate monetary returns and long-term planetary welfare, acting as a compass guiding us towards a future that is not just profitable but also just, inclusive, and environmentally conscious. Here are some specific examples of the benefits of investing in sustainable projects:

- A study by the Rocky Mountain Institute found that investing in renewable energy could create 2.4 million jobs in the United States by 2030.
- A study by the International Renewable Energy Agency found that investing in energy efficiency could create 13.8 million jobs in the global economy by 2030.
- A study by the World Bank found that investing in climate-smart agriculture could lift 100 million people out of poverty by 2030.

By investing in sustainable projects, we can help to create a better future for ourselves and for generations to come.

6.5 Case Studies

Orsted: From Fossil Fuels to Renewable Energy

Background: Orsted, formerly known as DONG Energy (Danish Oil and Natural Gas), was one of Northern Europe's most fossil fuel-intensive utilities. However, in the past decade, they undertook a transformative journey to become a global leader in offshore wind energy.

Investment Strategy:

- **Transition Focus:** Shifted their core business from oil and gas exploration to renewable energy, mainly offshore wind.
- **Asset Divestment:** Sold their upstream oil and gas business to focus resources on renewables.
- **Innovation:** Invested in R&D to improve offshore wind technology, making it more efficient and cost-effective.

Outcome: By 2020, Orsted had reduced its carbon emissions by over 80% compared to 2006. The company aims to be carbon neutral in its energy generation and operations by 2025, and by 2040, they intend to tackle emissions from its entire carbon footprint.

TerraCycle: Pioneering Waste into Value

Background: TerraCycle, founded in 2001, is a U.S.-based company known for turning hard-to-recycle waste into usable products. They have positioned themselves as leaders in the circular economy movement.

Investment Strategy:

- **Partnerships:** Collaborated with major consumer brands like Colgate, Procter & Gamble, and Nestle to design waste-reduction initiatives.
- **Diverse Revenue Streams:** Apart from recycling services, they launched the 'Loop' initiative—a shopping platform where products are sold in reusable packaging.
- **Global Expansion:** Started in the U.S. but expanded operations to over 20 countries, addressing waste issues on a global scale.

Outcome: TerraCycle has successfully diverted millions of pounds of waste from landfills and incinerators. Their innovative models have garnered significant media attention and made them a favored partner for brands looking to enhance their sustainability credentials.

Lessons Learned

Orsted:

- **Flexibility is Key:** Even companies rooted in traditional energy sectors can pivot their business models when aligned with a clear vision and strategy.
- **Holistic Transition:** Merely investing in renewables is not enough. A successful transition requires divesting from unsustainable sectors and continually measuring and adjusting for impacts.

TerraCycle:

- **Innovation Drives Value:** By looking at waste as a resource, TerraCycle created new value streams that are both profitable and environmentally beneficial.
- **Collaboration Amplifies Impact:** Working with established brands allowed TerraCycle to scale its impact and bring circular economy concepts into mainstream consumer consciousness.

Both case studies underscore that sustainable investment is not just a philanthropic endeavor—it can be a lucrative strategy when aligned with market demand, innovation, and collaboration.

6.6 Challenges and Mitigations

Sustainable investing is investing in companies and projects that meet certain environmental, social, and governance (ESG) criteria. The goal of sustainable investing is to generate financial returns while also having a positive impact on the environment and society.

Sustainable investment is undoubtedly the need of the hour. However, as with any transformative approach, it presents a unique set of challenges. Understanding these challenges and implementing strategic mitigations can pave the way for smoother transitions and more successful outcomes. Some of the challenges include:

- **Initial Capital Costs:** Sustainable solutions, especially in sectors like renewable energy, often require substantial initial investments. It may deter businesses, especially smaller ones, from adopting sustainable measures due to the perceived high upfront costs.
- **Lack of data and transparency.** A lack of standardized data and reporting on ESG factors makes it difficult for investors to assess the sustainability of companies and projects.
- **Greenwashing.** Some companies may make misleading claims about their sustainability practices, making it difficult for investors to distinguish between truly sustainable investments and those not.
- **Higher costs.** Sustainable investments may be more expensive than traditional investments due to the additional costs of ESG research and due diligence.
- **Lower returns.** There is evidence that sustainable investments may generate lower returns than traditional investments. However, this is a complex issue, and the evidence is mixed.
- **Lack of Standardised Reporting**: In the domain of sustainable investment, a prevalent and pressing challenge is the lack of standardized reporting. As investors and stakeholders become increasingly keen on understanding the environmental, social, and governance (ESG) aspects of businesses, the need for standardized, transparent, and comparable reporting metrics becomes paramount. However, the current landscape is fragmented, leading to many challenges and implications.
- **Market and Regulatory Uncertainty**: The sustainable investment landscape is evolving, leading to potential policy shifts and market changes. Investors might hesitate to invest if they believe the regulatory environment might change unfavorably.
- **Knowledge and Skill Gaps**: A significant learning curve involves understanding sustainable investment's nuances. Without the right expertise, investors and companies might make sub-optimal decisions.
- **Technological Barriers**: Some sustainable solutions are still in developmental phases or may not be as efficient as their traditional counterparts. This might slow down the adoption rate of certain sustainable technologies.
- **Geopolitical Risks**: In certain regions, investing in sustainability might come with geopolitical risks, such as regulatory changes, civil unrest, or trade restrictions. Investors might forgo sustainable opportunities in these regions due to perceived risks.
- **Sociocultural Resistance**: In some communities or regions, there might be resistance to sustainable practices due to cultural beliefs or lack of awareness. This can slow down the implementation of sustainable projects in such areas.

- **Short-term Financial Focus**: The traditional investment model focuses on short-term returns, while many sustainable investments are long-term endeavors. There's a potential conflict between short-term profitability and long-term sustainable goals.
- **Supply Chain Complexity**: Ensuring sustainability throughout the supply chain can be challenging, especially in global operations. Companies might struggle to ensure that every part of their supply chain adheres to sustainability standards.

There are a number of mitigations to these challenges, including:

- **The development of standardized ESG data and reporting standards.** This would make it easier for investors to assess the sustainability of companies and projects.
- **The creation of more rigorous ESG disclosure requirements.** This would help to prevent greenwashing.
- **The development of more affordable ESG investment products.** This would make sustainable investing more accessible to a broader range of investors.
- **The evidence that sustainable investments can generate higher returns over the long term.** This would encourage more investors to invest in sustainable investments.

Despite the challenges, sustainable investing is a growing trend. More and more investors are looking to invest in companies and projects that positively impact the environment and society. As the market for sustainable investments grows, the challenges are likely to be addressed and the benefits of sustainable investing will become more apparent.

6.7 Conclusion

While these challenges are significant, they are not insurmountable. Addressing each of these challenges requires a combined effort from businesses, governments, investors, and civil society. The potential rewards—both in terms of financial returns and positive societal impact—justify the collective endeavor to overcome these obstacles. Through collaborative action, continuous learning, and innovation, the global community can ensure that sustainable investment becomes the norm, rather than the exception.

As we traverse through the twenty-first century, a notable shift is discernible in the global financial landscape. Once seen as a niche or risky proposition, sustainable investment has carved its rightful place at the heart of the global investment strategy. The journey we have explored—from understanding the importance of sustainable investment to dissecting the mechanics of renewable energy and circular economy projects to delving into the multifaceted influencers of investment decisions—demonstrates the depth and breadth of this new paradigm. But what does all this signify for our collective future, and how does this reshape our vision for global economies and ecosystems?

6.8 A Reckoning of Priorities

The twenty-first century brought with it an acute realization. The methods that fuelled our rapid industrial and technological progress in the last century are untenable. Fossil fuels, wasteful manufacturing processes, and the single-use consumer culture have depleted the planet's resources and disrupted its delicate ecological balance. Climate change, pollution, and biodiversity loss are no longer theoretical threats but pressing realities. In this scenario, sustainable investment is not merely a response; it is a reckoning—a recalibration of global priorities towards long-term survival and prosperity.

6.9 The Economics of Sustainability

Historically, one of the critical counterarguments against aggressive sustainable transitions was the perceived economic trade-off. There was a prevalent notion that sustainability came at the expense of profitability. However, as highlighted by our case studies on companies like Orsted and TerraCycle, and many others globally, this could not be further from the truth.

The economics of sustainability are straightforward. Renewable energy sources, for instance, offer cheaper, long-term alternatives to fossil fuels after overcoming initial capital costs. When implemented correctly, the circular economy conserves resources and opens avenues for innovation, job creation, and market diversification. Simply put, sustainability has become a catalyst for economic value creation instead of being a cost.

6.10 The Evolving Investor Landscape

Modern investors, both institutional and individual, are progressively discerning. They are not just chasing short-term financial returns; they are evaluating the broader impact of their investments. Stakeholders are pressuring companies to adopt ESG (Environmental, Social, Governance) metrics, and shareholder activism is at an all-time high. The rise of green bonds, ESG funds, and direct sustainable investment avenues signifies an evolving investor consciousness that balances economic returns with global impact.

Moreover, a new generation of consumers and investors, notably millennials and Gen Z, are emerging as influential drivers of sustainable practices. For them, an investment or purchase's ethical, environmental, and social dimensions are often as significant, if not more, than the financial dimension.

6.11 Challenges and Opportunities Ahead

No major paradigm shift is devoid of challenges, and the same holds for sustainable investment. Technological barriers in some sectors, existing infrastructural inertia, regulatory inconsistencies across regions, and sometimes even socio-cultural resistance can impede the pace of sustainable transitions.

However, history has shown that challenges often sow the seeds of innovation. The intermittency issue of renewable energy has catalyzed advancements in energy storage technologies. The challenge of waste in the circular economy has given birth to pioneering recycling and upcycling solutions. In every challenge, there lies an investment opportunity.

6.12 The Road Ahead

At this pivotal juncture, the path forward is one of collaboration, innovation, and steadfast commitment. Policymakers, business leaders, investors, researchers, and the broader public have roles to play.

Policymakers must frame regulations and incentives that facilitate sustainable transitions and penalize unsustainable practices. Business leaders must imbibe sustainability at the core of their strategy, ensuring it permeates every facet of their operations—from supply chains to customer engagement. Investors, both big and small, must continue to champion and prioritize sustainable ventures, signalling the market's direction. Researchers and innovators have the onus of pushing the boundaries, making sustainable solutions more accessible, efficient, and cost-effective. And lastly, the public, as consumers and citizens, can drive demand and shape discourse, ensuring sustainability remains at the forefront.

Sustainable investment is more than a financial strategy; it is a collective vision for a future where prosperity coexists harmoniously with planetary well-being. The trajectory is set, the momentum is building, and every stakeholder has a part to play in scripting this promising narrative for the future. The onus is on us, as a global community, to ensure this vision translates into a tangible reality for future generations.

References

Okkonen, L., & Lehtonen, O. (2016). Socio-economic impacts of community wind power projects in Northern Scotland. *Renewable Energy, 85*, 826–833. https://doi.org/10.1016/j.renene.2015.07.047

Yazar, M., Hestad, D., Mangalagiu, D., Saysel, A. K., Ma, Y., & Thornton, T. F. (2020). From urban sustainability transformations to green gentrification: Urban renewal in Gaziosmanpaşa, Istanbul. *Climatic Change, 160*(4), 637–653. https://doi.org/10.1007/s10584-019-02509-3

Chapter 7
Policy Framework for Sustainable Energy Transition and the Circular Economy

Abstract This chapter delves into the interconnection of sustainable energy transition and the circular economy within the realm of policymaking. With growing environmental concerns, there is an urgent call for a consolidated policy framework that weaves these two concepts into a unified blueprint for a green future. Initially, the chapter concisely elucidates the sustainable energy transition, emphasizing the shift from nonrenewable to renewable resources and the circular economy, which advocates for resource optimization through reuse, recycling, and regeneration. At the core of the discussion are the essential elements of a viable policy framework, ranging from vision-setting and incentive structures to legislation and public education. Moreover, the chapter identifies potential economic, technological, and sociocultural challenges in the path of policy implementation. Through case studies, such as Denmark's leadership in wind energy and Japan's pioneering efforts in the circular economy, readers gain insights into the real-world applications of these policies. The overarching theme posits that for a truly sustainable future, a harmonious integration of energy transition and circular economy principles is paramount in policy frameworks.

Keywords Sustainable energy transition · Circular economy · Policy framework · Resource optimization · Renewable resources · Green future

7.1 Introduction

As global concerns about climate change, resource depletion, and environmental degradation intensify, there is a growing need to reorient our energy and resource management strategies. At the heart of this paradigm shift are two intertwined concepts: the transition to sustainable energy and the circular economy. The sustainable energy transition entails moving from finite, polluting energy sources to renewable, cleaner alternatives. On the other hand, the circular economy seeks to redefine the conventional 'take-make-waste' industrial model, promoting a system where resources are reused, recycled, and regenerated. Bringing these two concepts to

V. Kandpal et al., *Sustainable Energy Transition*, Circular Economy and Sustainability, https://doi.org/10.1007/978-3-031-52943-6_7

fruition necessitates a robust policy framework to guide and streamline efforts across sectors and borders (Li et al., 2023). A policy framework for sustainable energy transition and the circular economy entails the formulation and execution of policies that promote a transition from conventional, linear patterns of consumption and energy utilization to more sustainable, efficient, and renewable approaches.

The world is facing several serious environmental challenges, including climate change, air pollution, and resource depletion. These challenges can only be addressed through a fundamental transformation of our energy and economic systems. A transition to a cleaner, more efficient, and more equitable energy system is a sustainable energy transition. This transition will necessitate significant investments in renewable energy, energy efficiency, and other low-carbon technologies. It will also necessitate modifications to our transportation, construction, and industrial sectors. A circular economy is an economic system that seeks to eliminate waste and pollution by extending the lifespan of materials. This can be achieved through reuse, repair, and recycling. A circular economy can help reduce our reliance on finite resources and mitigate climate change. The dual challenge of transitioning to sustainable energy sources and adopting a circular economy requires an integrated policy approach. Such a framework ensures harmonized objectives, maximized synergies, and reduced contradictions between the two paradigms (Hariram et al., 2023).

The COVID-19 pandemic has posed substantial obstacles to the process of transitioning to alternative energy sources. Global concerns have been voiced regarding the excessive focus on economic recovery at the expense of advancements in energy transition. There is an increasing demand for a "green" recovery program that aims to revive the economy without sacrificing the environment. Nevertheless, there has been a lack of thorough scholarly examination regarding the consequences of COVID-19 on the worldwide shift towards sustainable energy (Tian et al., 2021). The implementation of both the sustainable energy transition and the circular economy is crucial to addressing climate change and attaining a sustainable future. A comprehensive policy framework has the potential to facilitate these changes, guaranteeing their seamless implementation, fair distribution of benefits, and optimal outcomes. This chapter explores the diverse policy components and techniques that can effectively support these dual transitions.

Policy Framework for Sustainable Energy Transition and the Circular Economy would cover various essential domains:

- Policies that prioritize the shift towards renewable energy sources, including solar, wind, and hydro power. These measures may encompass inducements for the advancement of renewable energy, policies to gradually eliminate fossil fuels, and backing for research and breakthroughs in clean energy technologies.
- Strategies aimed at optimizing energy use in industrial, architectural, and transportation sectors. This may entail establishing criteria for energy-efficient appliances, advocating for environmentally-friendly construction methods, and promoting the utilization of public and low-emission transportation.
- Promoting the utilization of items through their reuse, repair, and recycling in order to decrease the generation of trash. This may entail implementing regula-

tions that facilitate the creation of products with extended lifespans, encourage recycling and composting, and discourage the use of single-use plastics and other materials that cannot be recycled.
- Economic incentives and regulation encompass various financial stimuli, like as subsidies, tax breaks, or grants, that are provided to firms and consumers as a means to encourage the adoption of sustainable practices. In addition, implementing regulatory mechanisms such as carbon pricing or emissions trading schemes can effectively incorporate the environmental expenses.
- Promoting cooperation among government, industry, and civil society to guarantee a comprehensive and inclusive approach. This may entail conducting public discussions, forging alliances with corporations, and engaging in collaboration with international organizations.
- Initiatives aimed at instructing the general public and industry leaders on the advantages of sustainable energy and circular economy practices, encompassing the ways in which individuals and corporations may make a positive impact.
- Monitoring and Evaluation involves the establishment of quantifiable measures and frameworks to monitor the progress made towards achieving sustainability objectives. It also entails the adaptation of policies in response to data and comments received.

7.2 Objectives of the Policy Framework

I. **Transition to Sustainable Energy Systems:** The transition towards sustainable energy systems is widely recognized as a key undertaking in contemporary times. This need arises from the urgent need to address climate change, safeguard energy stability, and foster economic advancement. This document provides a comprehensive examination of the transition process, emphasizing its significance, inherent difficulties, and viable approaches.

   - **Decarbonization**: Achieve a significant reduction in carbon emissions by promoting the shift from fossil fuels to renewable energy sources. Decarbonization is the act of diminishing carbon dioxide (CO2) emissions, specifically those caused by human actions, in order to ultimately establish a low-carbon economy. It is crucial in the effort to combat climate change, as carbon dioxide (CO2) is a significant greenhouse gas that contributes to the phenomenon of global warming. Decarbonization is a multifaceted task that necessitates synchronized efforts across all areas of the economy and society. It entails more than simply substituting outdated technology with modern ones; it also necessitates substantial transformations in societal functioning and human lifestyles.
   - **Energy Security**: Energy security encompasses the dependable accessibility of energy resources at a reasonable cost. It has a vital impact on the stability and prosperity of a nation. Energy security encompasses not only the guaran-

tee of a consistent energy supply, but also the imperative of ensuring that the energy provided is both sustainable, cost-effective, and ecologically conscious. With the global shift towards a more sustainable energy future, the notion of energy security is becoming more closely linked to the advancement of renewable energy sources and the mitigation of carbon emissions. It ensures a stable, reliable supply of energy by diversifying energy sources and reducing dependency on imports.

II. **Technological Advancement**: Technological advancement encompasses the ongoing development and introduction of new and improved technologies that consistently reshape our lifestyles, occupations, and interactions with our surroundings. This process includes a broad spectrum of disciplines and has multiple effects:

(a) Advancements in Information and Communication Technology (ICT), such as the internet, mobile technology, and artificial intelligence, have fundamentally transformed the way we communicate and share information. This has had a profound impact on various sectors, including education, healthcare, business, and government.
(b) Artificial Intelligence (AI) and Machine Learning (ML) are advancing rapidly, enhancing data processing, decision-making, and automation in areas such as healthcare, finance, manufacturing, and transportation.
(c) Advancements in solar, wind, hydroelectric, and geothermal energy are enhancing the efficiency and cost-effectiveness of renewable sources, facilitating the shift towards a more sustainable energy future.
(d) Biotechnology and genomics are making significant progress in the areas of medicine, agriculture, and environmental conservation. This progress includes the development of customized medicine, genetically modified crops, and sustainable materials derived from biological sources.
(e) Nanotechnology refers to the manipulation of matter at the atomic or molecular level. It finds applications in several fields such as health, electronics, and materials science, resulting in the development of products that possess unique qualities and functions.
(f) Advancements technologies and autonomous vehicles, high-speed rail, and aeronautical technology are revolutionizing global mobility for both people and cargo.
(g) Progress in this domain is not only broadening our comprehension of the cosmos but also yielding practical utilities such as satellite-based communication, navigation, and earth observation.
(h) Robotics are progressively employed in industrial, healthcare, and service sectors, enhancing efficiency, safety, and precision.
(i) The field of Material Science focuses on the advancement of novel materials such as graphene and aerogels, which are facilitating the emergence of innovative applications in the domains of electronics, construction, and energy storage.

(j) Environmental technology plays a crucial role in safeguarding the environment and promoting sustainable development through advancements in waste management, water purification, and pollution control.
(k) Progress in medical devices, telemedicine, and health informatics is enhancing the accuracy of diagnosis, the effectiveness of therapy, and the quality of patient care.
(l) Cybersecurity is of utmost importance as our dependence on digital technology increases, necessitating innovations in order to safeguard data and systems from cyber assaults.

The progress of technology has both advantages and disadvantages. On one hand, it provides significant advantages, but on the other hand, it poses issues such as the displacement of jobs caused by automation, worries over privacy, and the existence of a gap in access to digital resources. Striking a balance between these advantages and difficulties is crucial in utilizing technology for an improved future.

III. **Promotion of Circular Economy Principles:** Advocating for circular economy concepts entails transitioning from the conventional linear economic framework of "extract, produce, discard" to a more environmentally sound approach that eliminates waste, prolongs the lifespan of products and resources, and restores natural systems. This method not only has positive effects on the environment but also has the potential to generate new business prospects and contribute to economic expansion.

- Promoting the creation of products that are durable, can be easily repaired, and can be recycled throughout their lifecycle. This entails creating goods that are readily disassembled for the purpose of repair or recycling, as well as utilizing materials that can be sustainably procured and repurposed.
- Enacting tactics to optimize resource utilization and minimize waste generation. This can encompass enhancing production methodologies, minimizing material use, and repurposing waste materials.
- Extended Producer Responsibility (EPR) refers to the implementation of policies that hold producers accountable for the complete life cycle of their products, which includes managing the products at the end of their useful life. This incentivizes enterprises to develop items that are more recyclable or have a reduced ecological footprint.
- Circular Supply Chains: Establishing supply chains that facilitate the repurposing and reutilization of commodities. This may entail utilizing recycled materials in the manufacturing process, restoring previously used products, or creating mechanisms for product return post-consumption.
- Informing consumers about the advantages of circular economy principles and motivating them to actively partake in circular behaviors, such as engaging in recycling, purchasing goods crafted from recycled materials, or favoring services over items.

- Regulatory frameworks and incentives can be established to encourage circular economy practices. This might involve implementing legislation that provide tax benefits to enterprises that adopt sustainable practices, as well as imposing penalties on those that generate excessive waste.
- Investing in research and development of novel technologies and methodologies that facilitate circular economy practices, such as innovative recycling technologies or materials that possess more sustainability.
- Promoting cooperation among enterprises, governments, and non-profit organizations to cultivate circular economy solutions. This may encompass collaborations between public and commercial entities, coalitions within industries, and cooperation on a global scale.
- Promoting the adoption of business models that enable circular activities, such as product-as-a-service models, wherein consumers lease things rather than owning them.
- Green Infrastructure and Urban Planning involve the creation of physical structures and urban areas that facilitate circular economy principles. This includes the establishment of recycling and composting facilities, as well as the design of cities that promote resource sharing and reuse.

Advancing circular economy concepts necessitates a fundamental transformation in the approach to product creation, utilization, and recycling. This endeavor entails the cooperation of diverse stakeholders, such as governments, enterprises, and consumers. This method not only mitigates the environmental impact but also provides economic advantages through novel business models and employment prospects.

IV. **Economic and social objectives:** Economic and social objectives are essential elements of a country's development aspirations. They frequently include a diverse array of objectives and strategies aimed at enhancing the quality of life and welfare of the populace. Achieving a balance between these aims can be difficult, as they may require making compromises. Primary economic and social goals encompass:

- Economic growth is typically quantified by the rise in a nation's Gross Domestic Product (GDP). Economic growth is vital in order to allocate the necessary resources for enhancing living conditions and alleviating poverty.
- Generating employment possibilities for the expanding labor force. This entails not just augmenting the quantity of employment opportunities but also enhancing the caliber of work — guaranteeing that occupations are remunerative, stable, and satisfying.
- The act of regulating the rate of inflation in order to maintain stability in prices. Significant inflation can gradually diminish the ability to buy goods and services, particularly for individuals with lower incomes, whereas deflation can result in a decline in economic productivity.
- The objective is to decrease economic gaps and poverty levels by implementing measures such as progressive taxation, social assistance programs, and policies that ensure equal opportunities for everyone.

- Social welfare and security entail the creation of a protective system to support and safeguard the marginalized segments of society by means of healthcare, education, housing, and social security initiatives.
- Environmental sustainability refers to the practice of ensuring that economic growth is not achieved at the detriment of the environment. This entails advocating for sustainable practices, renewable energy, and conservation endeavors.
- Allocating resources towards the development of education and healthcare systems to enhance the intellectual and physical capabilities of the population, which is essential for sustainable economic progress and overall societal welfare.
- Striving for a consistent economic climate characterized by minimal fluctuations in economic growth, employment levels, and pricing. The maintenance of this stability is essential for the purpose of strategic planning and investment over an extended period of time.
- Advocating for balanced regional development to ensure fair and equal development across various parts of the country, hence mitigating regional inequities and imbalances.
- Inclusive Growth refers to the objective of ensuring equitable distribution of the advantages resulting from economic growth across all segments of society, including marginalized and disadvantaged populations.
- Infrastructure Development refers to the process of constructing and upkeeping critical services, such as transportation, communication, energy, and water, that are crucial for economic productivity and enhancing the overall standard of living.
- Fostering research and development (R&D) and innovation to stimulate future expansion and tackle societal issues.

Achieving a balance between these objectives necessitates meticulous policy planning and coordination. Frequently, there are compromises to be made; for instance, measures that promote swift economic expansion might occasionally result in environmental deterioration or heightened disparity. Therefore, it is essential to adopt a comprehensive approach that takes into account the interconnectedness of different goals in order to achieve sustainable development.

V. **Environmental Protection and Restoration:** Environmental conservation and remediation encompass a wide range of tactics and measures designed to conserve natural resources, maintain biodiversity, and rectify ecological harm caused by human actions. This strategy is crucial for preserving the ecological well-being of the earth and guaranteeing the viability of living conditions for future generations.

    Environmental conservation and rehabilitation necessitate a comprehensive strategy that encompasses governmental regulations, corporate accountability, and individual engagement. This is a worldwide endeavor that acknowledges the fact that environmental concerns transcend national boundaries and that the well-being of the Earth impacts every individual.

- **Biodiversity Preservation**: Reduce the environmental impact of energy production and waste, thus aiding in the preservation of ecosystems and biodiversity.
- **Pollution Reduction**: Minimize the release of pollutants, toxins, and other harmful substances in air, water, and soil.
- **Climate Resilience**: Enhance the resilience of communities and ecosystems against climate change impacts through sustainable energy and resource practices.

VI. **Strengthened governance and collaboration**

- **Stakeholder Engagement**: Foster a participatory approach where businesses, civil society, local communities, and other stakeholders play an active role in shaping and implementing policies.
- **International Collaboration**: Engage in collaborative efforts with other nations to share knowledge, technologies, and best practices in sustainable energy and the circular economy.
- **Transparency and Accountability**: Establish clear benchmarks, monitoring mechanisms, and regular reporting to ensure policies are effectively implemented and goals are met.

By clearly articulating these objectives, the policy framework provides a roadmap that stakeholders at all levels—from government agencies to individual consumers—can rally behind. The transition to a sustainable energy system and circular economy requires a collective effort, and these objectives serve as guiding stars for the journey ahead (Romero-Castro et al., 2023).

7.3 Overview of the Importance of Sustainable Energy Transition and the Circular Economy

With the global consensus on the imperativeness of addressing climate change, there is an increased focus on reshaping our energy and consumption narratives. Two transformative concepts, the sustainable energy transition and the circular economy, are emerging as crucial pathways to this end.

7.3.1 Sustainable Energy Transition: Rethinking Energy

A. **The Evolution of Energy Consumption**
Historically, human civilization has relied predominantly on nonrenewable energy sources. From the coal-driven Industrial Revolution to today's oil-centric economies, our development trajectory has leaned heavily on finite resources.

B. **The Impetus for Change**

1. **Environmental Imperative**: Conventional energy sources emit vast quantities of greenhouse gases. The resultant global warming is linked to extreme weather events, rising sea levels, and biodiversity loss.
2. **Economic Considerations**: Volatile oil prices and geopolitical tensions in resource-rich regions have underscored the fragility of current energy systems.
3. **Technological Feasibility**: Advancements in renewable technology, particularly solar and wind, are making these options increasingly efficient and cost competitive.

C. **Benefits of Transition**

1. **Climate Mitigation**: A swift energy transition is pivotal to achieving the targets set under international agreements such as the Paris Accord.
2. **Energy Independence**: Localized renewable sources decentralize energy production, curtailing reliance on foreign oil and enhancing national security.
3. **Innovation and Employment**: The renewable sector is rapidly becoming a hub for tech innovations, offering substantial job opportunities.

7.4 Circular Economy: Redefining Consumption

A. **The Linear Model's Downfall**
For decades, our consumption pattern has been linear: extract resources, produce, consume, and then dispose. This 'once-and-done' approach is untenable in a world with finite resources **and growing populations.**

B. **Principles of a Circular Economy**

1. **Design for Longevity**: Products are designed to last, reducing the frequency of replacements and waste.
2. **Recycle, reuse, repurpose**: Rather than discarding, products or their components are reintroduced into the production cycle.
3. **Systemic Thinking**: This is not just about recycling more. It is a complete systemic shift in production, consumption, and waste management.

C. **Circular Advantage**

1. **Resource Efficiency**: With the planet's resources under strain, using them judiciously ensures that they last for future generations.
2. **Economic Opportunities**: Businesses can unlock new revenue streams, from product-as-a-service models to markets for recycled materials.
3. **Environmental Health**: Reduced waste translates to fewer landfills, less pollution, and a cleaner environment.

7.5 Synergizing the Dual Transitions

Although distinct, the transition to a sustainable energy source and the circular economy are inextricably linked. A circular model necessitates energy, which must be pure. In contrast, renewable energy systems have material requirements as well, including solar panels, wind turbine blades, and batteries. Circular principles, such as recycling and prolonged product lifespans, play an essential role in this context.

7.5.1 Global Endeavors and Success Stories

Countries and corporations worldwide exemplify how these twin transitions can be achieved:

1. **Scandinavia's Renewable Push**: With the support of strong policies, countries like Denmark and Sweden are close to producing 100% renewable electricity.
2. **Japan's Circular Initiatives**: With policies such as the "3Rs" (Reduce, Reuse, Recycle), Japan is pioneering in waste management and circular manufacturing.

The sustainable energy transition and the circular economy are not just environmental imperatives; they are the bedrock for a stable, prosperous future. As resource constraints tighten and environmental challenges mount, these intertwined strategies offer a beacon of hope and a roadmap to sustainability.

In our rapidly evolving world, they signify not just the 'right thing to do' but the most logical, beneficial path forward for both people and the planet.

7.5.2 Global Energy Landscape and Challenges

A confluence of technological, environmental, and socioeconomic factors has caused significant changes in the global energy landscape. While new opportunities emerge, these shifts also bring forth significant challenges.

Historical Context

Historically, the world has heavily relied on fossil fuels—coal, oil, and natural gas—as primary energy sources. The Industrial Revolution marked the beginning of this intense fossil fuel consumption, powering economies and shaping modern civilization.

Current Energy Mix

As of the last few decades:

1. **Fossil Fuels**: Fossil fuels still dominate the global energy mix, with coal serving many power plants, oil powering transportation, and natural gas being used for both electricity and heating.
2. **Nuclear Energy**: It has a significant share in some countries, offering a low carbon but sometimes controversial energy source.
3. **Renewables**: Rapidly growing, especially wind, solar, and hydroelectric power, as costs decrease, and efficiency improves.

Drivers of Change
Several factors are reshaping the energy landscape:

1. **Climate Change**: As the effects of global warming become more evident, there is an imperative need to reduce greenhouse gas emissions, compelling nations to investigate alternative, healthier energy sources.
2. **Technological Advancements**: Innovations in renewable energy technologies, battery storage, and grid management make alternative energy sources more viable.
3. **Economic Factors**: Fluctuating fossil fuel prices, combined with decreasing costs of renewables, are altering economic calculations.
4. **Geopolitical Considerations**: Energy resources play a significant role in geopolitics, with nations striving for energy independence and security.

Challenges in the Global Energy Landscape

1. **Infrastructure Legacy**: Much of the current infrastructure is tailored for fossil fuels. Transitioning requires substantial investments in new infrastructure and potential stranded assets.
2. **Intermittency of Renewables**: Solar and wind energy are intermittent, demanding innovations in energy storage and grid management.
3. **Scale and Accessibility**: While renewables are growing, the scale required to meet global energy demands is immense. Additionally, ensuring energy access to underserved regions remains a challenge.
4. **Economic Impacts**: Transitioning away from fossil fuels can have socioeconomic implications, including job losses in traditional energy sectors.
5. **Regulatory and Policy Hurdles**: Inconsistent policies, regulations, and subsidies can hinder the adoption of cleaner energy sources.
6. **Environmental and Social Concerns**: Even green energy sources have environmental implications, such as the impact of hydroelectric dams on aquatic ecosystems or land use for large-scale solar farms.

The global energy landscape is at a pivotal juncture, with the urgency to transition to sustainable sources juxtaposed against the challenges of scale, technology, and socioeconomics. Understanding this intricate landscape is vital for stakeholders, from policymakers to consumers, as the choices made today will shape the planet's environmental and economic future for future generations.

7.6 Overview of Resource Consumption and Waste

From the dawn of human civilization to the present day, resource consumption and waste generation have been intertwined with our societal evolution. Our relationship with resources has changed profoundly, reflecting our technological advancements, cultural shifts, and economic expansions.

Early Human Societies

Hunter-Gatherers

- Resource consumption was directly linked to immediate needs.
- Materials such as stone, bone, and wood were used to craft tools.
- Waste was predominantly organic and biodegradable, leaving minimal long-term environmental impact.

Agricultural Revolution:

- Settled farming communities began to form, leading to increased resource consumption, including water, land, and primitive tools.
- The surplus of agricultural produce allowed populations to grow and the first settlements to develop.
- Waste generation increased, but it was mainly organic and often repurposed or returned to the land.

Ancient Civilizations

Mesopotamia, the Indus Valley, Egypt, and China:

- Advanced agricultural practices led to surplus food, supporting large urban centers.
- The introduction of metallurgy meant increased mining and resource consumption.
- The creation of long-lasting products such as pottery and metal objects resulted in more durable waste.
- Sewage and waste management systems, such as the drains of the Indus Valley cities or Roman aqueducts, began to emerge.

Medieval Era

- Urban centers continued to grow across Asia and Europe, driving resource consumption.
- Innovations such as water mills increased resource extraction and processing.
- Craftsmanship led to diverse products and, consequently, varied waste.
- Many cities face sanitation challenges without organized waste management, contributing to health crises such as plagues.

Industrial Revolution

- The transition from agrarian societies to industrial urban centers in the eighteenth and nineteenth centuries marked a dramatic increase in resource consumption.

- Coal became the primary energy source, leading to massive mining operations.
- Manufactured goods, from textiles to machinery, increased exponentially.
- Waste became a significant challenge. Cities were often choked with smoke, sewage, and industrial waste.
- The beginning of man-made pollutants introduced new challenges for ecosystems.

Twentieth Century to Present

- The discovery and widespread use of oil changed the global resource landscape, leading to the rise of the petrochemical industry and plastic production.
- Post-WWII, consumerism took off, especially in Western countries, leading to the 'throwaway culture.'
- The scale of resource extraction, from deforestation to overfishing, reached unprecedented levels.
- Waste, especially nonbiodegradable plastics, began accumulating in landfills and oceans.
- The late twentieth and early twenty-first century saw a growing awareness of sustainability. Efforts to reduce, reuse, and recycle have gained traction, but the challenge of waste, especially in a globalized world, remains daunting.

Historically, our relationship with resource consumption and waste mirrored our technological and socioeconomic advancements. As we have evolved, our footprint on the planet has expanded significantly. Recognizing our past patterns can offer insights into current challenges and guide us toward a more sustainable future.

7.7 Necessity for an Integrated Policy Framework: Bridging Sustainable Energy Transition and the Circular Economy

As society inches toward embracing both sustainable energy transition and circular economy models, the importance of a unified policy approach becomes evident. Singular, siloed policies may lead to isolated achievements, but the larger goal of sustainability demands holistic, interconnected strategies. An integrated policy framework, therefore, emerges not as a choice but as an imperative.

Interdependence of Energy and Consumption
To grasp the necessity of an integrated policy framework, one must first recognize the interdependence of energy production and consumption models. Our methods of producing energy and consuming resources are intrinsically linked.

1. **Material Needs for Renewable Energy**: Building renewable energy infrastructure, be it solar panels, wind turbines, or battery storage systems, requires significant material input. The lifecycle management of these materials can benefit from circular principles, thereby reducing waste and optimizing resource use.

2. **Energy Needs of Circular Processes**: Processes in the circular economy, such as recycling and remanufacturing, are energy intensive. To ensure that these processes are genuinely sustainable, they must be powered by renewable sources.

Coherent Vision, Holistic Results

By integrating policies for sustainable energy and a circular economy, governments and organizations can set a coherent vision and direction, leading to the following:

1. **Consistency in Decision-Making**: An integrated policy framework ensures that decisions made in one sector do not adversely impact the other.
2. **Optimized Resource Allocation**: Unified policies can guide investments and incentives to areas of maximum impact, ensuring efficient use of public funds.
3. **Synergistic Solutions**: Cross-sectoral strategies can lead to innovative solutions that might not be clear when looking at sectors in isolation.

Addressing Overlap and Gaps

When policy frameworks for the energy transition and circular economy are designed separately, there is a risk of overlapping efforts or, worse, gaps in policy coverage.

1. **Avoiding Duplication**: An integrated approach can prevent repetitive initiatives, saving time, resources, and funds.
2. **Ensuring Comprehensive Coverage**: With a unified view, policymakers can ensure that no critical area falls between the cracks due to a segmented approach.

Engaging Diverse Stakeholders

Both energy transition and circular economy models involve many stakeholders, from businesses and local communities to international partners.

1. **Unified Messaging**: A unified policy conveys a consistent message to all stakeholders, reducing confusion and ensuring that everyone is aligned with the overarching objectives.
2. **Collaborative Efforts**: By bringing diverse entities under a shared framework, there is a better chance for collaborative solutions, shared responsibilities, and pooling of resources.

Economic and Social Resilience

The world is increasingly interconnected. Economic or environmental disruptions in one sector can ripple across others. An integrated policy framework offers:

1. **Buffer Against Volatility**: With multiple sectors working in tandem, if one faces challenges, the others can provide stability, ensuring economic and social resilience.
2. **Enhanced Economic Opportunities**: With aligned strategies, businesses can identify cross-sectoral opportunities, leading to innovative products, services, and job creation.

In an era of global challenges, isolated efforts can no longer suffice. An integrated policy framework for sustainable energy transition and the circular economy is not just a strategic choice—it is a necessity to ensure that our efforts are harmonized, effective, and future-proof. By recognizing the intertwined nature of energy and consumption and acting upon it through unified policies, we stand a better chance of creating a sustainable and resilient future.

7.8 Sustainable Energy Transition

Sustainable Energy Transition is the transition from nonrenewable energy sources like coal and hydrocarbons to more sustainable alternatives like wind, solar, and hydroelectric power. The need to combat climate change, improve energy security, and promote technological innovation are driving forces behind the global push for a sustainable energy transition. As the negative effects of fossil fuel consumption become more evident, the transition to cleaner and more sustainable energy sources becomes necessary for ecological, social, and economic reasons.

The sustainable energy transition is the systemic shift from a global energy system dominated by fossil fuels to one based on renewable and sustainable energy sources. This transformation must ensure economic viability, social inclusion, and resilience against potential disruptions, in addition to reducing environmental impacts.

7.8.1 Drivers for Current Energy Transition

1. **Environmental Concerns**: The overwhelming evidence of climate change, primarily driven by carbon emissions from fossil fuels, underscores the need for cleaner energy solutions.
2. **Technological Advancements**: Innovations in renewable energy technologies, battery storage, and grid management make alternative energy sources increasingly viable.
3. **Economic Factors**: As the costs of renewable energy technologies fall, their economic competitiveness against fossil fuels rises.
4. **Geopolitical and Energy Security**: A diversified energy mix reduces dependencies on politically volatile regions and enhances energy security.
5. **Public Demand and Awareness**: As awareness of environmental challenges grows, public demand for cleaner energy sources escalates.

7.8.2 Key Technologies and Approaches

1. **Renewable Energy**: The adoption of wind, solar, hydro, and geothermal sources is essential.

2. **Energy Efficiency**: Improving energy efficiency in homes, industries, and transportation reduces energy consumption.
3. **Electrification of Transportation**: Transitioning from gasoline and diesel-powered vehicles to electric ones minimizes emissions from the transportation sector.
4. **Grid Modernization**: Adapting to a decentralized energy production system, integrating smart grids, and improving storage capabilities.
5. **Policy and Regulatory Support**: Governments play a crucial role by setting favorable policies, standards, and incentives to support the transition.

7.8.3 Challenges in the Transition

1. **Infrastructure Overhaul**: Many existing infrastructures are tailored for fossil fuels and need significant modifications or replacements.
2. **Economic and Job Impacts**: Some regions and industries heavily reliant on fossil fuels may face economic hardships and job losses.
3. **Intermittency and Reliability**: Renewable sources such as solar and wind are intermittent, necessitating robust storage solutions and backup systems.
4. **Financial Barriers**: While costs are decreasing, the initial investments required for renewable technologies can be substantial.
5. **Cultural and Behavioral Changes**: Achieving a sustainable energy transition also requires shifts in societal behaviors and preferences.

7.8.4 The Role of Stakeholders

1. **Governments** can guide the transition through policy frameworks, financial incentives, and public investments.
2. **Industry and Private Sector**: Drive technological innovations, investments, and scaling of sustainable solutions.
3. **Civil Society and NGOs**: Play watchdog roles, raise awareness, and facilitate community-level initiatives.
4. **General Public**: Their consumption choices, public demand, and voting power can accelerate the transition.

The sustainable energy transition is not just a technical shift in energy sources but a comprehensive transformation in the way societies produce, distribute, and consume energy. Embracing this transition is paramount for the health of the planet and its inhabitants, ensuring a resilient and prosperous future for all.

7.9 Circular Economy

A circular economy is an economic system that reduces pollution by reusing, recycling, and regenerating resources in a closed-loop cycle, thereby minimizing the need for untapped resources and environmental impacts.

7.10 Synergy

Both the sustainable energy transition and the circular economy champion the reduction of waste, the promotion of efficiency, and the preservation of our planet's natural resources. Together, they have a unified vision for a green and sustainable future.

7.11 Key Elements of the Policy Framework

7.11.1 Vision and Goals

- **Clear Vision**: Establish a long-term vision for a sustainable and circular future.
- **Measurable Goals**: Set concrete, time-bound targets for energy transition and waste reduction.

7.11.2 Incentive Structures

- **Tax Incentives**: Provide tax breaks for companies and individuals investing in sustainable energy.
- **Grants and Subsidies**: Allocate funds for research and development in clean energy and circular economy innovations.

7.11.3 Legislation and Regulation

- **Mandates**: Set mandatory renewable energy adoption targets for utilities.
- **Standards**: Establish standards for sustainable design, production, and waste management.

7.11.4 Public Awareness and Education

- **Campaigns**: Promote the benefits of sustainable energy and the circular economy through public campaigns.
- **Education**: Integrate these principles into school curriculums and vocational training programs.

7.11.5 Collaboration and Partnership

- **Public-private partnerships**: Engage businesses, NGOs, and governments in cooperative ventures.
- **International Cooperation**: Foster collaboration between nations to share knowledge, technology, and best practices.

7.12 Implementation Challenges

7.12.1 Economic

- **Transition Costs**: Addressing short-term economic disruptions and costs associated with the transition.
- **Market Dynamics**: Navigating volatile energy markets and ensuring competitive pricing.

7.12.2 Technological

- **Innovation Requirements**: The need for continued technological advancements in sustainable energy and circular economy methods.
- **Infrastructure**: Updating and adapting current infrastructure to accommodate modern technologies.

7.12.3 Sociocultural

- **Public Resistance**: Addressing public skepticism or resistance to innovative technologies or practices.
- **Cultural Norms**: Changing deeply ingrained consumerist behaviors that contradict circular economy principles.

7.13 Case Studies

7.13.1 Denmark: A Leader in Wind Energy

Denmark leads the globe in wind energy. In 2022, 49.3% of Denmark's electricity generation came from wind power, making it the country's primary source of electricity. Denmark has the world's greatest wind power capacity per capita. Numerous factors have contributed to Denmark's achievement in the wind energy industry. These include:

- **A strong political commitment to renewable energy**: Denmark has a long history of supporting renewable energy and has set ambitious targets for the use of wind power.
- **A favorable policy environment**: Denmark has put in place several policies that have helped to promote wind energy, such as feed-in tariffs and tax breaks.
- **A skilled workforce**: Denmark has a strong engineering and manufacturing sector, which has helped to develop and build wind turbines.
- **A favorable wind resource:** Denmark has a strong wind resource, with average wind speeds of 10–12 meters per second.

Denmark's success in wind energy has made it a global leader in the industry. Danish companies are major players in the design, manufacturing, and installation of wind turbines, and Denmark is a major exporter of wind power technology. With a strong political commitment, a favorable policy environment, and a skilled workforce, it is possible to achieve a high level of renewable energy penetration, as demonstrated by Denmark's experience. Denmark's accomplishments can serve as an example for other nations looking to transition to sustainable energy sources. Here are some of the specific policies that have helped Denmark to develop its wind energy industry:

- **Feed-in tariffs**: Feed-in tariffs are a policy that guarantees a certain price for renewable energy, such as wind power. This makes it more attractive for businesses to invest in wind energy projects.
- **Tax breaks:** Tax breaks can be given to businesses that invest in renewable energy projects. This can help to reduce the cost of wind energy and make it more competitive with other sources of energy.
- **Research and development funding**: Governments can fund research and development into new wind energy technologies. This can help to improve the efficiency and cost-effectiveness of wind power.
- **Public–private partnerships**: Governments can partner with businesses to develop wind energy projects. This can help to share the risks and costs of these projects.

Denmark is an excellent example of how a country can achieve a high level of renewable energy penetration. The policies that have been put in place in Denmark have helped to create a favorable environment for wind energy development and have made Denmark a global leader in the industry.

7.13.2 *Japan: Pioneering the Circular Economy*

Japan is a pioneer in the circular economy. The country has a long history of resource efficiency and waste reduction and has recently actively promoted the circular economy. Here are some of the things that Japan is doing to promote the circular economy:

- **Setting ambitious targets:** Japan has set a target of recycling 60% of all municipal waste by 2030. The country is also working to reduce its reliance on virgin resources by increasing the use of recycled materials.
- **Enacting legislation:** Japan has enacted several laws promoting the circular economy, such as the Waste Management Act and the Recycling Promotion Act. These laws set standards for waste management and recycling and provide financial incentives for businesses to adopt circular practices.
- **Investing in research and development:** Japan is investing in research and development into modern technologies that can help to create a circular economy. This includes research into new materials, recycling technologies, and waste-to-energy technologies.
- **Promoting public awareness:** Japan promotes public awareness of the circular economy through education and awareness campaigns. The country is also working to create a circular economy culture, encouraging people to reduce, reuse, and recycle.

Japan's efforts to promote the circular economy were recognized by the World Economic Forum, which named Japan a "Circular Economy Pioneer" in 2020.

Here are some of the specific examples of Japan's circular economy initiatives:

- **The "3R" campaign**: Japan has a long-running campaign to promote the 3Rs of waste management: reduce, reuse, and recycle. The campaign has successfully raised public awareness of the importance of waste reduction and recycling.
- **The Eco-Town program**: The Eco-Town program is a government initiative to promote sustainable development in urban areas. The program provides financial assistance to municipalities that implement circular economy initiatives.
- **The Zero Waste Challenge**: Municipalities compete in the Zero Waste Challenge to see who can attain the highest recycling rate. The competition has increased awareness of the significance of recycling and refuse reduction.
- **The Circular Economy Partnership**: The Circular Economy Partnership is a public–private partnership that brings together businesses, government agencies, and other organizations to promote the circular economy. The association has developed many initiatives to promote circular practices, such as a product stewardship scheme for batteries and electronics.

Japan's efforts to promote the circular economy are still in their initial stages, but the country is making considerable progress. Japan is a model for other countries that are looking to transition to a more sustainable economy.

7.14 Conclusion

A unified policy framework that integrates the sustainable energy transition and circular economy is not merely advantageous for addressing the challenges of our time; it is essential. Nations can pave the way for a more sustainable, efficient, and regenerative future by grasping the crucial components of such a framework and the obstacles to its implementation. Transitioning to a sustainable energy system and a circular economy is a difficult and complex endeavor. However, it is necessary if we are to resolve the environmental challenges we face today. The policy framework can be vital in making this transition a reality. The policy framework must be comprehensive and ambitious, and it must address a wide range of issues, such as:

- Setting standards for energy efficiency, renewable energy, and waste management
- Providing incentives for investment in sustainable technologies
- Promoting research and development in modern technologies
- Educating the public about sustainable practices
- Creating a supportive regulatory environment

The policy framework must also be flexible enough to adapt to the changing needs of society and the economy. It must be fair and implemented in a way that does not disadvantage any group of people. The transition to a sustainable energy system and a circular economy is a long-term goal. However, by putting in place the right policies, we can make this goal a reality.

Here are some specific policy recommendations that could be included in the conclusion:

- Establish ambitious goals for renewable energy and energy efficiency.
- Provide financial incentives for businesses to invest in sustainable technologies.
- Promote research and development in modern technologies that can help to create a circular economy.
- Educate the public about the importance of sustainable practices.
- Create a supportive regulatory environment for sustainable energy and circular economy initiatives.

The policy framework for a sustainable energy transition and a circular economy is a crucial component of the solution to the environmental problems of our time. We can create a more sustainable future for ourselves and future generations by implementing the proper policies. Financing the energy transition entails not just substituting fossil fuels with renewable energy sources, but also establishing a future that is economically, socially, and environmentally sustainable. Through the dedication and collaboration of all stakeholders, Indonesia has the potential to accomplish a prosperous energy transition, significantly contribute to addressing the climate issue, and actualize the aspiration of an eco-friendly energy future for forthcoming generations.

References

Hariram, N. P., Mekha, K. B., Suganthan, V., & Sudhakar, K. (2023). Sustainalism: An integrated socio-economic-environmental model to address sustainable development and sustainability. *Sustainability, 15*(13). https://doi.org/10.3390/su151310682

Li, S., Zhang, L., Liu, X., & Zhu, C. (2023). Collaborative operation optimization and benefit-sharing strategy of rural hybrid renewable energy systems based on a circular economy: A Nash bargaining model. *Energy Conversion and Management, 283*, 116918. https://doi.org/10.1016/j.enconman.2023.116918

Romero-Castro, N., Ángeles López-Cabarcos, M., Miramontes-Viña, V., & Ribeiro-Soriano, D. (2023). Sustainable energy transition and circular economy: The heterogeneity of potential investors in rural community renewable energy projects. *Environment, Development and Sustainability*, 0123456789. https://doi.org/10.1007/s10668-022-02898-z

Tian, J., Yu, L., Xue, R., Zhuang, S., & Shan, Y. (2021, January). Global low-carbon energy transition in the post-COVID-19 era. *Applied Energy, 307*, 118205. https://doi.org/10.1016/j.apenergy.2021.118205

Chapter 8
Corporate Social Responsibility (C.S.R.) and E.S.G. Reporting: Redefining Business in the Twenty-First Century

Abstract As the twenty-first century unfolds, the traditional objectives of businesses, centered primarily on profit maximization, are being reevaluated. The emerging consensus acknowledges the need for companies to play a constructive role in addressing some of the pressing challenges facing society and the environment. This shift is encapsulated in the concepts of C.S.R. and E.S.G. Originating from philanthropic endeavors, C.S.R. now includes a vast spectrum of ethical practices from labor to environmental considerations. E.S.G., with its quantifiable metrics, offers an empirical approach to assessing a company's performance in these areas. Although distinct, C.S.R. and E.S.G. were intertwined, presenting qualitative intentions and quantitative measurements. While the rise of E.S.G. reporting marks a shift towards transparency, challenges such as lack of standardization and greenwashing persist. Looking forward, regulatory mandates, technological integration, and shifting stakeholder expectations, especially from younger generations, will shape the future of C.S.R. and E.S.G. Businesses have now emerged as responsible global entities, intertwining success with planetary and social well-being.

Keywords Corporate Social Responsibility (C.S.R.) · Environmental · Social · And governance (E.S.G.) reporting · Stakeholder considerations · Ethical practices · Quantifiable metrics · Standardization · Greenwashing · Regulatory mandates · Technological integration · Stakeholder expectations

8.1 Introduction

Corporate Social Responsibility (C.S.R.) and Environmental, Social, and Governance (E.S.G.) reporting reflect the transformation of corporate ethics and operational frameworks in the twenty-first century. Historically, profit maximization has been the primary objective of the corporate world. However, contemporary discourse, supported by extensive research, emphasizes that businesses have a broader societal obligation that extends beyond shareholders to all stakeholders (Freeman et al. 2010). From early philanthropic efforts by industrial magnates to

V. Kandpal et al., *Sustainable Energy Transition*, Circular Economy and Sustainability, https://doi.org/10.1007/978-3-031-52943-6_8

today's nuanced, data-driven approaches to sustainability and governance, the trajectory of C.S.R. and E.S.G. demonstrates the evolution of the corporate conscience. These frameworks and practices have changed the business landscape from one dominated by profit-driven pursuits to one that seeks a balance between profit, people, and the planet. This chapter contextualizes C.S.R. and E.S.G.'s evolution, significance, challenges, and prospective future within contemporary business imperatives. The discussions are grounded in literature, integrating historical context, empirical findings, and prospective insights.

Defining C.S.R. and E.S.G. Reporting

Corporate Social Responsibility (C.S.R.): C.S.R. refers to a company's commitment to operating ethically and responsibly, considering the impact of its actions on society, the environment, and various stakeholders. It encompasses a broad range of initiatives, including philanthropy, community engagement, employee well-being, and environmental sustainability. Corporate social responsibility (C.S.R.) is a business model that enables corporations to be socially accountable to themselves, their stakeholders, and the public. It demonstrates a company's dedication to contributing to sustainable development by providing economic, social, and environmental benefits to all stakeholders (Carroll, 1999; Dahlsrud, 2008; Lee et al., 2019; Wang & Chen, 2022). C.S.R. is the commitment of businesses to contribute to sustainable economic development by working with employees, their families, the local community, and society at large to improve their lives in ways that are good for business and growth (Hajji, 2009).

Corporate social responsibility is a business model that integrates self-regulation into the business's operations, ensuring its active compliance with the spirit of ethical standards, international norms, and the law.

Components of C.S.R.

Economic Responsibilities: Ensuring Business Viability and Broader Socio-Economic Impact: Economic responsibilities refer to the obligations and duties of businesses to produce goods and services that meet the needs of consumers and generate a return on investment for their shareholders. While profitability is at the core of these responsibilities, it is crucial to understand that in a modern context, economic responsibilities also encompass how businesses impact broader economic systems, including job creation, economic growth, and equitable wealth distribution. These include the fundamental duties of producing society-desired products and services and selling them for a profit (Friedman, 1970).

Core Aspects of Economic Responsibilities

1. **Profit Generation:** The primary economic responsibility of a business is to be profitable, ensuring sustainability, growth, and shareholder returns.
2. **Value Creation:** Businesses should offer products and services that provide real value to customers, meeting or surpassing their needs and expectations.

3. **Job Creation and Fair Compensation:** Businesses have the responsibility to create jobs and ensure that employees receive fair compensation and benefits.
4. **Economic Growth Contribution:** Through operations, expansion, and investments, businesses contribute to the G.D.P. of a region or country.
5. **Supply Chain Management:** Ensuring ethical and economic practices throughout the supply chain, from sourcing raw materials to delivering the end product to consumers.

Importance of Economic Responsibilities

1. **Business Sustainability:** Without profitability and sound economic practices, a business cannot sustain its operations.
2. **Stakeholder Trust:** Shareholders, investors, and employees place their trust in businesses to make economically sound decisions.
3. **Social Stability:** By creating jobs and fostering economic growth, businesses play a pivotal role in ensuring social stability.
4. **Innovation and Development:** Economic success allows businesses to invest in research and development, leading to innovation.

Challenges in Meeting Economic Responsibilities

1. **Short-termism:** A focus on short-term financial gains can sometimes overshadow long-term economic sustainability and broader societal impacts.
2. **Economic Fluctuations:** Global and local economic downturns can challenge a business's ability to meet its economic responsibilities.
3. **Balancing Stakeholder Interests:** The interests of shareholders, employees, suppliers, and the broader community might sometimes conflict, challenging businesses to find a balance.
4. **Rapid Technological Changes:** With the fast pace of technological innovation, businesses must continually adapt, which can have significant economic implications.

Strategies for Fulfilling Economic Responsibilities

1. **Diversification:** Diversifying product lines, markets, or investments to reduce risks related to economic downturns in specific sectors or regions.
2. **Long-term Planning:** Adopting a long-term perspective in strategy formulation, ensuring sustainable economic benefits.
3. **Ethical Practices:** Emphasizing ethical practices, which often lead to better economic outcomes eventually.
4. **Stakeholder Engagement:** Regularly engaging with stakeholders to understand their perspectives and align business strategies accordingly.
5. **Continuous Learning and Adaptability:** Investing in training and development to adapt to changing economic scenarios and technological advancements.

While economic responsibilities fundamentally pertain to profitability and value creation, it is essential to recognize that businesses operate within a broader socio-economic fabric. Their economic decisions and actions can have ripple effects, impacting not just their bottom line but also the well-being of communities, regions, and even entire economies. Balancing immediate economic interests with long-term sustainability and broader societal impacts is the challenge and duty of modern businesses.

8.2 Legal Responsibilities: The Backbone of Ethical and Sustainable Business Practices

Legal responsibilities refer to the obligations that businesses must operate within the parameters set by laws and regulations. These responsibilities ensure that businesses act in a manner that respects rights, protects consumers, and maintains the larger socioeconomic order. In an era of globalized commerce and heightened societal awareness, the importance of legal responsibilities cannot be overstated (Carroll, 1991).

Core Aspects of Legal Responsibilities

1. **Compliance:** At its most basic level, legal responsibility involves adhering to local, national, and international laws governing business operations.
2. **Transparency:** Businesses must provide accurate and timely information to regulators, shareholders, and other stakeholders as mandated by laws.
3. **Protection of Stakeholder Rights:** Laws are often framed to protect the rights of stakeholders such as consumers, employees, and the community. Businesses have a duty to uphold these rights.
4. **Liability:** If a business fails to meet its legal responsibilities, it can be held liable, leading to penalties, sanctions, or legal action.

Importance of Legal Responsibilities

1. **Trust and Credibility:** Businesses that consistently operate within the confines of the law build trust among consumers, investors, and other stakeholders.
2. **Level playing field:** Laws create a standardized framework within which all businesses operate, ensuring fair competition.
3. **Risk Mitigation:** Compliance with legal standards can help businesses preempt potential issues, reducing legal risks and associated costs.
4. **Societal Order:** Laws reflect the values and norms of a society. When businesses respect these, they contribute to social harmony and order.

Challenges in Meeting Legal Responsibilities

1. **Dynamic Regulatory Landscape:** With the fast-paced changes in technology, global commerce, and societal values, laws and regulations also evolve, posing challenges for businesses to stay updated.
2. **Interpreting Ambiguities:** Legal provisions can sometimes be open to interpretation, leading to uncertainties in compliance.

3. **Global Operations:** For multinational corporations, adhering to the legal standards of multiple jurisdictions can be complex.
4. **Cost Implications:** Ensuring full compliance, especially in heavily regulated industries, can entail significant costs.

Strategies for Fulfilling Legal Responsibilities

1. **Regular Training:** Conducting regular legal compliance training for employees, especially for those in decision-making roles.
2. **Legal Audit:** Periodically reviewing and auditing business operations to ensure that they align with prevailing legal standards.
3. **Engaging Legal Counsel:** Retaining or having access to legal experts who can provide guidance on compliance and legal challenges.
4. **Stakeholder Engagement:** Keep open channels of communication with stakeholders, such as regulators and community representatives, to better understand legal expectations.
5. **Investing in Compliance Infrastructure:** This includes software and systems that can help in tracking, reporting, and ensuring compliance with legal requirements.

Legal responsibilities function as both a safeguard and a compass for businesses. They deter malpractice and guide businesses toward actions that respect societal values and rights. As the business landscape becomes more intricate and interconnected, the role of legal responsibilities in ensuring ethical and sustainable business practices grows ever more crucial.

8.3 Ethical Responsibilities: Guiding Principles Beyond Legal Obligations

Ethical responsibilities refer to the moral obligations that businesses have, which often go beyond legal compliance. These responsibilities reflect the expectations that society has of businesses based on values, principles, and norms. Addressing ethical responsibilities means making decisions that, while not always legally needed, are considered right, fair, and just. Fulfilling societal expectations not codified in law, such as fair employment practices (Crane & Matten, 2007).

Core Aspects of Ethical Responsibilities

1. **Integrity:** Businesses should act honestly and transparently in all dealings, fostering trust with stakeholders.
2. **Fairness:** Ensure that decisions and practices are just, equitable, and non-discriminatory.
3. **Respect for Stakeholder Interests:** Considering and valuing the rights and interests of all stakeholders, not just shareholders.

4. **Sustainability:** Making decisions that protect the interests of future generations, not just present needs.
5. **Transparency:** Openly sharing relevant information with stakeholders, even if not legally mandated.

Importance of Ethical Responsibilities

1. **Reputation Management:** Ethical behavior is a significant determinant of a business's reputation and public perception.
2. **Sustainable Growth:** Ethical practices often lead to more sustainable business models and long-term success.
3. **Employee Satisfaction:** Organizations that operate ethically often see higher morale and employee retention rates.
4. **Stakeholder Trust:** Ethical conduct fosters trust and loyalty among consumers, investors, and other stakeholders.
5. **Risk Mitigation:** Ethical practices can help identify and mitigate potential risks before they escalate.

Challenges in Meeting Ethical Responsibilities

1. **Subjectivity:** What is considered ethical can sometimes be subjective and vary across cultures or individuals.
2. **Conflicting Interests:** Ethical responsibilities might at times conflict with short-term profitability or stakeholder interests.
3. **Economic Pressures:** In challenging economic times, maintaining ethical standards can become more difficult.
4. **Lack of Clear Guidelines:** Unlike legal responsibilities, ethical boundaries might not always be clear-cut.

Strategies for Fulfilling Ethical Responsibilities

1. **Ethical Training:** Offering regular training sessions to employees on the company's ethical standards and decision-making frameworks.
2. **Clear Ethical Guidelines:** Developing and publicizing a code of ethics that provides guidance to all members of the organization.
3. **Stakeholder Engagement:** Regular dialogue with stakeholders to understand their ethical concerns and expectations.
4. **Ethical Leadership:** Leaders must exemplify ethical behavior, setting a tone at the top that permeates throughout the organization.
5. **Whistleblower Mechanisms:** Establishing channels for employees and stakeholders to safely report unethical behaviors.

Ethical responsibilities, while intangible and often subjective, form the moral compass of a business. They reflect the business's values and principles and play a crucial role in determining its societal role and reputation. As consumers and stakeholders become increasingly conscious of ethical standards, businesses that prioritize these responsibilities will not only foster trust and loyalty but also ensure their long-term viability and success.

8.4 Philanthropic Responsibilities: Businesses' Commitment to the Greater Good

Philanthropic responsibilities pertain to the voluntary actions businesses take to improve the quality of life for their communities and society at large. Beyond economic, legal, and ethical obligations, these responsibilities reflect a company's commitment to making a positive impact and contributing to the welfare of its larger ecosystem. Voluntary activities that contribute to a community's quality of life include charitable donations (Porter & Kramer, 2002).

Core Aspects of Philanthropic Responsibilities

1. **Community Development:** Engaging in activities that support local communities, such as building infrastructure, supporting education, or promoting health and wellness.
2. **Charitable Giving:** Donating funds, products, or services to charitable organizations or causes.
3. **Volunteering:** Encouraging employees to give their time and skills to community projects or charitable organizations.
4. **Supporting Social Causes:** Backing issues of societal importance, such as environmental conservation, hunger alleviation, or disease eradication.
5. **Promoting Cultural and Artistic Endeavors:** Supporting the arts, culture, and heritage for societal enrichment.

Importance of Philanthropic Responsibilities

1. **Strengthened Brand Image:** Companies that engage in philanthropy often earn a positive reputation and increase goodwill from the public.
2. **Employee Morale and Retention:** Employees feel proud to be part of organizations that make a positive societal impact, leading to higher engagement and retention.
3. **Building Stronger Communities:** Businesses, especially local ones, benefit from thriving communities, creating a virtuous cycle of mutual support.
4. **Long-term Business Sustainability:** By addressing societal challenges, businesses can also ensure a more stable and sustainable future for their operations.
5. **Stakeholder Relations:** Philanthropy can foster better relationships with various stakeholders, including customers, investors, and local governments.

Challenges in Addressing Philanthropic Responsibilities

1. **Resource Allocation:** Balancing between business needs and philanthropic activities can be challenging, especially for smaller enterprises.
2. **Determining Impact:** Measuring the true impact of philanthropic endeavors is not always straightforward.
3. **Public Skepticism:** Sometimes, philanthropic actions can be seen as mere P.R. exercises, leading to skepticism.

4. **Navigating Politics and Sensitivities:** Philanthropic choices can sometimes inadvertently take on political dimensions or touch on sensitive issues.

Strategies for Fulfilling Philanthropic Responsibilities

1. **Align with Core Values:** Philanthropic activities should align with the company's core values and mission for genuine impact.
2. **Stakeholder Engagement:** Engaging with community members and other stakeholders can guide meaningful and effective philanthropic initiatives.
3. **Collaboration:** Partnering with N.G.O.s, local governments, or other organizations can amplify the impact of philanthropic activities.
4. **Transparency:** Being transparent about philanthropic initiatives, their costs, and their impacts can build trust with stakeholders.
5. **Set Clear Objectives:** Establishing clear goals and regularly assessing the outcomes of philanthropic activities can ensure continued relevance and impact.

Philanthropic responsibilities signify a business's dedication to contributing positively to society and its broader environment. While these actions are voluntary, they resonate deeply with stakeholders and reflect the soul of a business. In a world increasingly focused on holistic success and shared value, philanthropic responsibilities carve a path for businesses to forge deeper connections, not just as economic entities but as vital societal pillars.

8.5 Environmental Management: A Key Driver for Sustainable Business and Development

Environmental management refers to the systematic approach and practices used by individuals, businesses, and governments to protect the environment and ensure sustainable development. In the context of businesses, it relates to making decisions that ensure operational efficiency and reduce the environmental footprint.

Importance of Environmental Management

1. **Conservation of Resources:** Effective management ensures the efficient use of natural resources, ensuring their availability for future generations.
2. **Regulatory Compliance:** Many governments now impose environmental regulations that companies must comply with to operate legally.
3. **Reputation Management:** As consumer awareness about environmental issues grows, businesses that prioritize eco-friendly practices are viewed more favorably.
4. **Cost Savings:** By focusing on resource efficiency and waste reduction, companies can realize substantial cost savings in the long run.
5. **Biodiversity Conservation:** Through responsible practices, businesses can ensure that they do not harm the delicate balance of ecosystems.

Key Components of Environmental Management

1. **Environmental Auditing:** Regular checks to ensure that business practices align with environmental goals and regulatory requirements.
2. **Waste Management:** Proper disposal and, where possible, recycling, or reusing waste materials.
3. **Energy Efficiency:** Using resources such as electricity, gas, and water efficiently and investing in renewable energy sources.
4. **Pollution Control:** Implementing measures to prevent the release of harmful substances into the air, water, or land.
5. **Sustainable Procurement:** Sourcing products and services that have been produced sustainably.
6. **Training and Education:** Ensure that employees understand the importance of environmental issues and are trained to act responsibly.

Challenges in Implementing Environmental Management

1. **Initial Costs:** The initial investment required for green technology or sustainable infrastructure can be high.
2. **Complex Regulations:** Varying and constantly evolving regulations can make compliance challenging.
3. **Resistance to Change:** Employees and stakeholders might be resistant to changes in processes or technologies.
4. **Global Variations:** Multinational corporations may struggle with diverse environmental standards and cultural perceptions about sustainability in different countries.

Strategies for Effective Environmental Management

1. **Stakeholder Engagement:** Engage with stakeholders, including employees, suppliers, and local communities, to gain insights and support for environmental initiatives.
2. **Continuous Monitoring:** Regularly evaluate the impact of business operations on the environment and adjust strategies accordingly.
3. **Adopting Technology:** Use technology such as A.I. and IoT for better monitoring and management of environmental factors.
4. **Setting Clear Targets:** Define clear, measurable environmental goals and track progress regularly.
5. **Integration with Business Strategy:** Make environmental management an integral part of the overall business strategy rather than a separate initiative.

In the face of climate change and other pressing environmental concerns, environmental management is not just a business imperative but also a societal responsibility. Companies that integrate sound environmental management practices not only benefit in terms of cost savings and reputation but also contribute to a more sustainable and healthier planet.

8.6 Community Involvement: An Essential Pillar of Modern Organizations

Community involvement, sometimes referred to as community engagement or corporate citizenship, is the participation and commitment of individuals or organizations in activities that positively impact the well-being of local communities. For businesses, this involvement translates to initiatives that go beyond profit-making, emphasizing a genuine contribution to the betterment of society.

Benefits of Community Involvement

1. **Builds Trust and Reputation:** Organizations that are actively involved in community initiatives are viewed more favorably, earning them goodwill and trust.
2. **Enhanced Employee Morale:** Employees feel proud to be associated with companies that give back, resulting in higher job satisfaction and retention rates.
3. **Strengthened Community Ties:** Active involvement fosters relationships with community leaders, N.G.O.s, and local residents, which can prove advantageous in various ways for businesses.
4. **Enhanced Corporate Image:** Community involvement often results in positive media coverage, which can help in improving a company's public image.
5. **Networking Opportunities:** Community activities often involve interactions with other businesses and stakeholders, leading to potential collaborations or partnerships.

Forms of Community Involvement

1. **Philanthropy:** Direct financial contributions or donations to local charities, N.G.O.s, or causes.
2. **Volunteering:** Employees dedicate their time and skills to local causes, whether it is renovating a school or planting trees.
3. **Partnerships with N.G.O.s:** Collaborative projects where businesses and N.G.O.s join hands to address specific community needs.
4. **Sponsoring Local Events:** Supporting community events, such as local sports, festivals, or cultural events.
5. **Educational initiatives:** Offering scholarships, internships, or mentorship programs to local students.
6. **Local Sourcing:** Procuring goods and services from local suppliers to boost the local economy.

Challenges to Effective Community Involvement

1. **Misalignment of Objectives:** The goals of the business might not always align seamlessly with the needs of the community.
2. **Insufficient Resources:** Especially for smaller businesses, dedicating resources for community projects can be challenging.

3. **Community Skepticism:** Initial community engagement efforts might be met with skepticism, especially if the company's past actions were not community centric.
4. **Measuring Impact:** Determining the tangible and intangible impact of community involvement can be complex.

Best Practices for Community Involvement

1. **Stakeholder Engagement:** Engage with community representatives to understand their needs before initiating any project.
2. **Transparency:** Be open about the objectives, outcomes, and any challenges faced during community projects.
3. **Long-term Commitment:** Instead of one-off projects, projects aim for sustained involvement to create a lasting impact.
4. **Inclusivity:** Ensure the involvement of diverse groups from the community, ensuring a wider reach and impact.
5. **Employee Involvement:** Encourage employees to actively participate and even lead community initiatives.

Community involvement is no longer an optional C.S.R. activity for businesses; it is becoming an integral part of their identity. In an age where consumers, employees, and stakeholders are increasingly conscious of the societal roles of corporations, community involvement acts as a bridge, forging deeper connections and fostering mutual growth.

8.7 Ethical Marketing Practice: Elevating Integrity in Promotions

Ethical marketing involves making honest claims and helping consumers make informed decisions based on accuracy, fairness, and respect. It stands in stark contrast to deceptive advertising or business practices that may mislead consumers or promote harmful behaviors. At its core, ethical marketing enriches the relationship between businesses and consumers, fostering trust and long-term loyalty.

Principles of Ethical Marketing

1. **Honesty:** Always present products and services truthfully. Misleading advertisements or exaggerations erode trust and can lead to legal repercussions.
2. **Transparency:** Clearly, state all pertinent details of a product, service, or promotion, including pricing, limitations, terms, and any associated risks.
3. **Respect for Consumer Rights:** Understand and adhere to consumer rights in all marketing efforts. This includes respecting their privacy, honoring opt-out requests promptly, and not making intrusive or unwelcome pitches.

4. **Avoidance of Harm:** Avoid marketing practices that might be harmful or exploit vulnerable groups. This includes not promoting products or services that can be detrimental to public health, safety, or well-being.
5. **Fairness:** Treat customers and competitors fairly. Avoid practices that can be considered discriminatory or biased.
6. **Accountability:** Be ready to accept responsibility for marketing campaigns and, if necessary, rectify any negative outcomes.

Implementing Ethical Marketing Practices

1. **Comprehensive Training:** Employees, especially those in marketing and sales roles, should be trained in ethical considerations, helping them recognize potential pitfalls and make principled decisions.
2. **Clear Guidelines:** Create a detailed marketing ethics policy. This document should offer guidance on how to oversee various scenarios, from online advertising to data collection.
3. **Ongoing Monitoring:** Regularly review marketing campaigns to ensure they align with ethical standards. If concerns arise, address them promptly.
4. **Open Communication:** Foster an organizational culture where team members feel comfortable raising ethical concerns without fear of retaliation.
5. **Engage with Feedback:** Listen to feedback from consumers, peers, and other stakeholders. Their perspective can offer valuable insights into the ethical implications of marketing campaigns.

Benefits of Ethical Marketing

1. **Builds Trust:** Ethical marketing builds trust between the company and its customers, leading to long-term loyalty and brand advocacy.
2. **Avoids Legal Issues:** Abiding by ethical practices minimizes the risk of lawsuits, fines, or public relations disasters related to deceptive advertising or other unscrupulous behaviors.
3. **Enhances Brand Image:** Companies known for their ethical approach are often viewed favorably by consumers, setting them apart from competitors.
4. **Internal Morale:** Employees prefer working for ethical companies. Ethical marketing practices can boost morale and retention and even attract top talent.
5. **Long-term Profitability:** While ethical marketing might sometimes involve higher short-term costs or foregone opportunities, it can lead to greater profitability overall due to enhanced customer loyalty and fewer legal complications.

Ethical marketing is not just a "nice to have" aspect of business; it is a fundamental component of long-term success in the modern marketplace. Beyond just avoiding deceptive practices, it emphasizes creating genuine value for consumers, fostering trust, and building enduring relationships. In an age of instant communication and empowered consumers, ethical marketing practices are more crucial than ever.

8.8 Stakeholder Engagement: Stakeholder Engagement: Building Strong Relationships for Mutual Success

Stakeholder engagement refers to the process by which an organization involves individuals, groups, or entities who can be affected by, or can affect, the outcome of its actions. This engagement is more than just communication; it is about building relationships, fostering trust, and collaborating to achieve shared goals.

Who are Stakeholders?
Stakeholders can be internal or external

1. **Internal stakeholders** include employees, managers, board members, and shareholders.
2. **External stakeholders** encompass customers, suppliers, creditors, regulators, N.G.O.s, community groups, and sometimes even competitors.

Importance of Stakeholder Engagement

1. **Risk Management:** Identifying and addressing stakeholder concerns can prevent potential challenges or disputes.
2. **Informed Decision Making:** Engaging stakeholders can provide valuable insights, leading to better-informed decisions.
3. **Enhanced Reputation:** Positive stakeholder relationships boost an organization's reputation, fostering goodwill and trust.
4. **Long-term Success:** Engaged stakeholders can contribute to an organization's long-term sustainability by offering support, resources, or expertise.
5. **Responsiveness:** Active stakeholder engagement ensures that an organization is responsive to changing societal needs and expectations.

Principles of Stakeholder Engagement

1. **Transparency:** Openly share information, including organizational intentions, plans, and performance.
2. **Inclusivity:** Ensure that all relevant stakeholder groups are included in engagement activities.
3. **Timeliness:** Engage stakeholders early and regularly to prevent potential misunderstandings or conflicts.
4. **Two-way Communication:** Encourage dialogue, allowing stakeholders to share feedback and concerns.
5. **Respect:** Recognize the value of each stakeholder's perspective, treating them with dignity and respect.

Effective Strategies for Stakeholder Engagement

1. **Stakeholder Mapping:** Identify and prioritize stakeholders based on their relevance, influence, and interest.
2. **Engagement Plan:** Develop a structured plan detailing how, when, and why stakeholders will be engaged.

3. **Feedback Mechanisms:** Implement tools or platforms for stakeholders to share feedback, concerns, or suggestions.
4. **Regular Reviews:** Periodically evaluate engagement efforts, adapting strategies based on feedback and changing circumstances.
5. **Collaborative Initiatives:** Partner with stakeholders on projects or initiatives, fostering deeper engagement and mutual benefit.

Challenges in Stakeholder Engagement

1. **Diverse Interests:** Different stakeholders might have conflicting priorities or interests.
2. **Resource Constraints:** Effective engagement can be resource intensive in terms of time, personnel, and finances.
3. **Communication Barriers:** Differences in language, culture, or technical understanding can hinder communication.
4. **Trust Issues:** Previous negative experiences can result in skepticism or mistrust among stakeholders.

In the interconnected world of the twenty-first century, no organization operates in isolation. Engaging with stakeholders is not merely a corporate responsibility; it is a strategic imperative. When done effectively, it can lead to enhanced trust, collaborative successes, and long-term sustainability. As businesses navigate the complexities of the modern landscape, those that prioritize stakeholder engagement are more likely to thrive, innovate, and build lasting legacies.

8.9 Environmental, Social, and Governance (E.S.G.) Reporting

E.S.G. reporting pertains to the disclosure of a company's operations and performance in three critical areas—Environmental, Social, and Governance. It provides measurable criteria for assessing how a corporation manages its impact on the environment, interacts with people, and governs itself (Kotsantonis et al., 2016). E.S.G. reporting evaluates a company's performance based on environmental, social, and governance factors. Environmental factors include climate change, resource use, and pollution. Social factors address labor practices, employee diversity, and community relations. Governance factors involve the company's leadership, transparency, and adherence to ethical business practices.

Key Components

1. **Environmental:** This reflects a company's energy use, waste, pollution, natural resource conservation, and animal treatment. It provides a metric to assess a company's impact and sustainability practices (Clark et al., 2015).

2. **Social:** This encompasses relationships with employees, suppliers, customers, and communities. It includes areas such as labor practices, diversity, and inclusion (Aguinis & Glavas, 2012).
3. **Governance:** This pertains to a company's leadership, executive pay, internal controls, shareholder rights, and business ethics (Aguilera et al., 2007).

While C.S.R. primarily concerns qualitative approaches and strategic alignment with societal values, E.S.G. reporting adds a quantitative dimension by offering concrete, measurable criteria.

8.10 Redefining Business in the Twenty-First Century

- **Stakeholder Engagement:** With the rise of C.S.R. and E.S.G. practices, businesses are no longer solely accountable to shareholders; they are also responsible to stakeholders—those affected by a company's operations, including communities, employees, and the environment.
- **Long-term Value Creation:** Rather than focusing solely on short-term profitability, there is an increasing emphasis on creating sustainable long-term value for all stakeholders.
- **Risk Management:** Recognizing and addressing E.S.G. risks can protect a company from reputational, legal, and financial pitfalls.
- **Access to Capital:** Many investors are now keen on investing in companies with strong E.S.G. performance, believing that these companies will have superior financial performance in the long run.
- **Competitive Advantage:** Companies that engage in C.S.R. and uphold E.S.G. standards are often viewed more favorably by consumers, potentially leading to increased brand loyalty and market share.
- **Regulatory Compliance:** As governments and international bodies become more aware of sustainability and ethical issues, they implement stricter regulations and standards. C.S.R. and E.S.G. practices ensure that companies are prepared to meet or exceed these regulations.
- **Attraction and Retention of Talent:** Modern workers, especially millennials and Gen Z, prioritize working for companies that align with their values.

C.S.R. and E.S.G. reporting has fundamentally shifted the definition of success in business. They urge companies to envision success in terms of financial gains and in a broader context, including societal and environmental impacts. This comprehensive approach to business will likely dominate the corporate landscape throughout the twenty-first century, compelling companies to innovate, adapt, and evolve in ways that benefit their bottom line and the world at large.

8.11 Importance of Corporate Social Responsibility (C.S.R.) and Environmental, Social, and Governance (E.S.G.) Reporting

Risk Management and Resilience

- **C.S.R.:** By actively engaging in C.S.R. practices, corporations can foresee potential risks, especially in their supply chains or operational areas. They can preemptively address issues such as labor disputes, environmental accidents, or community relations, thereby protecting their brand and ensuring continuity (Godfrey, 2005).
- **E.S.G. Reporting:** Transparency in E.S.G. factors allow corporations to recognize vulnerabilities within their governance structures, environmental impact, or social interactions and address these vulnerabilities before they escalate into significant threats (Lee & Faff, 2009).

Enhanced Brand Value and Reputation

- **C.S.R.:** Positive C.S.R. initiatives can significantly bolster a company's brand image, making it more appealing to consumers, increasingly considering companies' ethical and social stands before making purchasing decisions (Sen & Bhattacharya, 2001).
- **E.S.G. Reporting:** The disclosure of E.S.G. metrics demonstrates a company's commitment to transparency, which can foster trust among investors, customers, and other stakeholders, thereby enhancing corporate reputation (Eccles et al., 2014).

Financial Performance and Investor Attraction

- **C.S.R.:** There is growing evidence that firms that commit to C.S.R. can achieve better financial performance due to improved operational efficiency, customer loyalty, and brand differentiation (Orlitzky et al., 2003).
- **E.S.G. Reporting:** Investors are increasingly relying on E.S.G. reports to assess the long-term viability of firms. Robust E.S.G. performance often correlates with better financial returns, lower volatility, and reduced systemic risks (Khan et al., 2016).

Employee Satisfaction and Talent Attraction

- **C.S.R.:** Companies with strong C.S.R. initiatives often report higher levels of employee satisfaction, reduced turnover, and an easier time attracting top talent, given the increasing preference of the workforce towards ethically responsible companies (Turban & Greening, 1997).
- **E.S.G. Reporting:** Transparency in governance and social initiatives can make a company more attractive to potential employees who prioritize these factors in their job search (Berrone et al., 2017).

Long-Term Sustainability and Competitive Advantage

- **C.S.R.:** C.S.R. strategies that align with a company's core business can lead to sustainable competitive advantages by fostering innovation, improving relationships with stakeholders, and ensuring long-term viability in the market (Porter & Kramer, 2006).
- **E.S.G. Reporting:** E.S.G. metrics provide tangible data on a company's sustainability practices. By continuously monitoring and improving these metrics, companies can ensure long-term sustainability in a world facing various environmental and social challenges (Clark et al., 2015).

Challenges of C.S.R. and E.S.G. Reporting

C.S.R. (Corporate Social Responsibility) and E.S.G. (Environmental, Social, and Governance) reporting come with unique challenges, despite being essential components of contemporary business operations. Here are the main challenges organizations face:

1. **Lack of Standardization**: There are numerous standards, frameworks, and guidelines for C.S.R. and E.S.G. reporting, leading to inconsistencies. While some entities use the Global Reporting Initiative (G.R.I.) standards, others might use the Sustainability Accounting Standards Board (SASB) standards or other guidelines, making comparisons difficult.
2. **Data Collection and Verification**: Reliable data collection can be cumbersome, especially for multinational corporations with operations across various jurisdictions. Another significant challenge is ensuring accurate, consistent, and verifiable data.
3. **Greenwashing**: Some companies may exaggerate or falsely report their environmental or social efforts, a deceptive practice called greenwashing. This can mislead stakeholders and undermine the true purpose of C.S.R. and E.S.G. reporting.
4. **Short-term vs Long-term Perspective**: The focus on quarterly financial results can make it challenging for companies to prioritize and invest in long-term sustainability initiatives. This can lead to an underrepresentation of genuine long-term E.S.G. risks and opportunities in reporting.
5. **Subjectivity in Reporting**: While certain E.S.G. metrics, such as carbon emissions, are quantifiable, others, such as social impact or governance practices, can be subjective and difficult to measure.
6. **Integration with Financial Reporting**: Integrating C.S.R. and E.S.G. metrics with traditional financial reporting can be complex. It requires new systems, processes, and expertise to ensure seamless integration.
7. **Resource Constraints**: Particularly for small and medium enterprises (S.M.E.s), allocating resources for comprehensive E.S.G. reporting can be challenging due to limited expertise and financial constraints.
8. **Stakeholder Engagement**: Engaging a wide array of stakeholders, understanding their expectations, and then addressing those in the report can be a significant challenge.

9. **Evolving Expectations**: As societal expectations change, what is considered best practice in E.S.G. reporting can shift, requiring companies to continually adjust their strategies and reporting mechanisms.
10. **Legal and Regulatory Risks**: As governments implement more regulations concerning E.S.G. disclosures, companies face potential legal risks if they do not adhere to these regulations or if their reports are found to be misleading.
11. **Complexity of Global Operations**: For multinational corporations, adhering to different regulatory environments, understanding diverse stakeholder expectations, and dealing with cultural differences can add complexity to E.S.G. reporting.
12. **Balancing Transparency and Competitive Risks**: While stakeholders demand increased transparency, companies must be careful not to reveal sensitive or proprietary information that competitors could exploit.

While C.S.R. and E.S.G. reporting are vital for the modern business landscape, their challenges require a strategic approach. Organizations must prioritize transparency, invest in necessary infrastructure, and continuously dialogue with stakeholders. Overcoming these challenges not only improves the credibility of the reports but also helps businesses realize the actual benefits of sustainable practices.

8.12 Benefits of Corporate Social Responsibility (C.S.R.) and Environmental, Social, and Governance (E.S.G.) Reporting

The advantages of C.S.R. and E.S.G. reporting span a wide range of domains, from financial to societal. Integrating these practices into a corporate framework yield both tangible and intangible rewards. Here is a detailed look:

Financial Returns and Market Performance

- **C.S.R.:** Studies have indicated that firms with strong C.S.R. profiles often experience a positive correlation with financial performance. This is attributed to increased operational efficiencies, enhanced brand loyalty, and better stakeholder relationships (Margolis et al., 2009).
- **E.S.G. Reporting:** Firms with high E.S.G. scores tend to perform better in the stock market and have lower capital constraints. Additionally, they are often associated with reduced tail risks (Dhaliwal et al., 2010).

Stakeholder and Investor Relations

- **C.S.R.:** Companies engaged in C.S.R. activities cultivate deeper trust and commitment from their stakeholders, resulting in stronger relationships and collaborations (Bhattacharya & Sen, 2004).
- **E.S.G. Reporting:** E.S.G. disclosure provides transparency that investors highly value. Companies that report their E.S.G. metrics attract more long-term and sustainability-focused investors (Eccles et al., 2014).

Enhanced Corporate Reputation and Brand Loyalty

- **C.S.R.:** C.S.R. initiatives, especially those aligned with core business strategies, can bolster corporate reputation, leading to increased loyalty and a potential market share gain (Fombrun et al., 2000).
- **E.S.G. Reporting:** Transparent E.S.G. reports display a company's commitment to ethical practices, which can differentiate it in the market, leading to enhanced credibility and brand strength (Eccles et al., 2014).

Operational Efficiency and Innovation

- **C.S.R.:** Embracing C.S.R. can lead to innovations in products, services, and processes. For instance, focusing on environmental sustainability can drive a company to innovate in reducing waste, thus saving costs (Nidumolu et al., 2009).
- **E.S.G. Reporting:** Continuous monitoring and reporting push companies to optimize operations, reduce resource consumption, and innovate in areas identified as inefficient or harmful (Goss & Roberts, 2011).

Employee Engagement and Talent Retention

- **C.S.R.:** Companies actively engaged in C.S.R. typically report higher employee morale and satisfaction. Such companies tend to attract and retain top talent, especially among the younger generations who prioritize ethical considerations in their workplaces (Turban & Greening, 1997).
- **E.S.G. Reporting:** Organizations that publicly commit to E.S.G. principles tend to attract employees who identify with these values, leading to better organizational alignment, reduced turnover, and a more engaged workforce (Berrone et al., 2017).

Risk Mitigation

- **C.S.R.:** Proactively addressing societal and environmental concerns helps companies anticipate and manage risks, especially those associated with regulatory sanctions, reputational damages, or supply chain disruptions (Godfrey, 2005).
- **E.S.G. Reporting:** Regular E.S.G. assessments provide companies with insights into potential vulnerabilities in their operations, enabling them to address risks before they escalate into significant issues (Lee & Faff, 2009).

Both C.S.R. and E.S.G. reporting brings significant benefits to corporations. While C.S.R. fosters societal development, brand reputation, employee satisfaction, and long-term strategic alignment, E.S.G. reporting enhances investment attraction, performance measurement, stakeholder trust, risk management, and global alignment. Together, they represent a comprehensive approach to responsible corporate governance, reflecting a firm's commitment to shareholders and a broader set of stakeholders and global societal goals.

Role of Technology

Technology has an integral role in the execution, monitoring, and enhancement of C.S.R. and E.S.G. initiatives. From the initial stages of planning to reporting and communicating outcomes, advancements in technology have revolutionized the way businesses approach these responsibilities. Here is a deeper dive into the role of technology in C.S.R. and E.S.G.:

1. Data Collection and Analysis

- **Big Data and Analytics:** Large datasets relating to environmental impacts, employee welfare, and community engagements can be collected and analyzed to draw insights, trends, and predictions. This aids companies in understanding their effects and in making informed decisions.
- **Sensors and IoT:** Internet of Things (IoT) devices, such as environmental sensors, can monitor real-time emissions, resource usage, and waste generation. These real-time data allow companies to make immediate changes and improvements.

2. Supply Chain Transparency:

- **Blockchain:** Blockchain technology ensures traceability and transparency in supply chains. Companies can use it to prove the ethical sourcing of materials, ensuring that products are not linked to environmental degradation or unethical labor practices.
- **Artificial Intelligence (A.I.):** A.I. can predict supply chain disruptions, ensuring that sustainable and ethical suppliers are available and reliable.

3. Employee Engagement and Welfare:

- **Remote Working Tools:** The rise of remote work technology, such as video conferencing and collaboration platforms, aids in work-life balance, reducing carbon footprints from commuting, and can be part of a broader C.S.R. strategy related to employee welfare.
- **Wellness Apps and Platforms:** Companies can provide digital tools and platforms to support employee mental health, physical wellness, and continuous learning.

4. Stakeholder Engagement:

- **Virtual Reality (V.R.) and Augmented Reality (A.R.):** V.R. and A.R. can be used for immersive experiences, helping stakeholders understand projects or initiatives more deeply. For example, a company can use V.R. to show the positive impact of its community development projects.

5. Environmental Conservation:

- **Smart Grids:** Technology-driven smart grids enable efficient energy use, reducing waste and ensuring sustainable energy consumption.
- **Remote Sensing:** Drones and satellites can monitor deforestation, land degradation, or the progress of reforestation projects.

6. Reporting and Communication

- **Digital Platforms:** Interactive digital reports with real-time data, infographics, and multimedia elements can make E.S.G. reports more engaging and transparent. Stakeholders can access up-to-date information at any time.
- **Social Media:** Allows companies to communicate C.S.R. and E.S.G. initiatives directly to consumers and stakeholders. It also provides a platform for feedback, ensuring that companies remain accountable.

7. Education and Training

- **E-learning Platforms:** Digital platforms can train employees and stakeholders on the importance of C.S.R. and E.S.G., ensuring that everyone is aligned with the company's values and goals.

Technology acts as a powerful enabler for C.S.R. and E.S.G. initiatives, ensuring that strategies are not just theoretical but actionable, measurable, and communicable. As technological advancements continue to evolve, they will further intertwine with the world of C.S.R. and E.S.G., leading to more robust, impactful, and efficient initiatives.

8.13 Drivers behind the Rise of C.S.R. And E.S.G.

The prominence of Corporate Social Responsibility (C.S.R.) and Environmental, Social, and Governance (E.S.G.) metrics in modern business practices did not emerge in a vacuum. Instead, a range of factors have driven their rise and made them essential components of corporate strategy. Here are the principal drivers behind their ascent:

1. **Stakeholder Awareness and Demand:**
    - **Informed Consumers:** The digital age has empowered consumers with a plethora of information. Consumers now frequently choose brands aligned with their values and that demonstrate a commitment to positive societal and environmental impact.
    - **Investor Pressure:** E.S.G. metrics offer insights into the long-term viability and risk profile of a company. Institutional investors increasingly incorporate E.S.G. factors into their investment decisions, pressuring companies to improve their performance in these areas.
2. **Global Challenges and Collective Action:**
    - **Environmental Concerns:** Issues such as climate change, biodiversity loss, and pollution have highlighted the need for businesses to play their part in environmental stewardship.

- **Social Equity:** As socioeconomic disparities grow globally, there is a call for businesses to contribute to more equitable societies, providing fair wages and ensuring no exploitation within supply chains.

3. **Regulatory and Policy Evolution:**
   - As the global community recognizes the importance of sustainable development, governments and international bodies have established regulations and guidelines urging or mandating businesses to adopt C.S.R. and E.S.G. practices.
   - Initiatives such as the Paris Agreement have a cascading effect, leading to national policies that impact business operations.

4. **Risk Management:**
   - **Reputation:** In an age of instantaneous communication, a company's missteps can rapidly become public, damaging its reputation. Strong C.S.R. and E.S.G. practices can help mitigate such risks.
   - **Operational:** Companies that ignore E.S.G. factors may face operational risks, including supply chain disruptions due to unchecked environmental impacts or social unrest.

5. **Competitive Advantage and Profitability:**
   - As consumers increasingly align their spending with their values, businesses that lead in C.S.R. and E.S.G. can gain a competitive edge.
   - In the long term, sustainable practices can lead to cost savings, resilience, and access to new markets, thereby positively impacting profitability.

6. **Employee Attraction and Retention:**
   - The modern workforce, especially younger generations, seeks meaningful employment. Companies with robust C.S.R. and E.S.G. practices tend to attract and retain talent more effectively.

7. **Technological Advancements:**
   - Digital platforms and tools enable greater transparency, traceability, and reporting. Technologies such as blockchain, for instance, can help trace ethical sourcing, while big data can assist in monitoring and reporting E.S.G. metrics.

8. **Civil Society and Activism:**
   - N.G.O.s, community groups, and activists play a significant role in highlighting areas that need attention and holding corporations accountable. Their influence has been pivotal in pushing businesses towards better C.S.R. and E.S.G. practices.

9. **Collaborative initiatives:**
   - Partnerships and collaborative efforts, such as the United Nations Global Compact, provide platforms for businesses to share best practices, set industry standards, and collectively work towards sustainable goals.

10. **Evolution of Business Paradigms:**

- The twenty-first-century business paradigm has shifted from the sole focus on shareholder value to a broader stakeholder approach. There is a growing recognition that businesses thrive when they operate sustainably and ethically, benefiting all stakeholders in the process.

The rise of C.S.R. and E.S.G. is a response to evolving global challenges, stakeholder expectations, and a shifting understanding of what constitutes business success. It reflects the global business community's recognition of its role in shaping a sustainable and equitable future.

8.14 Evolving Trends and Future Outlook

Corporate social responsibility (C.S.R.) and Environmental, Social, and Governance (E.S.G.) reporting are vital to contemporary business practices. These concepts illustrate a company's commitment to social and environmental responsibilities and are intricately connected to a firm's financial performance and brand reputation.

Evolving Trends

- **Integration of E.S.G. Metrics in Investment Decisions**: E.S.G. criteria are increasingly used to guide investment decisions. Investors now recognize the long-term value of sustainable business practices, leading to a shift in investment strategies favoring companies with strong E.S.G. profiles.
- **Regulatory Developments**: Governments around the world are implementing regulations to enforce transparency in C.S.R. and E.S.G. reporting. This includes mandatory disclosure requirements and standardized frameworks for reporting.
- **Technology Integration**: With the advancement of A.I., machine learning, and big data analytics, companies are using sophisticated tools to measure and report on C.S.R. and E.S.G. efforts.
- **Stakeholder Engagement**: There is an increased focus on engaging with stakeholders, including employees, customers, and communities. This is reflected in more detailed reporting, aiming for greater transparency and accountability.

  - **Focus on Social and Ethical Factors**: As social issues such as diversity, inclusion, and human rights are gaining prominence, companies are now actively reporting on these aspects under the umbrella of E.S.G.
  - **Climate Impact and Sustainability**: The global focus on climate change and sustainability drives companies to develop strategies for reducing emissions and resource use, aligning with global goals such as the Paris Agreement.
  - **Alignment with Sustainable Development Goals (S.D.G.s)**: More companies are aligning their C.S.R. and E.S.G. reporting with the United Nations' S.D.G.s, aiming to contribute to global efforts in tackling various social, economic, and environmental challenges.

Future Outlook

- **Increased Standardization**: The future holds further standardization in reporting guidelines, allowing for greater comparability between companies and industries.
- **Broader Scope**: The scope of C.S.R. and E.S.G. will continue to expand, incorporating emerging issues such as data privacy, cybersecurity, and circular economy practices.
- **Interconnected Reporting**: An integrated approach to reporting that brings financial, environmental, social, and governance data into a cohesive framework will likely become more common.
- **Active Participation in Policy Making**: Corporations may play a more active role in shaping public policies related to social and environmental issues, collaborating with governments and N.G.O.s.
- **Greater Scrutiny and Accountability**: With increased access to information, investors and the general public will demand greater transparency and authenticity in reporting, leading to higher accountability standards.
- **Focus on Impact Measurement**: As organizations aim for tangible and meaningful impacts, the tools, and methods to measure these impacts will evolve, providing deeper insight into the effectiveness of C.S.R. and E.S.G. initiatives.
- **Emphasis on Resilience and Risk Management**: The future may see a shift towards a focus on resilience, with E.S.G. strategies that drive positive impact and mitigate business risks related to social and environmental factors.

The evolving trends and future outlook of C.S.R. and E.S.G. reporting reflect the growing importance of responsible business practices in the global economy. The trends highlight a shift towards greater transparency, standardization, engagement, and strategic alignment with global goals. The future of C.S.R. and E.S.G. reporting will be more dynamic and integral to overall business strategy, demanding innovative approaches and continuous adaptation to emerging global challenges. Corporations that effectively embrace these practices will likely position themselves as leaders in their respective industries and contribute positively to broader societal goals.

8.15 Practical Steps for Companies to Adopt C.S.R. And E.S.G. Reporting

Adopting C.S.R. and E.S.G. reporting is crucial for modern businesses seeking to achieve sustainable success and meet the expectations of a wide array of stakeholders. Here is a step-by-step guide for companies:

1. Top-Level Commitment

- **Board and C.E.O. Alignment**: Ensure that the top leadership understands the importance of C.S.R. and E.S.G. and is committed to integrating them into the company's strategy.

- **Allocate Resources**: Dedicate a budget, manpower, and other necessary resources to develop and execute C.S.R. initiatives and E.S.G. reporting.

2. Internal Assessment and Baseline Setting

- **SWOT Analysis**: Conduct a SWOT (Strengths, Weaknesses, Opportunities, Threats) analysis specific to C.S.R. and E.S.G. factors.
- **Current State Analysis**: Assess the company's current C.S.R. initiatives and E.S.G. performance to set a baseline.

3. Stakeholder Consultation

- **Engage Stakeholders**: Speak to employees, shareholders, suppliers, communities, and customers to understand their concerns and expectations.
- **Feedback Loop**: Create mechanisms for continuous feedback from stakeholders.

4. Define Materiality

- **Identify Key Issues**: From the feedback and analysis, determine which social, environmental, and governance issues are most relevant to the business and its stakeholders.
- **Prioritize Issues**: Not all issues will have equal weight. Prioritize based on potential impact on the business and importance to stakeholders.

5. Set Clear Objectives and KPIs

- **Align with Global Standards**: Consider aligning objectives with global standards and goals, such as the U.N.'s Sustainable Development Goals (S.D.G.s).
- **Establish KPIs**: Set clear, measurable key performance indicators (KPIs) for each objective to track progress.

6. Integration into Business Strategy

- **Cross-Departmental Collaboration**: Ensure that C.S.R. and E.S.G. strategies are not isolated but integrated across departments such as finance, H.R., operations, and marketing.
- **Embedded in Corporate Culture**: Incorporate C.S.R. and E.S.G. values into the company's mission, vision, and core values.

7. Regular Monitoring and Evaluation

- **Review Progress**: Regularly evaluate performance against the set KPIs.
- **Adjust Strategy**: If certain initiatives are not working or new issues emerge, be ready to adjust the strategy.

8. Transparent Reporting

- **Choose a Reporting Framework**: Use standardized frameworks such as the Global Reporting Initiative (G.R.I.), Sustainability Accounting Standards Board (SASB), or Task Force on Climate-related Financial Disclosures (TCFD) for guidance.

- **Annual Sustainability Report**: Publish an annual sustainability or C.S.R. report detailing the company's initiatives, performance, challenges, and future plans.
- **Third-party Verification**: Consider getting your report audited or verified by an external party for additional credibility.

9. Continuous Improvement

- **Stay Updated**: As C.S.R. and E.S.G. standards evolve, ensure you are up to date with the latest best practices and guidelines.
- **Innovate**: Look for new ways, technologies, or partnerships to enhance your C.S.R. initiatives and E.S.G. performance.

10. Celebrate and Communicate Success

- **Share Achievements**: Celebrate milestones and successes internally and externally.
- **Engage in Thought Leadership**: Participate in forums, conferences, and workshops to share your insights, challenges, and lessons learned.

Remember that the goal of C.S.R. and E.S.G. reporting is not just compliance but also to genuinely embed responsible and sustainable practices into the core of the business. It is a journey of continuous growth, reflection, and improvement. Embracing this journey will not only benefit the business and its stakeholders but also contribute positively to global challenges and societal well-being.

8.16 Case Studies

1. Patagonia

Background Founded in 1973 by Yvon Chouinard, Patagonia is an American company that sells outdoor clothing and gear. It is widely recognized for its commitment to sustainability and social responsibility.

Key C.S.R. & E.S.G. Initiatives

- **1% for the planet:** Since 1985, Patagonia has pledged to donate 1% of its sales to preserving and restoring the natural environment.
- **Worn Wear Program:** This initiative encourages customers to buy used Patagonia items or trade in their old ones, emphasizing the importance of reducing waste and promoting a circular economy.
- **Supply Chain Transparency:** Patagonia ensures that its supply chain is ethical, even going as far as disclosing the factories it works with to maintain accountability.

- **Fair Trade Certified:** Many of Patagonia's products are Fair Trade Certified, meaning that they pay a premium for every item that goes directly to the workers.

Outcomes & Impact

- Patagonia's dedication to sustainability and ethical practices has bolstered its brand image and loyalty among consumers who value these principles.
- Their environmental efforts, combined with their business model, have often been cited as leading examples of how companies can be profitable while prioritizing the planet.

2. Unilever

Background Unilever is a British-Dutch multinational consumer goods company co-headquartered in London and Rotterdam. It is one of the world's largest and oldest consumer goods companies, with products available in over 190 countries.

Key C.S.R. & E.S.G. Initiatives

- **Unilever Sustainable Living Plan:** Launched in 2010, this plan set out ambitious goals to decouple their growth from their environmental impact while increasing their positive social impact. The plan focused on improving health and hygiene, reducing environmental impact, and enhancing livelihoods.
- **Supply Chain:** Unilever is committed to sustainably sourcing 100% of its agricultural raw materials.
- **Plastic Pledge:** Unilever has committed to making all its plastic packaging recyclable, reusable, or compostable by 2025.

Outcomes and Impact

- As of recent reports, Unilever has witnessed growth in its "Sustainable Living" brands at a rate faster than the rest of the business.
- Their commitment to sustainability has made them attractive to socially conscious investors and consumers alike.
- The company's efforts towards sustainable sourcing have resulted in strengthened relationships with suppliers and have created resilience against potential disruptions.

Patagonia and Unilever highlight that businesses can remain profitable while embracing ethical and sustainable practices. Their commitment to C.S.R. and E.S.G. is not merely about compliance or marketing; it is woven into their core business strategies. As companies face increased scrutiny from consumers, regulators, and investors, these case studies are powerful examples of the convergence of doing good and doing well.

8.17 Conclusion

C.S.R. and E.S.G. reporting has become essential to modern business practices, shaping how companies operate, innovate, and interact with stakeholders. As expectations and regulations continue to evolve, companies must embrace these practices as compliance measures and as opportunities to drive sustainable growth and positively impact society and the environment. The business landscape of the twenty-first century is enduring a radical transformation. Organizations can no longer view success exclusively through the lens of financial benefit. Profitability, societal improvement, and environmental stewardship comprise the modern success metric for corporations.

The rise of C.S.R. and E.S.G. is not merely a fad but a reflection of a global society that expects more from its businesses. Investors and consumers are voting with their capital and purchasing power, putting pressure on businesses to align their strategies with broader societal values. As we have seen, companies that adopt these paradigms stand to gain reputation and brand loyalty and discover new development and innovation opportunities. In addition, businesses are equipped with tools that enhance their C.S.R. and E.S.G. initiatives by integrating advanced technologies. These technologies assure transparency, facilitate improved stakeholder engagement, and allow for increased resource utilization. However, along with these advancements come obstacles. Greenwashing, ambiguous metrics, and the possibility of superficial compliance remind us of the importance of genuine commitment and ongoing scrutiny. Standardized reporting and persistent stakeholder engagement will ensure that corporations maintain accountability.

The business landscape of the twenty-first century is not merely characterized by innovations in technology, global market dynamics, or evolving consumer preferences. At its core, it is being redefined by a deep-seated recognition of businesses' roles as integral stakeholders in a larger societal and environmental context. Corporate Social Responsibility (C.S.R.) and Environmental, Social, and Governance (E.S.G.) Reporting have emerged as potent manifestations of this paradigm shift.

Businesses today are expected to look beyond profitability as the sole measure of success. The four pillars of C.S.R. – economic, legal, ethical, and philanthropic responsibilities – delineate a comprehensive framework for companies to engage conscientiously with society. Simultaneously, E.S.G. reporting provides stakeholders with a transparent lens through which to view how businesses impact the world and where their true values lie.

This transformation in the business ethos is driven by a confluence of factors:

1. **Informed Stakeholders:** Empowered by information and connectivity, stakeholders – from consumers to investors – demand transparency, ethics, and genuine commitment to societal well-being.
2. **Global Challenges:** Issues such as climate change, inequality, and resource depletion require collective action, with businesses playing a pivotal role.

3. **Holistic Success Metrics:** Financial performance, although crucial, is now one of the many yardsticks for business success. E.S.G. metrics underscore the importance of broader impacts.
4. **Long-term Viability:** Companies are realizing that addressing social and environmental challenges is not just altruistic; it ensures long-term sustainability and resilience.

While the journey towards fully integrating C.S.R. and E.S.G. principles is fraught with challenges from navigating complex regulations to balancing stakeholder interests, the direction is clear. The businesses of the twenty-first century are expected to be more than economic entities. They are partners in crafting a sustainable, equitable, and prosperous future. It is worth reflecting on the ancient adage, "To whom much is given, much is expected." Businesses, wielding considerable influence and resources, have much given to them. In this new era, society's expectations are clear: operate responsibly, transparently, and with a sincere commitment to the greater good. The future of business is not just about products, services, or innovations. It is about purpose, values, and impact.

8.18 Reflection on the Paradigm Shift in Twenty-First-Century Business

The twenty-first century has ushered in an era of rapid change and transformation, particularly for businesses worldwide. Reflecting on this paradigm shift, several key dimensions emerge:

- **From Profit to Purpose**: Traditional business models were singularly focused on profit generation. However, today's businesses recognize that long-term profitability is intricately tied to purpose. Corporate social responsibility (C.S.R.) is no longer a peripheral activity; it is central to a company's mission, with businesses now expected to pursue profit while also considering the planet and people.
- **Stakeholder Capitalism**: The twenty-first century has transitioned from shareholder capitalism, which emphasizes shareholder returns, to stakeholder capitalism. Businesses now recognize the interconnectedness of their relationships with all stakeholders, from employees and customers to communities and the environment.
- **Digital Transformation**: The rise of the internet, artificial intelligence, big data, and other digital technologies has revolutionized operations, marketing, and communication. This has enabled global outreach but has also introduced challenges such as data privacy concerns.
- **Transparency and Accountability**: With the proliferation of information through digital channels, businesses are under increasing scrutiny. There is a demand for greater transparency in operations, supply chains, and especially in how businesses tackle ethical challenges.

- **Sustainability and Climate Consciousness**: Climate change and sustainability are no longer topics confined to activists. They are central to business strategies, with companies focusing on sustainable practices, renewable energy, and circular economies to ensure long-term viability.
- **Diversity, Equity, and Inclusion (D.E.I.)**: There is a growing recognition of the importance of D.E.I. in the workplace. A diverse workforce drives innovation and reflects global customer bases, and companies are striving to ensure representation and inclusivity.
- **Redefining Value and Success**: The metrics for success have evolved. Nonfinancial metrics related to environmental, social, and governance (E.S.G.) factors have become as crucial as financial KPIs. These provide a holistic view of a company's performance and its societal impact.
- **Globalization vs. Localization**: While the twentieth century was marked by rapid globalization, recent years have seen a resurgence in the importance of local contexts. Businesses are now navigating a balance between global operations and local sensitivities.
- **Agility and Resilience**: Given the unpredictability of global events, from financial crises to pandemics such as COVID-19, businesses have learned the importance of agility. The ability to pivot, adapt, and evolve has become a critical competency.
- **The Role of Business in Societal Challenges**: Today's businesses are expected to play a role in addressing societal challenges, including battling pandemics, ensuring education, or fighting against systemic inequalities.

The twenty-first-century paradigm shift is not merely a change; it is a metamorphosis. Businesses are not just economic entities; they are social entities intertwined with the fabric of society. The future will belong to those organizations that recognize these evolving expectations and embed them in their D.N.A., achieving a harmonious balance between profit and purpose.

The Imperative for Businesses to Embrace Both C.S.R. and E.S.G. for Sustainable Success

The landscape of modern business is marked by dynamic change and increasing stakeholder scrutiny. In this environment, merely focusing on profit margins and shareholder returns no longer suffices. Both Corporate Social Responsibility (C.S.R.) and Environmental, Social, and Governance (E.S.G.) criteria have emerged as essential frameworks for businesses aiming for sustainable success. Here is why embracing both is imperative:

- **Stakeholder Expectations**: Modern stakeholders, including investors, consumers, and employees, demand more than just financial returns. They expect companies to have a broader purpose, one that positively impacts society and the environment. By integrating C.S.R. and E.S.G. into business strategies, companies can meet these elevated expectations.

- **Risk Management**: Addressing E.S.G. factors helps businesses identify and mitigate various risks, from environmental hazards to social unrest. Proper E.S.G. management can prevent regulatory fines, boycotts, and reputational damage, which can have significant financial implications.
- **Long-term Viability**: Sustainability is not just about protecting the environment; it is about ensuring that the business itself is sustainable in the long run. By focusing on C.S.R. and E.S.G., companies can anticipate future challenges, innovate solutions, and ensure that they remain viable and competitive in a changing world.
- **Attracting Investment**: The investment community is increasingly valuing E.S.G. performance. Sustainable and responsible investment has grown exponentially, with many institutional investors and funds integrating E.S.G. factors into their investment decisions. Businesses that prioritize E.S.G. are more likely to attract and retain these investors.
- **Enhanced Reputation**: A strong C.S.R. and E.S.G. record enhances corporate reputation, which in turn can lead to increased customer loyalty, attracting top talent, and even commanding premium prices for products and services.
- **Operational Efficiency**: Many C.S.R. and E.S.G. initiatives, such as reducing waste or energy consumption, also lead to operational efficiency and cost savings. For instance, transitioning to renewable energy sources can result in long-term savings.
- **Innovation and New Opportunities**: Engaging in C.S.R. and E.S.G. can drive innovation. As businesses seek sustainable solutions, they may discover new products, services, or markets, fostering growth and diversification.
- **Employee Engagement and Retention**: Modern employees, especially younger generations, want to work for companies that align with their values. A strong commitment to C.S.R. and E.S.G. can attract, engage, and retain top talent, leading to increased productivity and reduced turnover costs.
- **Regulatory Preparedness**: Governments and regulatory bodies worldwide are increasingly instituting policies related to environmental protection, social welfare, and corporate governance. By proactively integrating E.S.G. criteria, companies can stay ahead of the regulatory curve, ensuring compliance and avoiding potential penalties.
- **Moral Imperative**: Beyond the tangible benefits, there is a moral imperative. As influential entities, businesses have the power and resources to make a positive impact. Engaging in meaningful C.S.R. and E.S.G. practices reflects corporate ethics and values.

The integration of C.S.R. and E.S.G. is not just a trend but a fundamental shift in understanding the role of business in the twenty-first century. Companies that understand and act upon this imperative will not only achieve sustainable success but also play a pivotal role in shaping a more equitable and sustainable world.

8.19 Call to Action for Businesses, Investors, and Stakeholders

As we navigate the challenges and opportunities of the twenty-first century, the responsibility to drive sustainable and equitable growth is shared among businesses, investors, and stakeholders. Here is a call to action for each:

For Businesses

- **Integrate Purpose**: Move beyond profit as the sole driving force. Integrate a higher purpose that prioritizes planet, people, and societal well-being alongside profit.
- **Embed E.S.G. into Strategy**: Integrate E.S.G. factors into your business strategy, ensuring that decisions consider environmental, social, and governance impacts.
- **Innovate for Sustainability**: Invest in research and development that promotes sustainable products, services, and solutions. Seek to redefine industries with a focus on long-term viability.
- **Transparency and Reporting**: Regularly report on your C.S.R. and E.S.G. initiatives, ensuring that stakeholders are informed about both successes and challenges. Use standardized metrics where possible.
- **Collaborate**: Join forces with other businesses, nonprofits, governments, and communities to scale impact. Collective action often brings about systemic change faster.

For Investors

- **Value E.S.G. Performance**: Prioritize businesses that show strong E.S.G. performance, ensuring that investments drive positive change.
- **Engage and Influence**: Use your influence to encourage businesses to adopt sustainable practices. Engage in active dialogues, pushing for transparency and improvements.
- **Educate and Advocate**: Promote the importance of sustainable investment among your peers and networks. Share research and insights that underscore the long-term value of E.S.G. integration.
- **Diversify Portfolio with Impact Investing**: Allocate a portion of your portfolio to impact investments, which offer both financial returns and measurable societal impact.

For Stakeholders

- **Demand Accountability**: Whether you are a consumer, employee, or community member, demand businesses prioritize sustainability, equity, and ethics.
- **Stay Informed**: Educate yourself about the businesses you support. Understand their values, practices, and impact on the world.
- **Collaborate for Change**: Engage with businesses in constructive dialogues. Offer insights, feedback, and solutions that help them enhance their C.S.R. and E.S.G. efforts.

- **Support Ethical Choices**: Consumers choose products and services from companies that align with their values. This market-driven approach often catalyzes change more rapidly.
- **Amplify the Message**: Use your platform, whether it is in your local community, on social media, or within professional networks, to promote the importance of C.S.R. and E.S.G. Celebrate businesses doing it right and call out those falling short.

The journey towards a sustainable, equitable, and prosperous future requires collective action. Each of us, whether as a business leader, an investor, or a stakeholder, holds a piece of the puzzle. By committing to these calls to action, we can create a world where business is not just about wealth creation but also about ensuring well-being for all. Let us act now, ensuring that the legacy we leave behind is one of responsibility, resilience, and positive impact.

Declaration of Generative A.I. and AI-assisted Technologies in the Writing Process
During the preparation of this work, the author(s) used OpenAI to improve language and readability. After utilizing this tool/service, the author(s) reviewed and edited the content as necessary and take(s) full responsibility for the publication's content.

References

Aguilera, R. V., Rupp, D. E., Williams, C. A., & Ganapathi, J. (2007). Putting the S back in corporate social responsibility: A multilevel theory of social change in organizations. *Academy of Management Review, 32*(3), 836–863. https://doi.org/10.5465/amr.2007.25275678

Aguinis, H., & Glavas, A. (2012). What we know and don't know about corporate social responsibility a review and research agenda. *Journal of Management, 38*, 932–968. https://doi.org/10.1177/0149206311436079

Berrone, P., Fosfuri, A., & Gelabert, L. (2017). Does greenwashing pay off? understanding the relationship between environmental actions and environmental legitimacy. *Journal of Business Ethics, 144*(2), 363–379. https://doi.org/10.1007/s10551-015-2816-9

Bhattacharya, C. B., & Sen, S. (2004). Doing better at doing good: When, why, and how consumers respond to corporate social initiatives. *California Management Review, 47*(1), 9–24. https://doi.org/10.2307/41166284

Carroll, A. B. (1991). The pyramid of corporate social responsibility: Toward the moral management of organizational stakeholders. *Business Horizons, 34*(4), 39–48. https://doi.org/10.1016/0007-6813(91)90005-G

Carroll, A. B. (1999). Corporate social responsibility: Evolution of a definitional construct. *Business & Society, 38*(3), 268–295. https://doi.org/10.1177/000765039903800303

Clark, G. L., Feiner, A., & Viehs, M. (2015). *From the stockholder to the stakeholder: How sustainability can drive financial outperformance* (Issue March). http://ssrn.com/abstract=2508281HOWSUSTAINABILITYCANDRIVEFINANCIALOUTPERFORMANCE Electronic copy available at: https://ssrn.com/abstract=2508281

Crane, A., & Matten, D. (2007). *Business ethics: Managing corporate citizenship and sustainability in the age of globalization*. Oxford University Press.

Dahlsrud, A. (2008). How corporate social responsibility is defined: An analysis of 37 definitions. *Corporate Social Responsibility and Environmental Management, 15*(1), 1–13. https://doi.org/10.1002/csr.132

Dhaliwal, D., Oliver Zhen, L., Tsang, A., & Yang, G. (2010). Voluntary nonfinancial disclosure and the cost of equity capital: The initiation of corporate social responsibility reporting. *The Accounting Review, 86*(1), 59–100. https://doi.org/10.2139/ssrn.1687155

Eccles, R., Ioannou, I., & Serafeim, G. (2014). The impact of corporate sustainability on organizational processes and performance. *Management Science, 60*, 2835–2857. https://doi.org/10.1287/mnsc.2014.1984

Fombrun, C. J., Gardberg, N. A., & Sever, J. M. (2000). The reputation QuotientSM: A multi-stakeholder measure of corporate reputation. *Journal of Brand Management, 7*(4), 241–255. https://doi.org/10.1057/bm.2000.10

Freeman, R. E., Harrison, J. S., Wicks, A. C., Parmar, B. L., & de Colle, S. (2010). *Stakeholder theory: The state of the art*. Cambridge University Press. https://doi.org/10.1017/CBO9780511815768

Friedman, M. (1970). The social responsibility of business is to increase its profits. *New York Times Magazine*, 122–126.

Godfrey, P. (2005). The relationship between corporate philanthropy and shareholder wealth: A risk management perspective. *Academy of Management Review, 30*, 777–798. https://doi.org/10.5465/AMR.2005.18378878

Goss, A., & Roberts, G. S. (2011). The impact of corporate social responsibility on the cost of bank loans. *Journal of Banking & Finance, 35*(7), 1794–1810. https://doi.org/10.1016/j.jbankfin.2010.12.002

Hajji, S. (2009, December 7). The Spirit of socio-economic development. International petroleum technology conference. https://doi.org/10.2523/IPTC-14095-MS

Khan, M., Serafeim, G., & Yoon, A. (2016). Corporate sustainability: First evidence on materiality. *Accounting Review, 91*(6), 1697–1724. https://doi.org/10.2308/accr-51383

Kotsantonis, S., Pinney, C., & Serafeim, G. (2016). E.S.G. Integration in investment management: Myths and realities. *Journal of Applied Corporate Finance, 28*, 10–16. https://doi.org/10.1111/jacf.12169

Lee, D. D., & Faff, R. W. (2009). Corporate sustainability performance and idiosyncratic risk: A global perspective. *Financial Review, 44*(2), 213–237. https://doi.org/10.1111/j.1540-6288.2009.00216.x

Lee, H., Kim, S. Y., Kim, G., & Kang, H.-Y. (2019). Public preferences for corporate social responsibility activities in the pharmaceutical industry: Empirical evidence from Korea. *PLoS One, 14*(8), e0221321. https://doi.org/10.1371/journal.pone.0221321

Margolis, J., Elfenbein, H., & Walsh, J. (2009). Does it pay to be good...And does it matter? A meta-analysis of the relationship between corporate social and financial performance. *SSRN Electronic Journal*. https://doi.org/10.2139/ssrn.1866371

Nidumolu, R., Prahalad, C. K., & Rangaswami, M. R. (2009). Why sustainability is now the key driver of innovation. *Harvard Business Review*, 57–64.

Orlitzky, M., Schmidt, F. L., & Rynes, S. L. (2003). Corporate social and financial performance: A meta-analysis. *Organization Studies, 24*(3), 403–441. https://doi.org/10.1177/0170840603024003910

Porter, M. E., & Kramer, M. R. (2002). The competitive advantage of corporate philanthropy. *Harvard Business Review, 80*(12), 56–68.

Porter, M. E., & Kramer, M. R. (2006). Strategy and society: The link between competitive advantage and corporate social responsibility. *Harvard Business Review*, 78–83.

Sen, S., & Bhattacharya, C. B. (2001). Does doing good always lead to doing better? Consumer reactions to corporate social responsibility. *Journal of Marketing Research, 38*(2), 225–243. https://doi.org/10.1509/jmkr.38.2.225.18838

Turban, D., & Greening, D. (1997). Corporate social performance and organizational attractiveness to prospective employees. *Academy of Management Journal, 40*, 658–672. https://doi.org/10.2307/257057

Wang, H., & Chen, Z.-X. (2022). Corporate social responsibility and job applicant attraction: A moderated-mediation model. *PLoS One, 17*(3), e0260125. https://doi.org/10.1371/journal.pone.0260125

Chapter 9
Environmental Impact Assessment and Sustainable Energy Transition

Abstract Given the ongoing worldwide transition towards a sustainable energy trajectory, it is imperative to prioritize the advancement of technologies that facilitate the integration of this transition with ecological and socio-economic sustainability. The Environmental Impact Assessment (EIA) is recognized as a crucial instrument among the several mechanisms that are considered. This research investigates the intrinsic relationship between Environmental Impact Assessment (EIA) and the global shift towards sustainable energy. Environmental Impact Assessments (EIAs) offer a complete framework for the examination, anticipation, and alleviation of any environmental ramifications linked to energy developments. Multiple individuals and organizations are currently making concerted efforts to achieve a delicate equilibrium between the increasing worldwide need for energy and the pressing necessity to preserve the environment. This study examines the importance of the Environmental Impact Assessment (EIA) in evaluating greenhouse gas emissions, promoting long-term sustainability, and facilitating stakeholder participation. This remark underscores the importance of performing Environmental Impact Assessments (EIA) to assess not just the immediate impacts of energy projects but also their long-term viability within a constantly changing climate. The Environmental Impact Assessment (EIA) is not solely regarded as a legal impediment but rather as a crucial guiding principle in our endeavor to achieve a sustainable energy trajectory.

Keywords Sustainable energy · Environmental Impact Assessment (EIA) · Ecological sustainability · Socio-economic sustainability · Energy transition · Greenhouse gas emissions · Climate change

9.1 Introduction

Environmental Impact Assessment (EIA) is a valuable instrument utilized to ascertain the potential environmental, social, and economic consequences of a project before the final determination. The increasing global demand for sustainable energy

V. Kandpal et al., *Sustainable Energy Transition*, Circular Economy and Sustainability, https://doi.org/10.1007/978-3-031-52943-6_9

underscores the significance of the Environmental Impact Assessment (EIA) in guaranteeing the environmental friendliness and sustainability of energy transition initiatives. Environmental Impact Assessment (EIA) is a valuable instrument for advancing sustainability objectives, as it prevents the implementation of new projects that would result in adverse environmental consequences.

Within the realm of the energy industry, the Environmental Impact Assessment (EIA) serves as a valuable tool for evaluating the ecological consequences associated with both renewable energy initiatives, such as solar and wind farms, and non-renewable energy undertakings, such as coal-fired power plants (Sokka et al., 2016). EIA can also assess the environmental impacts of energy policies, such as carbon pricing or energy efficiency standards.

The principal objective of Environmental Impact Assessment (EIA) in the context of sustainability is to ascertain and assess the potential environmental ramifications linked to proposed projects and policies.

- Mitigate the negative environmental impacts of projects and regulations.
- Promote the progress of sustainable energy initiatives and the development of policies to facilitate their implementation.

The Environmental Impact Assessment (EIA) encompasses various dimensions. The endeavour requires substantial exertion; nevertheless, it fulfils a crucial role in safeguarding against environmentally undesirable consequences from novel energy initiatives and regulations. The following instances illustrate the utilisation of Environmental Impact Assessment (EIA) as a means to advance the cause of sustainable energy transition:

- National Environmental Policy Act (NEPA) mandates the implementation of Environmental Impact Assessments (EIA) for all federal undertakings within the United States that possess the potential to yield substantial environmental consequences. This measure has effectively mitigated the environmental impacts associated with energy projects, including but not limited to oil spills and air pollution.
- Within the European Union, the Strategic Environmental Assessment (SEA) Directive mandates the implementation of Environmental Impact Assessments (EIA) for any substantial plans and programmes that possess the potential to exert a noteworthy influence on the environment. This has contributed to the development of energy policies that aim to mitigate environmental concerns.
- In the context of China, the implementation of Environmental Impact Assessment (EIA) is mandatory for all significant energy-related initiatives. This measure has effectively mitigated the environmental consequences associated with energy generation in China, a nation that has the distinction of being the foremost contributor to global greenhouse gas emissions (Nolte, 2014).

EIA is a valuable tool for promoting sustainable energy transition. By identifying and assessing the environmental impacts of proposed projects and policies, EIA can help to ensure that new energy projects and procedures are developed in a way that minimises environmental impacts. In conjunction with the Environmental Impact

Assessment (EIA), various alternative techniques can aid the progression of a sustainable energy transition.

- Life Cycle Assessment (LCA) can be employed to evaluate the environmental effects of products and services across their complete life cycle.
- Carbon pricing is a mechanism that imposes a monetary value on carbon emissions, thereby incentivising both enterprises and individuals to curtail their emission levels.
- Energy efficiency standards mandate that businesses and structures adhere to regulations to reduce energy consumption.
- Renewable energy targets establish objectives for the proportion of energy derived from renewable sources.

These methods can be used to establish a holistic strategy for facilitating the transition towards sustainable energy. Through the utilisation of these technologies, it is possible to ascertain if our energy system exhibits qualities of cleanliness, efficiency, and sustainability.

9.2 Role of Environmental Impact Assessment (EIA) in Sustainable Energy Transition

Environmental Impact Assessment (EIA) has gained heightened significance as the world grapples with the dual challenge of climate change and surging energy demand. Here are the reasons why EIA is indispensable in this context:

Balancing Energy Needs with Environmental Protection

In the contemporary era of an increasingly interconnected global society, there exists a persistent and escalating demand for energy. As economies experience growth, society undergoes modernisation, and people expand, there is a corresponding increase in the demand for energy. Conversely, the profound ramifications of climate change, in conjunction with escalating environmental degradation, underscore the imperative nature of safeguarding the environment. The intricate task of reconciling the growing need for energy with the imperative of preserving the environment might be considered as one of the critical issues of our era.

- **Meeting Energy Demands**: As the global population grows and economies expand, the demand for energy is set to rise exponentially. Developing countries are in a phase of rapid industrialisation and urbanisation, necessitating more power.
- **Environmental Safeguard**: The majority of conventional sources of energy, particularly fossil fuels, possess substantial detrimental effects on the environment, ranging from the release of greenhouse gases to the degradation of habitats. EIA ensures that this surging energy demand is met in a manner that

minimally harms the environment. It serves as a check and balance system to ensure that economic growth does not come at an unsustainable environmental cost.

- **Transition to Renewable Energy:** Renewable energy sources, including solar, wind, and hydroelectric power, present a prospective trajectory towards sustainable development. In recent years, there has been a significant decline in the costs of renewable energy technologies, rendering them more economically viable and comparable to conventional energy sources. Ongoing research and development efforts contribute to advancing renewable technologies and storage solutions, thereby enhancing energy generation efficiency and ensuring its consistency and reliability.
- Nevertheless, it is imperative to acknowledge that even renewable sources of energy are not exempt from causing environmental consequences. The manufacturing process of solar panels entails the utilisation of potentially dangerous chemicals, wind turbines have been observed to impact bird and chiropteran populations, and the implementation of hydropower has the potential to disturb aquatic ecosystems. Therefore, it is imperative to achieve a harmonious equilibrium, especially in the presence of renewable energy sources.
- **Economic and Societal Growth**: The provision of reliable energy is a fundamental pillar of socio-economic progress. Electricity is a crucial source of energy that drives many sectors, such as industries, residential buildings, and transportation systems. Its utilisation has a direct positive relationship with the enhancement of living standards.

Achieving a harmonious balance between our energy requirements and environmental preservation entails more than simply identifying cleaner energy alternatives. It necessitates a comprehensive strategy that encompasses various dimensions, including technological advancements, governmental frameworks, societal engagement, and ecological considerations. In the face of impending problems, the prospect of innovation and adaptability is a source of optimism. By placing equal importance on meeting our energy demands and safeguarding the ecological well-being of our world, we may aspire to a future characterised by sustainable and environmentally friendly sources of power.

9.3 Mitigating Climate Change Impacts

The issue of climate change is characterised by the increasing presence of greenhouse gases (GHGs) in the atmosphere, which gives rise to complex and significant difficulties. Given the growing prominence of effects such as erratic weather patterns, escalating sea levels, and the depletion of biodiversity, it is crucial to investigate comprehensive approaches for mitigating these issues. This discourse extensively explores the diverse strategies available to mitigate the impacts of climate change.

A. **Transition to Renewable Energy.**

The transition towards renewable energy is often seen as a significant and transformative change in contemporary times. It signifies a departure from exhaustible and environmentally harmful fossil fuel resources towards cleaner and sustainable alternatives. The process of evolution described here is not solely driven by environmental factors but also by economic and societal considerations. This is particularly relevant in light of the global challenges posed by climate change and the need for energy security.

- **Solar and Wind**: Harnessing energy from the sun and wind represents one of the most promising pathways for reducing carbon emissions. With advances in technology, the costs of solar panels and wind turbines have decreased significantly, making them competitive alternatives to fossil fuels.
- **Hydropower and Geothermal**: Hydropower and geothermal energy are regarded as two of the most durable and dependable sources of renewable energy. Both of these methods harness the inherent mechanisms of the Earth, namely the movement of water and the thermal energy from its interior, to produce electrical power. The distinctive attributes of these entities present notable benefits, along with accompanying difficulties that necessitate resolution. Though already in use, these sectors have growth potential. They provide consistent, base-load power without emitting GHGs.
- **Bioenergy**: Bioenergy is a term used to describe the energy obtained from organic matter, commonly known as biomass. The aforementioned energy has the potential to be harnessed for power generation, heat production, or conversion into biofuels to facilitate transportation. The perception of plants and trees as renewable resources stems from their ability to regenerate and sequester carbon dioxide from the environment during their growth cycle. Although bioenergy shows potential as a renewable energy option, it is crucial to implement sustainable management practices to guarantee its environmental advantages. This involves obtaining energy from organic materials. When managed sustainably, bioenergy can be a carbon-neutral energy source.
- **Storage Solutions**: Transitioning to renewable energy necessitates advancements in energy storage solutions, such as batteries, to counteract the intermittent nature of renewables like solar and wind.

- The growing integration of renewable energy sources like solar and wind into the energy grid presents a challenge: how to ensure a consistent energy supply when the sun is not shining or the wind is not blowing. Energy storage solutions are critical for buffering these intermittent energy sources and ensuring a stable energy supply. These solutions vary in terms of scale, technology, cost, and application.

Battery Energy Storage Systems (BESS)

- **Lithium-ion Batteries**: Currently the dominant technology for both small-scale (like in electric vehicles) and large-scale applications due to their high energy density, decreasing costs, and efficiency.

- **Lead-Acid Batteries**: Traditional battery technology has been used for decades, especially for off-grid renewable systems.
- **Solid-State Batteries**: A promising emerging technology, they offer potential advantages in terms of safety and energy density over lithium-ion batteries.
- **Flow Batteries**: Instead of storing energy in the battery cell itself, they store it in external tanks. This design means they can be scaled easily by increasing tank size.

Thermal Energy Storage

- **Molten Salt Storage**: Used in concentrated solar power (CSP) plants, they store excess heat in molten salt, which can later be used to produce steam and generate electricity.
- **Ice/Water Storage**: For cooling applications, excess energy is used to freeze water. When cooling is needed, the ice is melted, providing a cooling effect.
- **Phase Change Materials (PCMs)**: PCMs absorb or release heat when they change phase (from solid to liquid or vice versa). They can be integrated into building materials to provide passive heating or cooling.

Mechanical Energy Storage

- **Pumped Hydro Storage**: When electricity demand is low, excess energy is used to pump water uphill to a reservoir. When demand is high, the water is released downhill, passing through turbines to generate electricity.
- **Compressed Air Energy Storage (CAES)**: Excess energy compresses air into underground caverns. When needed, the air is released, heated, and expanded in a turbine to produce electricity.
- **Flywheels**: These store energy in a rotating mass. When energy is needed, the flywheel's rotation is converted back into electricity.

Hydrogen and Power-to-Gas

- **Electrolysis**: Excess electricity is used to split water into hydrogen and oxygen. The hydrogen can be stored and later used in fuel cells to generate electricity.
- **Power-to-Gas**: The produced hydrogen can be combined with carbon dioxide to produce synthetic methane, which can be stored and used in existing natural gas infrastructure.

Supercapacitors: Supercapacitors store energy in an electric field (as opposed to a chemical reaction in batteries). They can charge and discharge much faster than batteries but typically store less energy.

As the transition to renewable energy accelerates, the role of energy storage solutions becomes increasingly critical. Innovations and advancements in storage technology will be paramount not only to buffer the intermittency of renewables but also to improve the resilience, efficiency, and sustainability of the global energy system.

The adoption of renewable energy sources encompasses more than simply technological and economic changes; it signifies a societal shift towards a future characterised by sustainability and fairness. Despite ongoing hurdles, the collaborative endeavours of researchers, governments, corporations, and the global populace facilitate the progression towards a more environmentally sustainable society.

9.4 Energy Efficiency Improvements

- **Building Retrofitting**: Modernising old buildings to meet current energy efficiency standards can result in significant energy savings. This includes better insulation, efficient heating and cooling systems, and energy-saving appliances.
- **Transport**: Developing and promoting the use of electric vehicles (EVs), enhancing public transportation, and promoting non-motorised transport can reduce GHG emissions from the transport sector.
- **Industry**: Modern, efficient machinery and processes can significantly reduce energy consumption in industries.

9.5 Carbon Capture and Storage (CCS)

CCS technology captures CO2 emissions at their source (like power plants) and stores them underground or underwater, preventing them from reaching the atmosphere. While promising, it is still an emerging field that requires further development to be widely adopted.

Reforestation and Afforestation
Trees function as "carbon sinks", absorbing more carbon dioxide than they emit.

- **Reforestation**: This involves planting trees in areas where they were previously cut down.
- **Afforestation**: This is the process of planting trees in areas where there were none before.

Both practices not only sequester carbon but also restore habitats and enhance biodiversity.

Sustainable Agriculture

- **Agroforestry**: Integrating trees into farming systems can improve soil health, reduce erosion, and sequester carbon.
- **No-till farming**: This method minimises soil disruption, helping retain carbon in the soil.
- **Crop rotation and cover crops**: These practices improve soil health, leading to better carbon storage and fewer GHG emissions.

Waste Management

- **Methane Recovery**: Landfills are significant sources of methane, a potent GHG. By capturing and utilising this methane, its emission is prevented.
- **Recycling and Composting**: These reduce the need for landfills, cutting down on methane emissions and reducing the energy required to produce new materials.

Policy and Economic Measures

- **Carbon Pricing**: This includes tools like carbon taxes or cap-and-trade systems that make emitting carbon dioxide more expensive, incentivising industries to reduce their emissions.
- **Regulations**: Governments can enforce laws and standards that mandate certain practices or set emission limits.
- **International Cooperation**: Climate change is a global issue, and international cooperation, as seen in agreements like the Paris Accord, can synchronise efforts and resources.

9.6 Public Awareness and Education

Educated communities are more likely to support and engage in mitigation measures. Society can be galvanised into action by emphasising the importance and urgency of climate change mitigation in educational curriculums and public awareness campaigns.

Mitigating the impacts of climate change requires a multi-pronged, comprehensive approach, blending technological, policy-based, and societal strategies. By acknowledging the urgency and adopting an initiative-taking stance, it is possible to not only combat the adverse effects of climate change but also chart a course toward a more sustainable and resilient future.

Carbon Footprint EIA evaluates the greenhouse gas emissions of proposed projects, ensuring that they align with national and international climate goals, such as those set by the Paris Agreement.

- **Adaptation Measures**: Climate change brings unpredictability, with more frequent extreme weather events and altered local climates. EIA can help assess how a project might fare in these varied conditions and recommend adaptation measures.

Holistic Evaluation of Renewable Energy Projects

- **Not All Renewables Are Equal**: While renewable energy sources are better for the climate than fossil fuels, they can have other environmental impacts. For example, large solar farms can disrupt local ecosystems, and hydropower projects can alter river dynamics.
- **In-depth Analysis**: EIA ensures that the full spectrum of potential impacts from renewable projects is considered, from habitat disruption to resource use.

Ensuring Long-term Sustainability

- **Beyond Immediate Impacts**: EIA does not just look at the immediate aftermath of project implementation. It assesses long-term impacts, ensuring that projects remain sustainable in changing environmental conditions.

- **Resource Depletion**: With increasing energy demand, there is a risk of over-exploiting resources. EIA can highlight unsustainable extraction rates and recommend conservation measures.

Public Involvement and Social Considerations

- **Stakeholder Engagement**: Climate change affects everyone, but not always equitably. EIA processes often involve public consultations, ensuring that vulnerable communities have a voice in projects that might exacerbate climate threats they face.
- **Social Justice**: EIA can help identify projects that might disproportionately impact disadvantaged communities, ensuring that the energy transition does not perpetuate environmental injustices.

Economic Considerations

- **Avoiding Stranded Assets**: As we transition to a low-carbon economy, assets rooted in the old energy paradigm might become obsolete. EIA can help prevent investments in projects that become economically unviable as climate policies tighten.
- **Highlighting Co-benefits**: EIA can reveal additional benefits of sustainable projects, such as job creation in renewable sectors or health benefits from reduced air pollution.

EIA, in the context of growing climate threats and soaring energy demand, is a tool of foresight. It ensures that the urgent need for energy does not push us to make short-sighted decisions that amplify climate risks. Instead, it steers us towards solutions that serve both present and future generations.

9.7 Decision-Making Tool

Environmental Impact Assessment (EIA) is a systematic process that identifies, evaluates, and interprets the potential adverse and beneficial environmental impacts of proposed projects, especially in the energy sector. It is a crucial tool to assist decision-makers in ensuring the sustainability and viability of these projects. Here are how the EIA functions in three primary areas: predicting impacts, suggesting alternatives, and planning for mitigation measures.

A. **Predicting Impacts**

- **Baseline Data Collection**: Before predicting impacts, the EIA establishes a baseline. This means understanding the current state of the environment—flora, fauna, water sources, air quality, and socio-economic conditions—where the project is proposed.
- **Modelling and Forecasting**: Advanced models are used to forecast changes based on the project's details. This could mean predicting emissions, water pollution, habitat disruption, and noise levels for an energy project.

- **Assessing Cumulative Impacts**: It is not just the impact of one project that is considered. EIA examines the cumulative effect of multiple regional projects or actions over time.
- **Socio-Economic Impact Prediction**: Beyond environmental concerns, the EIA predicts how a project might affect local communities. Will there be displacement? Could local economies be disrupted or boosted?

B. **Suggesting Alternatives**

- **Alternative Locations**: If a proposed site is deemed too sensitive or critical, the EIA can suggest alternative sites with less environmental harm.
- **Technology Alternatives**: The EIA can recommend different technologies relevant to energy projects. For instance, if a coal-fired power plant is proposed, could a solar or wind farm meet the same energy needs with fewer emissions?
- **Scale and Design Modifications**: Sometimes, a smaller project or a design tweak can significantly reduce environmental impacts. EIA can suggest such modifications.
- **No-Action Alternative**: An essential part of any EIA is considering what would happen if the project were pursued. This provides a comparative baseline and can sometimes be the most environmentally favorable option.

C. **Planning for Mitigation Measures**

- **Avoidance**: The first step in mitigation is seeing if negative impacts can be entirely avoided. Could the project be redesigned so that specific harm does not occur?
- **Minimisation**: If impacts cannot be avoided, how can they be minimised? This could involve technology changes, pollution controls, or timing adjustments (e.g., not operating machinery during certain animal mating seasons).
- **Rectification**: Sometimes, impacts can be rectified by restoring the environment post-project. An example is replanting trees after construction.
- **Compensation**: If impacts are inevitable and cannot be rectified, compensation comes into play. This could involve creating new habitats, supporting community projects, or financial compensation.
- **Monitoring and Adaptive Management**: Post-implementation, EIAs often recommend continuous monitoring to ensure that predicted impacts align with actual outcomes. If discrepancies arise, adaptive management strategies can be employed to rectify the situation.

EIA provides a roadmap for decision-makers. It does not just lay out the potential harms of an energy project but actively seeks solutions to ensure that our energy needs do not come at an undue environmental cost. By predicting impacts, suggesting alternatives, and meticulously planning for mitigation, EIA ensures that energy projects align with broader goals of environmental conservation and societal well-being.

9.8 Environmental Impacts of Traditional and Renewable Energy Sources

Traditional energy sources, such as fossil fuels (coal, oil, and natural gas), significantly impact the environment. They emit greenhouse gases, which contribute to climate change, and they also produce air pollution, which can cause respiratory and other health problems.

Renewable energy sources, such as solar, wind, hydro, and geothermal, have a much smaller environmental impact. They do not emit greenhouse gases or air pollutants, and they are a more sustainable option for generating electricity.

Fossil fuels have dominated the global energy landscape for centuries. As the threats of climate change and environmental degradation intensify, there has been a push towards renewable energy sources. Traditional and renewable energy sources have distinct environmental impacts, varying in magnitude and nature. Renewable energy sources also have some environmental effects. For example, solar panels can use hazardous materials in manufacturing, and wind turbines can kill birds. It is crucial to weigh the environmental impacts of both traditional and renewable energy sources when making decisions about energy production. Here is a more detailed look at the environmental effects of some conventional and renewable energy sources:

Traditional Energy Sources (Fossil Fuels)

1. **Air Pollution**: The combustion of fossil fuels is the primary source of several air pollutants.

- **Carbon dioxide (CO2)**: Representing the majority of greenhouse gas emissions, CO2 from burning fossil fuels significantly accelerates global warming. The Intergovernmental Panel on Climate Change (IPCC) states that CO2 is responsible for about three-quarters of the planet's anthropogenic greenhouse effect, contributing to the melting of polar ice, rising sea levels, and extreme weather events.
- **Other Pollutants**: Other than CO2, fossil fuel combustion releases sulphur dioxide (SO2) and nitrogen oxides (NOx). These pollutants contribute to smog formation, leading to respiratory illnesses and reducing visibility. Acid rain from SO2 and NOx mixing with atmospheric moisture deteriorates freshwater systems, harms vegetation, and corrodes infrastructure.

2. Water Pollution

- **Oil spills**: Events like the Deepwater Horizon disaster highlight the catastrophic environmental impact of oil spills. Oil spills decimate marine ecosystems, affecting everything from microorganisms to larger marine mammals. The repercussions linger for decades as oil compounds break down slowly.
- **Coal mining runoff**: Coal extraction, particularly mountain-top removal, poses significant water pollution risks. Runoff contaminated with heavy metals and toxins harms aquatic life and pollutes human water sources.

3. Land Degradation

- **Mining and Drilling**: Beyond immediate ecological disruption, coal mines and oil drilling sites are scars on landscapes that can take centuries to rehabilitate. They lead to habitat fragmentation, endangering flora and fauna.
- **Infrastructure**: Infrastructure for fossil fuels, including pipelines and refineries, further contributes to land degradation, often requiring deforestation and leading to potential leaks and spills.

4. Water Usage

- Traditional power plants, whether coal, gas, or nuclear, rely heavily on water for cooling. Excessive water withdrawal from rivers and lakes can reduce water availability for other uses and harm aquatic life by causing temperature fluctuations.

5. Waste

- The coal combustion waste by-product, coal ash, contains toxic heavy metals like arsenic, lead, and mercury. When improperly stored, these can leach into groundwater or spill into waterways.

9.9 Renewable Energy Sources

1. Hydropower

- **Ecosystem disruption**: Dams alter river ecosystems. They impede sediment flow, which can lead to downstream erosion and disrupt aquatic life cycles. Fish migration, particularly of species like salmon, is often adversely affected by dams.
- **Methane emissions**: Contrary to widespread belief, dams can be sources of greenhouse gas emissions. Rotting vegetation in reservoirs releases methane—a greenhouse gas far more potent than CO2 over short timescales.
- **Displacement**: Large dam projects often necessitate relocation of local communities, leading to socio-economic and cultural impacts.

2. Wind Power

- **Bird and bat mortality**: One of the most cited environmental impacts of wind farms is their effect on flying fauna. Collisions with turbine blades lead to bird and bat fatalities. However, with careful planning and placement, these impacts can be minimised.
- **Noise and aesthetics**: While subjective, the noise produced by wind turbines and their visual presence in landscapes can be points of contention.

3. Solar Power

- **Land use**: Large-scale solar installations can disrupt habitat, especially in desert ecosystems. It can also increase local temperatures, causing a heat island effect.
- **Water use**: Concentrated solar power systems, which use mirrors or lenses to focus sunlight, often require water for cooling, though less than fossil fuel plants.
- **Chemical usage**: Solar panel manufacturing uses hazardous chemicals. However, with proper handling and disposal, risks associated with these chemicals can be managed.

4. Geothermal

- **Subsurface impacts**: The drilling involved in geothermal energy extraction can lead to land subsidence in localised areas. Additionally, it might trigger minor seismic activities.
- **Emissions**: While their emissions are much lower than fossil fuels, geothermal plants release greenhouse gases like CO2 and, in some cases, low levels of toxic gases like hydrogen sulphide.

5. Biomass

- **Air pollution**: Although often touted as carbon-neutral, burning biomass does release CO2. The argument is that the plants, while growing, had absorbed an equivalent amount of CO2, hence negating the emissions. But biomass combustion also produces pollutants like particulate matter.
- **Land use**: Biomass cultivation can compete with land for food production. Also, monoculture plantations can reduce biodiversity.

The pressing urgency of environmental conservation and climate change mitigation demands a shift from traditional to renewable energy sources. However, it is essential to acknowledge that no energy source is without environmental impact. The goal is to understand these impacts thoroughly, mitigate them effectively, and make informed choices that balance energy needs, environmental health, and societal well-being. As technology progresses and we refine our methods, we move closer to an era of sustainable energy, minimising our ecological footprint.

9.10 Steps in Conducting EIA for Energy Projects

- **Screening**: Determining if the energy project requires an EIA.
- **Scoping**: Identifying the critical environmental issues to be addressed.
- **Impact Prediction**: Estimating the magnitude and significance of potential environmental changes.
- **Mitigation**: Proposing measures to prevent, reduce, or compensate for adverse impacts.

- **Reporting**: Preparing a document that presents the findings for decision-making.
- **Review and Monitoring**: Verifying the accuracy of the EIA predictions and effectiveness of mitigation measures.

9.11 Case Studies

- **Hydropower in the Amazon**: The impacts and mitigation strategies.
- **Wind Farms in Europe**: Addressing bird migration and aesthetic concerns.
- **Solar Farms in Desert Regions**: Ecological impacts and their resolutions.

9.12 Challenges and Future Directions

- **Dynamic Environmental Systems**: The difficulty of predicting impacts in ever-changing environments.
- **Interdisciplinary Approaches**: The need for experts from various fields to collaboratively assess impacts.
- **Public Participation**: Ensuring the public's concerns are heard and addressed in EIAs.

9.13 Sustainable Energy Transition: Beyond EIA

- **Life Cycle Assessment (LCA)**: A broader tool that considers impacts over the entire lifecycle of a product or service.
- **Systemic Approaches**: Looking at energy transitions in the context of broader socio-economic systems.
- **International Cooperation**: Collaborative efforts to share knowledge, technology, and resources for a global sustainable energy transition.

9.14 Conclusion

The need for a sustainable energy transition worldwide is of utmost importance, particularly in response to the increasing concerns about climate change and the growing strain on limited fossil fuel reserves. As demonstrated in the preceding sections of this scholarly article, Environmental Impact Assessments (EIAs) are of utmost importance in facilitating a successful and conscientious implementation of this shift.

Environmental Impact Assessments (EIAs) possess both positive and negative aspects. One perspective involves a critical evaluation of the potential negative consequences associated with energy projects, enabling decision-makers to make well-informed choices that extend beyond purely economic factors. On the contrary, they create opportunities for innovation by identifying regions that necessitate mitigation measures, thereby stimulating breakthroughs in energy deployment from both technological and strategic perspectives. These analyses prevent the unintentional emergence of additional environmental issues resulting from the implementation of renewable energy technologies such as hydropower, geothermal, bioenergy, and others. Moreover, with the increasing prominence of the shift towards renewable energy, the significance of energy storage systems is heightened. Energy storage plays a crucial role in maintaining a consistent, dependable, and uninterrupted energy provision, extending beyond its function as a mere safeguard against the intermittent nature of renewable energy sources. Nevertheless, these proposed remedies also have their own set of environmental consequences, hence emphasising the significance of conducting thorough Environmental Impact Assessments (EIAs).

The sustainable energy transition can be characterised as a comprehensive and intricate undertaking. Although there are numerous obstacles to overcome in the future, the utilisation of Environmental Impact Assessments (EIAs) offers a strategic plan that effectively manages the complex interplay between development and conservation. The indispensability of Environmental Impact Assessments (EIAs) lies in their ability to bring rigour and impartiality to the pursuit of a future characterised by the harmonious coexistence of energy sufficiency and environmental stewardship.

In light of the imminent global energy revolution, it is imperative to proceed with prudence, comprehensive understanding, and a deep reverence for the Earth and its ecological processes. When Environmental Impact Assessments (EIAs) are conducted with meticulousness and ethical standards, they guarantee that the pursuit of sustainability is not a blind leap but rather an illuminated path towards achieving a balanced and mutually beneficial relationship with our natural surroundings.

9.15 Implications

The significance of Environmental Impact Assessment (EIA) and Sustainable Energy Transition is emphasised in the chapter, highlighting the crucial role that EIAs play in facilitating a responsible and efficient worldwide transition to renewable energy sources. The consequences of this matter are extensive, as it underscores the urgent requirement for robust legal frameworks and global standards to guarantee the effectiveness and flexibility of Environmental Impact Assessments (EIAs) in various energy projects. In addition, the chapter argues that Environmental Impact Assessments (EIAs) have the potential to stimulate technological progress and foster innovation in sustainable energy solutions by identifying and addressing environmental obstacles. Therefore, in light of the complex challenge of reconciling energy needs with environmental preservation, the chapter underscores the significance of Environmental Impact Assessments (EIAs) as more than mere instruments but as indispensable guides for achieving a comprehensive and sustainable energy transition.

References

Nolte, K. (2014). Large-scale agricultural investments under poor land governance in Zambia. *Land Use Policy, 38*, 698–706. https://doi.org/10.1016/j.landusepol.2014.01.014

Sokka, L., Sinkko, T., Holma, A., Manninen, K., Pasanen, K., Rantala, M., & Leskinen, P. (2016). Environmental impacts of the national renewable energy targets – A case study from Finland. *Renewable and Sustainable Energy Reviews, 59*, 1599–1610. https://doi.org/10.1016/j.rser.2015.12.005

Chapter 10
Social and Governance Dimensions of Sustainable Energy Transition and the Circular Economy

Abstract This chapter examines the important roles played by social and governance factors in accomplishing a sustainable energy transition and implementing a circular economy. The essay begins by offering precise definitions of a sustainable energy transition and circular economy, highlighting their critical significance in achieving both environmental sustainability and economic progress. This discourse examines sociological aspects, including the impact of sustainable energy initiatives on employment creation, quality of life, and energy accessibility, particularly in impoverished communities. This chapter explores the difficulties related to governance, specifically in assessing the effectiveness of government policies and international cooperation in promoting renewable energy and energy efficiency. A comprehensive examination is undertaken to examine the principles of the circular economy and their significance in advancing sustainable resource management. The primary objective is to examine the correlation between circular economy practices and sustainable energy. In addition, the chapter assesses the societal ramifications and organizational structures necessary to maintain a circular economy. The book employs case studies and best practices to exemplify practical applications, analyzing the fundamental aspects that contribute to their efficacy. The chapter concludes by discussing the challenges and opportunities, emphasizing the importance of integrating social and governance considerations into sustainable energy and circular economy strategies to achieve enduring sustainability.

Keywords Sustainable energy transition · Circular economy · Environmental sustainability · Governance challenges · Social implications · Renewable energy · Energy efficiency · Policy harmonization · Resource management · Case studies

10.1 Overview

The twenty-first century has distinctive environmental and economic difficulties that necessitate inventive and sustainable solutions. The concepts of a sustainable energy transition and a circular economy are the main focuses of these solutions. A

V. Kandpal et al., *Sustainable Energy Transition*, Circular Economy and Sustainability, https://doi.org/10.1007/978-3-031-52943-6_10

sustainable energy transition entails the deliberate shift from conventional, finite energy sources such as fossil fuels to the utilization of renewable and ecologically sustainable sources such as solar, wind, and hydropower. This transition is essential for tackling global concerns such as climate change, air pollution, and energy security. The circular economy is an economic system that focuses on minimizing waste and maximizing resource efficiency. As firms strive to excel in a circular economy, it is beneficial to explicitly recognize and chart the interconnections between the circular economy, the supply chain, Industry 4.0, and technology transfer. This approach defies the conventional linear economy, which adheres to a production model of "extract, manufacture, discard" (Ren et al., 2023).

The demand for sustainable waste management techniques has increased among urban planners, stakeholders, industries, and business model developers, reflecting a growing environmental consciousness in communities and a recognition of human responsibility. However, the main obstacle to implementing sustainable waste management plans that promote responsible production and consumption, as well as the integration of the circular economy in urban, industrial, or environmentally compromised areas, is the lack of knowledge and awareness about the benefits of sustainability. Additionally, there is limited public participation and responsiveness to individual social, economic, and environmental responsibilities (Voukkali et al., 2023).

The concept of the circular economy aims to transform conventional production and economic growth models, which are traditionally linear systems involving the collection of raw materials, their conversion into consumer products, and their subsequent disposal as waste. Instead, circular dynamics are employed to establish a connection between resource utilization and waste generation, with the ultimate goal of reducing waste (Wang et al., 2020).

10.1.1 Importance in the Contemporary World

The significance of these concepts in contemporary culture cannot be exaggerated. The need to switch to sustainable energy sources is becoming more urgent due to the ongoing threat that climate change poses. The circular economy advocates for a more effective and accountable utilization of resources, facilitating this transition. These strategies are indispensable for attaining the United Nations Sustainable Development Goals (SDGs), specifically those pertaining to climate action, renewable energy, sustainable urban areas, responsible consumption, and economic advancement.

Sustainability is an essential characteristic of modern enterprises in the twenty-first century, and a management strategy may be used as a powerful tool to achieve it. Fourth-generation management emerged as a systems-oriented method for overseeing modern organizations prior to the emergence of Industry 4.0 and during a period of limited understanding of the circular economy (Hallioui et al., 2022).

10.1.2 The Interconnectedness Between Sustainable Energy and the Circular Economy Is a Topic of Discussion in This Context

The interdependence between sustainable energy and the circular economy is a crucial subject, as these notions are intricately linked and mutually supportive. Comprehending and utilizing this connection is crucial for attaining more sustainable and resilient economic and environmental results.

The correlation between the transition to sustainable energy and the circular economy is significant. Renewable energy technologies play a crucial role in fueling a circular economy, decreasing dependence on nonrenewable resources, and mitigating greenhouse gas emissions. On the other hand, applying circular economy principles in the energy sector enables the efficient utilization of energy and resources. Collaboration among several elements is essential for building a durable and robust economic framework (Dahchour et al., 2021).

10.1.3 Global Trends and Developments

There has been a significant worldwide trend in recent years toward the implementation of sustainable energy and circular economy methods. Currently, several countries are making substantial financial investments in renewable energy projects and enacting regulations to promote the efficient exploitation of resources. Furthermore, a growing number of enterprises are implementing circular economy ideas in their operations, resulting in a substantial influence on the involvement of the private sector. Environmental concerns and economic potential are significant driving factors behind the transition to sustainable practices, as they often lead to cost savings and innovative business methods.

10.1.4 Policy Frameworks and International Agreements

A plethora of international agreements and regulatory frameworks support the transition to sustainable energy and a circular economy. The Paris Agreement sets a worldwide framework with the objective of averting disastrous climate change by limiting the increase in global temperature to a threshold significantly lower than 2 degrees Celsius. Additionally, the European Union's Circular Economy Action Plan and various national laws from around the world make significant contributions to promoting the adoption and implementation of circular economy ideas.

10.1.5 Challenges and Barriers

Notwithstanding the advancements achieved, notable obstacles and impediments persist. The challenges encompass technological limitations, financial limitations, limited public awareness, and opposition from conventional industries. To surmount these obstacles, it is crucial for governmental institutions, corporations, and non-governmental organizations to cooperate and collaborate.

10.1.6 Roadmap for the Chapter

The objective of this chapter is to analyze these subjects comprehensively. The subsequent sections will examine the societal dimensions of a sustainable energy transition, investigating the impacts of this shift on several socioeconomic cohorts and the significance of public engagement. The analysis will also investigate the challenges associated with managing these changes, with a specific focus on the employment of policy tools, regulatory frameworks, and international cooperation. The subsequent chapter will address the circular economy by analyzing its core principles, social and governance aspects, and interrelations with sustainable energy. Concrete examples and actual case studies will be showcased to exemplify the practical implementation of these concepts in real-world scenarios. Finally, the chapter concludes by analyzing the perspective and strategies necessary to integrate social and governance elements into sustainable energy and circular economy policies, thus guaranteeing enduring sustainability.

The sustainable energy transition involves a substantial shift away from our reliance on fossil fuels, such as coal, oil, and natural gas, toward more sustainable and renewable energy sources, such as solar, wind, hydroelectric, and geothermal power. This change is essential for several reasons:

1. **Environmental Sustainability**: Renewable energy sources exhibit negligible to no emissions of greenhouse gases and pollutants, in contrast to fossil fuels**.** It is essential to adopt these sources to effectively address climate change, reduce air pollution, and preserve biodiversity.
2. **Economic development:** The utilization of sustainable energy sources has the capacity to stimulate economic growth by facilitating the emergence of innovative enterprises and generating job opportunities**.** Renewable energy technologies often necessitate more labor-intensive processes than conventional energy sources, hence creating potential for employment creation. Moreover, dedicating resources to the development of renewable energy infrastructure could lead to increased economic productivity and technological progress.
3. Energy security can be enhanced by relying on renewable energy sources. Unlike fossil fuels, renewable energy sources are not geographically limited and are less susceptible to geopolitical conflicts. Furthermore, they can be used locally.

4. **Long-term cost-effectiveness:** Despite the initial financial investment, adopting renewable energy technology provides substantial benefits in the future, including decreased operational costs and protection against volatile changes in fossil fuel prices.

The circular economy refers to an economic system that aims to minimize waste and maximize the use of resources by promoting the reuse, recycling, and regeneration of products and materials. The circular economy is an economic framework that prioritizes the elimination of waste and pollution, the continuous utilization of products and materials, and the restoration of natural systems (Tavanti, 2023). It differs from the conventional linear economy, which adheres to a 'take, make, dispose' framework. The circular economy holds significant importance for multiple reasons:

1. Resource efficiency is a key aspect of a circular economy where the emphasis is on maximizing the utilization of resources through recycling, reusing, and remanufacturing. This method aids in the preservation of natural resources and the mitigation of environmental deterioration.
2. **Waste Reduction:** The circular economy strives to close the loop by creating goods and processes that minimize waste and transform it into a valuable resource. It is imperative to minimize the environmental consequences of trash, such as the utilization of landfills and contamination of the environment.
3. **Economic advantages:** The circular economy has the potential to foster ingenuity and generate new prospects for businesses. Industries that prioritize recycling, refurbishing, and remanufacturing have the potential to expand into untapped markets and generate employment opportunities.
4. The circular economy plays a crucial role in environmental protection by minimizing waste and promoting the sustainable utilization of resources. It aids in the mitigation of greenhouse gas emissions and pollutants, thus supporting efforts to combat climate change and safeguard ecosystems.

A sustainable energy transition and a circular economy are both essential components for achieving environmental sustainability and promoting economic growth. These methods provide avenues toward a future that is more environmentally friendly, adaptable, and economically successful by tackling urgent and significant issues of our era.

10.1.7 The Interrelatedness of These Concepts and Their Significance in Attaining the United Nations Sustainable Development Goals (SDGs)

The concepts of sustainable energy transition and the circular economy are intricately connected and play a crucial role in achieving the United Nations Sustainable Development Goals (SDGs). The SDGs function as a worldwide call to eliminate poverty, protect the environment, and ensure universal peace and prosperity by

2030. The interrelationship between the shift toward sustainable energy and the circular economy and its importance in regard to the Sustainable Development Goals (SDGs) may be highlighted through many crucial factors:

1. **SDG 7 (cheap and clean energy):** The transition to sustainable energy is closely linked to Goal 7, which aims to ensure that everyone has access to cheap, reliable, sustainable, and modern energy. By adopting renewable energy sources and improving energy efficiency, we can effectively address the need for sustainable and eco-friendly energy solutions.
2. **SDG 11 (Sustainable Cities and Communities):** The principles of the circular economy improve the inclusiveness, safety, resilience, and sustainability of cities and human settlements, as stated in Goal 11 (Nwokolo et al., 2023). The implementation of circular economy techniques promotes the development of environmentally friendly urban environments through the efficient use of resources and reduction of waste.
3. **SDG 12 (Responsible Consumption and Production):** This study emphasizes the importance of utilizing sustainable energy sources and adopting circular economy principles to achieve Goal 12. This purpose focuses on ensuring the use of resources in an efficient manner, promoting the use of renewable energy, and reducing waste. These are all key components of the circular economy, which aims to establish sustainable consumption and production patterns (Nwokolo et al., 2023).
4. **SDG 13 (Climate Action):** The utilization of renewable energy sources is essential for tackling climate change (Goal 13) since it efficiently reduces greenhouse gas emissions. Moreover, the adoption of circular economy strategies plays a crucial role in tackling climate change by diminishing the production of waste and promoting the sustainable utilization of resources.
5. **SDG 8 (Decent Work and Economic Growth):** Shifting toward sustainable energy and circular economy models can promote economic growth and create new job opportunities in emerging industries such as renewable energy, recycling, and remanufacturing. This will greatly contribute to achieving Goal 8.
6. The ideas of a sustainable energy transition and a circular economy, which promote innovation and the development of infrastructure, are the driving forces behind SDG 9 (industry, innovation, and infrastructure). Investing in renewable energy technology and circular economy processes can promote industrial innovation and develop resilient infrastructure.
7. **Cross-cutting Impacts:** In addition to the stated objectives, the transition to sustainable energy and the adoption of circular economy principles have significant effects on several other Sustainable Development Goals (SDGs). These include improvements in health through pollution reduction, advancements in education by promoting sustainability knowledge, and the promotion of gender equality by empowering women through access to energy resources and sustainable economic activities.

The principles and practices of transitioning to sustainable energy and adopting a circular economy are intimately interconnected and mutually essential for achieving several Sustainable Development Goals. Their adoption can facilitate the creation of a more sustainable, egalitarian, and prosperous world for both current and future generations.

10.2 The Societal Implications of Shifting Towards Sustainable Energy

The transition to sustainable energy has significant ramifications for society, impacting multiple facets of everyday living, economic frameworks, and environmental well-being.

10.2.1 Impact on Various Social Groups

A sustainable energy transition has a substantial impact on several social groups. The potential of this initiative lies in its ability to generate employment opportunities and foster economic growth, thereby benefiting communities and improving overall quality of life. Nevertheless, this shift also presents difficulties, such as the possibility of employment reductions in conventional energy industries. To effectively address these repercussions, it is necessary to adopt a balanced approach that considers the requirements and capacities of all social groups, including those that are disadvantaged and excluded.

10.2.2 Ensuring Fair and Sustainable Access to Energy

Ensuring equitable and unbiased access to energy is a crucial element of the social aspect of sustainable energy. Energy poverty, which refers to the lack of sufficient access to modern energy services, is a significant issue in several locations worldwide. Utilizing sustainable energy sources helps alleviate this problem by providing economical, dependable, and eco-friendly energy, especially in underserved and rural areas. Access to these resources is vital for improving living conditions, enabling educational opportunities, and strengthening economic pursuits.

10.2.3 Role of Public Awareness and Education

Public awareness and education play vital roles in supporting the shift to sustainable energy. Spreading information to the public about the benefits of renewable energy and energy efficiency can create a positive atmosphere for the adoption of sustainable energy practices. Disseminating information through awareness campaigns, community programs, and educational activities can effectively elucidate the intricacies of renewable energy technology and foster its broad acceptance among local communities.

10.2.4 Job Creation and Skill Development

The transition to sustainable energy is a significant driver for the creation of job prospects in expanding sectors such as solar and wind energy, energy efficiency, and sustainable transportation. These industries require a wide range of skills, including expertise in renewable energy technology, project management, and policy development. It is essential to allocate resources to education and training programs to equip the workforce with the necessary skills for these growing, environmentally friendly career prospects.

10.2.5 Social Inclusion and Participation

Emphasizing social engagement and active participation is of utmost importance when shifting to renewable energy. This involves actively engaging communities in decision-making processes related to energy projects, taking into account their needs and preferences, and ensuring a fair distribution of the benefits of sustainable energy. Local community involvement can help to identify and address any social and environmental consequences related to renewable energy projects.

10.2.6 Tackling the Social Difficulties of Transition

While the sustainable energy transition offers several benefits, it also poses social challenges that need to be addressed. These responsibilities involve supervising the seamless shift for employees and communities dependent on the fossil fuel industry while also ensuring that the transition to renewable energy does not exacerbate existing inequalities. To effectively manage these challenges, it is imperative to establish social safety nets, retraining programs, and community involvement.

The social aspects of the sustainable energy transition involve a wide array of concerns, including fair energy availability, public consciousness, employment generation, and community harmony. To successfully transition to sustainable energy, it is crucial to thoroughly study these elements, ensuring that the energy is not only environmentally and economically viable but also socially inclusive and fair.

10.3 Issues of Governance in the Transition to Sustainable Energy

The shift towards sustainable energy entails intricate governance challenges, as it necessitates synchronized endeavors among several tiers of government, the commercial sector, and civil society.

10.3.1 Context

Governance in the context of sustainable energy encompasses the policies, regulations, and practices that guide and facilitate the transition from fossil fuels to renewable energy sources. Effective governance is crucial for ensuring that this transition is smooth and equitable and meets both environmental and developmental goals.

10.3.2 Role of Government Policies and Regulations

Governments play a pivotal role in shaping the energy landscape through policies and regulations. Instruments such as subsidies, tax incentives, and feed-in tariffs can encourage investment in renewable energy. However, creating a regulatory environment that balances the interests of different stakeholders while promoting sustainable practices is a complex challenge.

10.3.3 Sustainable Energy Governance Models

Varying governance structures, spanning from centralized to decentralized systems, influence the advancement and execution of sustainable energy solutions. Centralized governance can guarantee consistency and expansion, whereas decentralized approaches can empower local communities and stimulate inventive solutions that are customized to local needs.

10.3.4 Policy Instruments and their Efficacy

The effectiveness of policy instruments varies based on regional and national contexts. Subsidies and tax incentives might work well in some areas, whereas feed-in tariffs or renewable portfolio standards might be more effective in others. Assessing the impact of these instruments is essential for understanding their effectiveness in promoting sustainable energy.

10.3.5 International Cooperation and Policy Harmonization

International collaboration is crucial for addressing global concerns such as climate change. Agreements such as the Paris Agreement establish a structure for collaborative efforts. Nevertheless, achieving policy harmonization among nations continues to be a formidable task due to divergent economic interests, levels of development, and energy requirements.

10.3.6 Funding the Transition to Sustainable Energy

Securing financial support for sustainable energy initiatives poses a substantial obstacle to governance. Grants, loans, and green bonds are viable sources of funding, but the involvement of international financial organizations and development banks is also vital. Developing appealing investment prospects in renewable energy is crucial for facilitating the transition.

10.3.7 Regulatory Challenges and Market Dynamics

The incorporation of renewable energy into established energy markets poses regulatory obstacles. Carbon pricing and emissions trading systems can stimulate sustainable energy production, but their efficacy and fairness depend on meticulous design.

10.3.8 Stakeholder Engagement and Public–Private Partnerships

To achieve successful energy governance, it is crucial to include a wide range of stakeholders, including those in the commercial sector, civil society, and local communities. Public–private partnerships can leverage the benefits of both the public and private sectors to advance and accomplish sustainable energy goals.

10.3.9 Addressing the Energy Trilemma

The challenge of attaining equilibrium between cost-effectiveness, ecological sustainability, and the assurance of energy supply, commonly referred to as the energy trilemma, is a significant hurdle in governance. Policies must ensure that the transition to sustainable energy is economically efficient and dependable while also accomplishing environmental objectives.

10.3.10 Innovation and Technological Advancements

Moreover, governance frameworks should support and promote the advancement and use of cutting-edge energy technology. To achieve a sustainable energy transition, it is crucial to enact policies that encourage research and development, as well as the incorporation of advanced technologies such as smart grids and energy storage.

10.3.11 Transitioning from Fossil Fuels

Efficiently managing the decline of the fossil fuel industry and ensuring a just transition for affected communities and workers are vital responsibilities in governance. These policies require the adoption of measures that enable the acquisition of new skills, the expansion of economic activity, and the establishment of social welfare systems.

10.3.12 Environmental and Social Impact Governance

Incorporating environmental and social factors into energy governance is crucial. This entails conducting comprehensive evaluations of the ecological consequences and involving indigenous communities in the decision-making process.

10.3.13 Urban and Rural Governance

The challenges and strategies related to governance might significantly differ between urban and rural settings. It is crucial to adjust methods to these various scenarios to achieve efficient and comprehensive energy management.

10.3.14 Challenges in Policy Consistency and Long-Term Planning

The management of energy needs to address the issue of contradictory policies and the necessity for long-term, forward-thinking planning. Sustainable energy policies must be resilient to political volatility and align with long-lasting environmental and developmental goals.

10.3.15 Prospects for Future Energy Governance

Current developments in energy governance are characterized by an increasing focus on decentralization, digitalization, and cross-sectoral integration. Adapting governance structures to these changes will be essential for protecting energy policies in the long term.

An effective structure of governance is essential for facilitating the sustainable transition to renewable energy. It requires balancing multiple interests, adapting to new challenges, and fostering collaboration across industries and boundaries. The successful implementation of a seamless and just shift toward sustainable energy relies on the ability to effectively and fairly address these governance barriers.

10.4 The Circular Economy: A Fundamental Change in the Use of Resources

10.4.1 Overview of the Circular Economy

The circular economy is a paradigm shift from the traditional linear model of 'extract, produce, and dispose' to a system that eradicates waste, promotes the continuous use of products and materials, and revitalizes natural ecosystems. This method emphasizes the implementation of reusing, repairing, refurbishing, and recycling products and materials to construct a closed-loop system. The objective is to reduce resource use and minimize waste, pollution, and carbon emissions.

10.4.2 Fundamental Concepts of the Circular Economy

The circular economy is founded on three fundamental principles:

1. Implementing a strategy of waste and pollution reduction through the reconsideration and reconfiguration of products and processes can lead to their complete elimination rather than only addressing them as an aftereffect.

2. **Ensuring** the continuous utilization of products and resources entails creating designs that prioritize long-lastingness, reusability, remanufacturing, and recycling to maintain the circulation of materials within the economy.
3. The circular economy aims to optimize natural systems by reintroducing vital nutrients into the environment and utilizing renewable energy sources.

10.4.3 Linking the Circular Economy and Sustainable Energy

The circular economy and sustainable energy are intrinsically intertwined. Renewable energy sources are essential for powering a circular economy, reducing reliance on finite resources, and lowering greenhouse gas emissions. The implementation of circular economy practices, such as improving energy efficiency and utilizing waste-to-energy technology, plays a significant role in attaining sustainable energy goals.

Circular activity is significantly and favorably influenced by elements such as the quality of institutions, the productivity of resources, and entrepreneurship. As social protection levels rise, the proportion of the population embracing CE principles also increases, while EU countries exceed their capacity to recycle by consuming more raw materials (Hondroyiannis et al., 2023).

10.4.4 Implications for the Economy and the Environment

The implementation of a circular economy paradigm has resulted in significant economic and environmental benefits. From an economic perspective, this approach might lead to decreased costs, new opportunities for the organization, and overall economic growth. From an ecological perspective, climate change reduces the strain on natural resources, mitigates pollution and emissions of greenhouse gases, and enhances biodiversity and the well-being of ecosystems.

10.4.5 Policy and Regulatory Framework

Effective policies and regulations are essential for fostering a circular economy. These include incentives for eco-friendly design, regulations on waste management, and support for businesses specializing in material recycling and reuse. Governments play a crucial role in creating a favorable environment for the adoption of circular economy techniques.

10.4.6 *Business Models and Corporate Practices*

The circular economy offers innovative opportunities and economic strategies for organizations. These include product-as-a-service models, where companies retain ownership of products and offer them as services, as well as models focused on product durability, repairability, and recyclability.

10.4.6.1 Business Models

Renewable Energy Production and Supply
Companies investing in renewable energy sources like solar, wind, and hydro to produce and supply clean energy. This can include utility-scale power generation or smaller, distributed energy resources.

Energy-as-a-Service (EaaS)
Businesses offer energy solutions as a service rather than a product. This can include energy management, efficiency solutions, and renewable energy provisioning, reducing the upfront costs for customers.

Circular Economy Models
Implementing circular economy principles, where products are designed for durability, reuse, and recyclability, thereby minimizing waste and resource consumption.

Green Financing and Investment
Financial services companies offering green bonds, sustainability-linked loans, and other financial products that support environmentally sustainable projects.

Decentralized Energy Systems
Companies focusing on localized energy generation and storage systems, allowing communities and end-users to generate their own renewable energy.

Carbon Offset and Trading
Businesses engaging in carbon offsetting and trading, where they invest in environmental projects to compensate for their own carbon emissions.

10.4.6.2 Corporate Practices

- **Sustainability Reporting**:
- Corporations increasingly disclose their environmental impact and sustainability practices, often in line with global frameworks like the Global Reporting Initiative (GRI) or the Sustainability Accounting Standards Board (SASB).

- **Green Supply Chain Management**:
- Integrating environmental considerations into supply chain management, including sourcing from sustainable suppliers and optimizing logistics for lower emissions.
- **Energy Efficiency Improvements**:
- Investing in energy-efficient technologies and practices in corporate facilities, such as LED lighting, efficient heating and cooling systems, and energy management systems.
- **Employee Engagement and Training**:
- Encouraging sustainability practices among employees and providing training on environmental stewardship and energy conservation.
- **Corporate Social Responsibility (CSR) Initiatives**:
- Engaging in CSR projects that focus on environmental sustainability, such as tree planting, renewable energy installations in communities, or supporting environmental nonprofits.
- **Public-Private Partnerships**:
- Collaborating with governments and other organizations on sustainable energy projects and policy advocacy.

Innovation and R&D in Sustainable Technologies
Investing in research and development for new sustainable technologies, such as advanced renewable energy systems, battery storage, or carbon capture and utilization.

By adopting these models and practices, businesses not only contribute to the global effort to combat climate change but also often enhance their competitiveness, open new markets, and respond to the growing demand from consumers and investors for environmentally responsible operations. Businesses are embracing inventive models and practices that are in line with environmental objectives and also economically feasible, as they shift towards sustainable energy. These models not only contribute to the reduction of carbon footprints but also frequently result in cost savings and the creation of new market opportunities. Below are several fundamental business models and corporate practices in this domain:

10.4.7 Technological Innovation and Infrastructure

Incorporating technological advancements is crucial for facilitating a circular economy. Progress in the fields of materials science, digital technology, and product design is highly significant. Equally important is the establishment of the essential infrastructure for the collection, recycling, and reassembly of materials.

10.4.8 Challenges and Barriers

The adoption of a circular economy model faces several challenges, including technological barriers, market dynamics, consumer behavior, and the prevalence of traditional linear economic models. To overcome these challenges, coordinated effort is needed, requiring the collaboration of governments, businesses, and consumers.

The task of reconciling limited resources and the need for low-carbon practices with the need for material progress is substantial. The concept of a "circular economy," which emphasizes the reduction, reuse, recycling, and recovery of materials, offers a strategy for achieving global economic sustainability. Nevertheless, the existing recycling efforts fulfill only a mere 9% of the worldwide need for materials, underscoring a significant deficiency in scientific, policy, and technological aspects that must be resolved (Li & Xu, 2022).

As a result of increased urbanization and industrialization, significant quantities of municipal solid waste (MSW) are being collected on our planet. The development of waste has numerous adverse environmental consequences, which has led to it becoming a subject of dispute within the environmental community. Within the framework of the circular economy, global initiatives have been undertaken to implement a methodical management strategy along with an environmentally friendly treatment technology to optimize the utilization of municipal solid waste (MSW) (Ambaye et al., 2023).

10.4.9 Global and Local Initiatives

The circular economy has been demonstrated through a multitude of international and local initiatives, highlighting its significant potential. These projects encompass local initiatives focused on minimizing waste and utilizing sustainable materials in urban settings, as well as global partnerships aimed at mitigating plastic pollution in the sea. Both global and local initiatives have distinct but complementary functions in facilitating the shift towards sustainable energy. Global initiatives establish overarching objectives and promote the exchange of resources and information, but it is local initiatives that are essential for carrying out and adjusting to the unique circumstances of specific regions.

10.4.10 The Importance of Education and Public Awareness

Promoting education and cultivating public awareness are essential for creating a social environment that embraces the ideas of a circular economy. This entails incorporating these principles into the educational curriculum and organizing public awareness campaigns to promote the advantages of a circular economy.

10.4.11 Prospects for the Future

The circular economy concept is anticipated to have a crucial impact on advancing sustainable development in the future. The proper execution of this approach will rely on advancements in corporate frameworks, policy initiatives, and technological breakthroughs. The circular economy represents an innovative strategy for optimizing resource consumption that is in line with the objectives of sustainable development. By evaluating and restructuring our methods of manufacturing and utilization, we have the potential to establish a global environment that is environmentally sustainable, adaptable, and financially lucrative.

References

Ambaye, T. G., Djellabi, R., Vaccari, M., Prasad, S., Aminabhavi, T. M., & Rtimi, S. (2023). Emerging technologies and sustainable strategies for municipal solid waste valorization: Challenges of circular economy implementation. *Journal of Cleaner Production, 423*, 138708. https://doi.org/10.1016/j.jclepro.2023.138708

Dahchour, A., El Hajjaji, S., Dhiba, D., & Ghosh, S. K. (2021). Circular economy in Morocco: Status and perspectives. In S. K. Ghosh & S. K. Ghosh (Eds.), *Circular economy: Recent trends in global perspective* (pp. 29–67). Springer Nature Singapore. https://doi.org/10.1007/978-981-16-0913-8_2

Hallioui, A., Herrou, B., Santos, R. S., Katina, P. F., & Egbue, O. (2022). Systems-based approach to contemporary business management: An enabler of business sustainability in a context of industry 4.0, circular economy, competitiveness and diverse stakeholders. *Journal of Cleaner Production, 373*, 133819. https://doi.org/10.1016/j.jclepro.2022.133819

Hondroyiannis, G., Sardianou, E., Nikou, V., Evangelinos, K., & Nikolaou, I. (2023). Energy market dynamics and institutional sustainability: How affect the Europe's circular economy. *Circular Economy, 2*(3), 100048. https://doi.org/10.1016/j.cec.2023.100048

Li, J., & Xu, G. (2022). Circular economy towards zero waste and decarbonization. *Circular Economy, 1*(1), 100002. https://doi.org/10.1016/j.cec.2022.100002

Nwokolo, S. C., Singh, R., Khan, S., Kumar, A., & Luthra, S. (2023). Influencing the scale of Africa's energy transition. In S. C. Nwokolo, R. Singh, S. Khan, A. Kumar, & S. Luthra (Eds.), *Africa's path to net-zero: Exploring scenarios for a sustainable energy transition* (pp. 75–91). Springer Nature. https://doi.org/10.1007/978-3-031-44514-9_4

Ren, Y., Li, R., Wu, K.-J., & Tseng, M.-L. (2023). Discovering the systematic interlinkages among the circular economy, supply chain, industry 4.0, and technology transfer: A bibliometric analysis. *Cleaner and Responsible Consumption, 9*, 100123. https://doi.org/10.1016/j.clrc.2023.100123

Tavanti, M. (2023). Approaches to managing sustainability. In M. Tavanti (Ed.), *Developing sustainability in organizations: A values-based approach* (pp. 145–170). Springer International Publishing. https://doi.org/10.1007/978-3-031-36907-0_6

Voukkali, I., Papamichael, I., Economou, F., Loizia, P., Klontza, E., Lekkas, D. F., Naddeo, V., & Zorpas, A. A. (2023). Factors affecting social attitude and behavior for the transition towards a circular economy. *Sustainable Chemistry and Pharmacy, 36*, 101276. https://doi.org/10.1016/j.scp.2023.101276

Wang, H., Schandl, H., Wang, X., Ma, F., Yue, Q., Wang, G., Wang, Y., Wei, Y., Zhang, Z., & Zheng, R. (2020). Resources, Conservation & Recycling Measuring progress of China's circular economy. *Resources, Conservation & Recycling, 163*(11), 105070. https://doi.org/10.1016/j.resconrec.2020.105070

Chapter 11
Challenges and Opportunities for Sustainable Energy Transition and Circular Economy

Abstract Adopting sustainable energy and implementing a circular economy are crucial for tackling urgent issues such as environmental degradation, climate change, and resource shortages. This chapter presents a thorough examination of the various components associated with this transformation, highlighting both the inherent difficulties and new possibilities. We presented a comprehensive examination of the present state of global energy and resource use, highlighting the significant reliance on fossil fuels and the linear economic model characterized by the process of 'take-make-dispose'. The discussion emphasizes the role of these current practices in causing environmental problems, such as the rise in greenhouse gas emissions, pollution, and depletion of resources, thus emphasizing the immediate necessity for reform. The chapter primarily addresses the obstacles that hinder the progress of achieving a sustainable energy transition. The impediments to the extensive implementation of renewable energy encompass technological constraints, economic hurdles such as exorbitant upfront expenses and financial unpredictability, and societal and political challenges such as popular opposition and legislative stagnation. Nevertheless, the chapter also emphasizes the plethora of opportunities that stem from the transition to sustainable energy. The investigation of sustainable energy technologies, such as solar and wind power, is complemented by progress in energy storage and grid management. Furthermore, there is ongoing analysis of the possibility of creating business models, such as Energy as a Service (EaaS). The importance of the community's participation and the social consequences of the transition process are also acknowledged as a vital element. The chapter explores the societal consequences, encompassing concerns of unequal access to energy and the fair treatment of the environment. The organization supports the implementation of renewable energy projects at the community level, educational initiatives, and the integration of social innovation into commercial models. The chapter consolidates the knowledge acquired from the different issues examined, emphasizing that the pursuit of sustainable energy and a circular economy is complex yet crucial. The situation requires cooperation between governments, corporations, and communities, and highlights the importance of comprehensive, adaptable, and inclusive policies.

V. Kandpal et al., *Sustainable Energy Transition*, Circular Economy and Sustainability, https://doi.org/10.1007/978-3-031-52943-6_11

The chapter finishes by restating the crucial importance of creativity and flexibility in successfully navigating the hurdles and capitalizing on the opportunities presented throughout this transformation.

Keywords Sustainable energy transition · Circular economy · Renewable energy technologies · Environmental sustainability · Policy and regulatory framework · Social implications · Community engagement · Technological innovations · Economic challenges · Resource efficiency

11.1 Introduction

The advent of the twenty-first century has ushered in a time where the sustainability of our energy supplies and the efficiency of our economic systems have become vital global concerns. The increasing acknowledgment of climate change, environmental degradation, and the finite nature of our planet's resources has underscored the urgent need for a transition to sustainable energy and the adoption of a circular economy. The Intergovernmental Panel on Climate Change (IPCC) provides a lucid and disturbing evaluation of the dire consequences that will arise from continued reliance on fossil fuels and irresponsible use of resources. It highlights the urgent necessity for a global shift in our fundamental approach. Simultaneously, the World Energy Outlook reports consistently illustrate the growing need for energy and the environmental repercussions of our existing consumption patterns (IEA, 2020). This background serves as the basis for analyzing novel methods to attain sustainable energy and circular economy frameworks, both of which are crucial for tackling environmental crises and ensuring a livable future. The transition towards sustainable energy and the circular economy is not only a necessity for the environment, but also presents a significant socioeconomic prospect. This initiative guarantees to stimulate innovation, generate employment opportunities, and foster more streamlined and competitive economies. Nevertheless, this shift is accompanied by a multitude of difficulties, encompassing technological obstacles as well as economic and social impediments.

This chapter seeks to examine these issues while also emphasizing the significant opportunities that exist within this transformation. By exploring the complexities of these notions, we can gain a deeper comprehension of the way forward and the necessary measures to be taken by different stakeholders, such as governments, businesses, and communities. The pursuit of a sustainable future necessitates collective accountability, and by working together, we can effectively utilize the capabilities of sustainable energy and circular economy methodologies to establish a resilient, wealthy, and ecologically aware global community.

11.1.1 *Defining the Key Concept*

Sustainable Energy Transition This pertains to the transition from conventional energy systems reliant on fossil fuels to renewable and environmentally friendly energy sources. This transformation is crucial in mitigating greenhouse gas emissions, addressing climate change, and guaranteeing a secure and sustainable energy provision for the future.

Circular Economy The circular economy diverges from the conventional linear economy model of 'take-make-dispose' by prioritizing the perpetual utilization of resources. This concept advocates for waste reduction, product reuse, and material recycling, thus establishing a closed-loop system that minimizes environmental harm.

11.1.2 *Relevance in the Contemporary World*

The shift towards sustainable energy and circular economy models is widely acknowledged as both an environmental imperative and an economic and social advantage. Studies such as (Rockström et al., 2009) have provided thorough evidence that climate change presents significant risks to ecosystems and human societies, potentially jeopardizing their very existence. Simultaneously, the limited availability of resources has a significant impact on both the global economy and geopolitical stability. The adoption of sustainable energy transition and circular economy strategies offers effective means to address these threats, while also promoting economic growth and social development. They provide a detailed plan for a future in which the responsible management of the environment is in harmony with economic success and the overall welfare of society. When examining the need to transition to sustainable energy and a circular economy, it is crucial to explore how these changes not only tackle ecological concerns but also address wider societal and economic difficulties.

Environmental Imperatives

- **Combating Climate Change**: The foremost environmental priority is to alleviate climate change. The widespread utilization of fossil fuels has resulted in a substantial rise in greenhouse gasses, predominantly carbon dioxide (CO2), which is a major factor in the phenomenon of global warming. Adopting sustainable energy sources, such as solar and wind power, is essential for mitigating these emissions. The IPCC reports emphasize the imperative of this transition to limit the increase in global temperature to below 1.5 °C, a critical point beyond which the consequences of climate change become progressively more severe and irreversible.

- **Conservation of Resources**: The circular economy aspires to effectively preserve natural resources in response to the pressing need for conservation. Conventional linear consumption models result in the swift exhaustion of resources and the deterioration of the environment. A circular economy promotes the reuse, recycling, and reduced consumption of resources, leading to a more effective and sustainable utilization of resources. This approach helps to save biodiversity and natural environments.

Socio-Economic Imperatives

- **Economic Resilience and Job Creation**: Adopting sustainable energy and circular economy concepts can stimulate economic expansion and generate fresh employment prospects. As per a report published by the International Renewable Energy Agency (IRENA), the renewable energy industries are expected to generate a substantial number of employment opportunities, surpassing the number of jobs lost in conventional energy sectors. Likewise, the implementation of circular economy techniques, such as recycling and remanufacturing, creates opportunities for the development of fresh markets and business models, hence enhancing economic resilience.
- **Social Equity and Health**: This transformation also has significant social ramifications. The generation of energy from fossil fuels frequently has a greater impact on underprivileged populations, resulting in difficulties such as pollution and health issues. Renewable energy sources have the potential to mitigate these disparities by offering cleaner and more readily available energy alternatives. Furthermore, the implementation of a circular economy can enhance the standard of living by minimizing waste and pollution, resulting in the development of healthier and more sustainable societies (Hegab et al., 2023).
- **Global Cooperation and Stability**: The shift also relies on worldwide collaboration and stability. Energy plays a pivotal role in geopolitical dynamics, and transitioning to sustainable energy sources can decrease reliance on places abundant in fossil fuels, so fostering more stable international relations. The extensive implementation of circular economy principles can lead to a more equitable distribution of resources and a reduction in conflicts pertaining to natural resources.
- **Innovation and Technological Advancement**: Essentially, the transition to sustainable energy and circular economy serves as a spur for innovation. It promotes the growth of cutting-edge technology and economic models, propelling technical progress that can offer answers to several environmental and societal concerns.

The drive to transition to sustainable energy and circular economies is multifaceted, as it addresses crucial environmental concerns and offers significant socio-economic benefits. This transition is not merely a response to challenges, but also an opportunity to create a more equitable, sustainable, and prosperous future for all individuals.

11.2 The Current Landscape of Energy and Resource Use

Global Energy Consumption Patterns

- **Overview of Energy Use**: Fossil fuels, namely coal, oil, and natural gas, currently have a prominent position in the global energy sector, constituting a substantial majority of the world's energy consumption. As per the World Energy Outlook by the International Energy Agency (IEA), these resources remain the primary source of energy production, despite increasing investments in renewable sources.
- **Regional Disparities**: Energy consumption exhibits significant variation among different locations. Industrialized nations, characterized by their advanced economy, consume an imbalanced quantity of energy resources in comparison to undeveloped countries. This discrepancy not only signifies varying degrees of industrialization and economic advancement, but also highlights disparities in energy accessibility and efficiency.
- **Sectoral Energy Use**: An analysis of energy use by sector provides valuable insights. The industrial sector typically has the highest level of consumption, followed by the transportation and residential sectors. Every industry has distinct obstacles and advantages when it comes to shifting towards renewable energy sources.

The Linear Economy Model

- **Characteristics of the Linear Economy**: The linear economy, which is characterized by the 'take-make-dispose' model, is widespread in the majority of global economies. This paradigm is heavily dependent on the extraction of raw resources, the production of things, and the disposal of trash, typically without much consideration for resource conservation or recycling.
- **Environmental Impact**: The linear paradigm has resulted in substantial environmental deterioration. These factors encompass the exhaustion of resources, the destruction of habitats, and the contamination of the environment. Furthermore, it exacerbates the accumulation of refuse, as non-biodegradable items are being deposited in landfills and oceans.

Environmental Impact of Current Energy and Resource Use

- **Climate Change and Greenhouse Gas Emissions**: The utilization of fossil fuels for energy is a significant catalyst for the release of greenhouse gases, hence propelling climate change. The resultant increase in global temperatures has resulted in severe weather phenomena, elevation of sea levels, and disturbances in ecosystems.
- **Pollution and Health Risks**: In addition to greenhouse gasses, the combustion of fossil fuels emits chemicals that provide substantial health hazards. Respiratory and cardiovascular problems have been specifically associated with air pollution.

- **Resource Depletion and Biodiversity Loss**: The mining of fossil fuels and minerals for energy and production has resulted in the exhaustion of finite resources. The process of extraction frequently causes disturbances in ecosystems and results in a decline in biodiversity.

Current Trends in Renewable Energy and Sustainability Initiatives

- **Growth of Renewable Energy**: Renewable energy sources such as solar, wind, and hydroelectric power has experienced a substantial transition. The transition is propelled by technology innovations, declining expenses, and heightened consciousness of environmental concerns.
- **Energy Efficiency Measures**: Several nations are adopting strategies to enhance energy efficiency in buildings, transportation, and industry. These solutions not only lower energy usage but also diminish the release of greenhouse gases.
- **Policies and Global Agreements**: Several global agreements, like the Paris Agreement, seek to tackle climate change by mitigating carbon emissions. National policies are crucial in fostering the use of renewable energy and sustainable activities (Zakaria et al., 2023).

The present state of energy and resource utilization underscores the urgent requirement for a shift towards sustainable energy and circular economy models. Although transitioning from a fossil fuel-based and linear economic system presents significant obstacles, there are also emerging trends and regulations that are facilitating this shift. The transition towards renewable energy, in conjunction with initiatives to enhance energy efficiency and minimize waste, is facilitating the progression towards a more sustainable future. However, attaining this objective necessitates collaborative endeavors by governments, enterprises, and individuals on a global scale.

11.3 Circular Economy as a Solution

11.3.1 Principles of Circular Economy

- **Reduce, Reuse and Recycle**: The fundamental principles of the circular economy encompass waste reduction, resource reuse, and material recycling. This strategy starkly contrasts with the conventional linear economy's 'take-make-dispose' model, prioritizing instead the closure of the resource utilization loop.
- **Design for Sustainability**: The design phase of the circular economy is of utmost importance since it focuses on creating items that are durable, easily repairable, and capable of being recycled. This method aims to reduce waste and promote the utilization of sustainable materials.
- **Resource Efficiency**: The circular economy prioritizes the optimization of resource utilization. It entails the effective utilization of raw resources and energy, resulting in a decreased environmental footprint and improved economic efficiency.

11.3.2 Case Studies of Successful Circular Economy Implementations

- **Industrial Symbiosis**: One striking example is the town of Kalundborg in Denmark, where industrial symbiosis practices have been implemented. Here, waste products from one company become resource inputs for another, leading to significant reductions in waste and energy consumption.
- **Consumer Goods Recycling**: Companies like Patagonia in the apparel industry have adopted circular economy principles by recycling materials and offering repair services for their products, significantly extending their lifecycle.
- **Urban Circular Initiatives**: Cities like Amsterdam and Copenhagen are pioneering circular economy strategies at the urban level, focusing on sustainable urban planning, waste management, and local recycling programs.

11.3.3 Benefits of Integrating Circular Economy in Energy Sector

- **Reducing Resource Dependence**: By adopting circular economy principles, the energy sector can reduce its reliance on finite resources, such as fossil fuels, and minimize environmental impacts.
- **Enhancing Energy Efficiency**: Circular economy practices, like the use of waste-to-energy technologies, can enhance energy efficiency. Repurposing waste as an energy source not only disposes of waste responsibly but also generates additional energy resources.
- **Economic and Environmental Synergies**: The circular economy offers synergies that benefit both the economy and the environment. It can lead to cost savings, innovation in sustainable technologies, and job creation, while simultaneously reducing environmental footprints.

The circular economy presents a viable solution to the current environmental and economic challenges posed by the linear economic model. By embracing the principles of reducing, reusing, and recycling, and implementing them across various sectors, significant strides can be made towards sustainability. Success stories from around the world demonstrate the practicality and benefits of the circular economy. Its integration into the energy sector, in particular, holds great promise for reducing environmental impact while fostering economic growth. As we move forward, the circular economy should be a cornerstone of sustainable development strategies, requiring commitment and innovation from all sectors of society.

11.3.4 Opportunities in Technology and Innovation

Advances in Renewable Energy Technologies

- **Solar and Wind Power Innovations**: Technological advancements in solar photovoltaics (PV) and wind turbines are leading to higher efficiency and lower costs. Innovations such as perovskite solar cells and floating wind farms are expanding the potential applications of these renewable sources.
- **Bioenergy and Geothermal Developments**: Advances in bioenergy, including biofuels and biomass energy, offer solutions for both power generation and transportation fuels. Similarly, improvements in geothermal energy technologies are unlocking new possibilities for both heating and electricity generation.
- **Hydrogen as an Energy Carrier**: The development of green hydrogen, produced using renewable energy, is gaining attention as a potential game-changer for storing and transporting energy, especially in sectors hard to electrify like heavy industry and long-haul transportation.

11.4 Role of Digital Technologies

- **Artificial Intelligence and Machine Learning**: AI and ML are being increasingly used to optimize energy systems, from predictive maintenance of renewable energy installations to smart grid management and energy demand forecasting.
- **Internet of Things (IoT)**: IoT technologies enable more efficient energy management and integration of various renewable sources into the grid. Smart meters and home energy management systems illustrate the potential of IoT in enhancing energy efficiency and consumer engagement.
- **Blockchain for Energy Transactions**: Blockchain technology offers a novel approach to energy transactions, enabling peer-to-peer energy trading and transparent, secure tracking of renewable energy certificates, fostering decentralized energy markets.

11.4.1 Innovations in Energy Storage and Grid Management

- **Advancements in Battery Technology**: The development of advanced battery technologies, such as solid-state batteries, is crucial for storing renewable energy more efficiently and for longer durations, addressing the intermittency challenge of solar and wind power.
- **Smart Grids and Energy Management Systems**: Smart grids, which are connected with sensors and digital communication technologies, have the potential to dynamically control the supply and demand of energy. This ensures that the

grid remains stable even with a significant amount of renewable energy being used.
- **Microgrids and Distributed Energy Resources**: Microgrids and distributed energy resources (DERs) are gaining significance in enhancing the resilience and adaptability of the energy system, particularly in geographically isolated and disaster-prone regions.

11.4.2 Emerging Business Models and Market Dynamics

- **Energy as a Service (EaaS)**: The adoption of energy efficiency measures and renewable technology is encouraged by this approach, in which customers are charged for energy services rather than energy consumption.
- **Circular Economy in Technology**: The application of circular economy concepts in technology production and use prioritizes the reutilization and recycling of electronic components and batteries, hence minimizing waste and resource extraction.
- **Collaborative Consumption and Sharing Economy**: The emergence of sharing economy models in the energy sector, exemplified as community-shared solar power installations, signifies a transition towards more cooperative and distributed energy systems.

The technology and innovation in sustainable energy and circular economy are continuously changing, providing many chances to tackle environmental concerns and promote economic growth. The successful adoption of technical developments for a more sustainable future hinges on supportive legislation, market readiness, and public acceptability. In order to fully harness the potential of these technologies in promoting sustainable energy transition and circular economy practices, it is crucial to prioritize ongoing innovation and foster collaboration among governments, businesses, and communities.

11.5 Policy and Regulatory Framework

11.5.1 Renewable Energy Policies

- **Feed-in Tariffs (FiTs)**: These policies have played a crucial role in multiple nations by providing incentives for the widespread use of renewable energy. They achieve this by ensuring a stable and appealing price for electricity generated from renewable sources. The implementation of Feed-in Tariffs (FiTs) in nations such as Germany has been remarkable, resulting in a substantial surge in the production of renewable energy.

- **Renewable Portfolio Standards (RPS)**: RPS laws, by requiring a specific proportion of electricity to be generated from renewable sources, have stimulated investments in renewable energy in the United States and other countries. Nevertheless, the efficacy of the measures fluctuates depending on the enforcement procedures and specific objectives established.
- **Tax Incentives and Subsidies**: Several nations provide tax incentives or subsidies to encourage the adoption of renewable energy. Although these subsidies have contributed to the growth of the sector, there is a continuing discussion on the long-term viability and potential market imbalances caused by them.

11.5.2 Circular Economy Legislation

The Circular Economy Action Plan of the European Union exemplifies this, as it establishes high objectives for recycling and waste reduction, while also advocating for eco-design. This legislative approach is exerting an influence on product lifecycles, prompting producers to consider the complete lifecycle of their products when designing them.

11.5.3 International Agreements

The Paris Agreement has established an international framework for the mitigation of carbon emissions, exerting influence over domestic policies. The influence of this can be observed through the heightened dedication of nations to reduce emissions and shift towards renewable energy sources.

11.5.4 Recommendations for Policymakers to Foster Transition

- Integrating Sustainability into Economic Policies
- Policymakers are advised to perceive sustainable energy not just as an environmental endeavor but also as an essential component of economic policy. This entails harmonizing environmental objectives with economic growth strategies, guaranteeing that investments in infrastructure, such as, are both financially feasible and ecologically sustainable.
- Building Resilient and Flexible Regulatory Frameworks
- Regulations must be updated in accordance with the rapid progress of technology. Flexibility can be attained by implementing adaptive policymaking, which involves the constant evaluation and revision of policies in response to technological advancements and market input.

- Facilitating Public-Private Partnerships
- It is advisable for policies to promote collaborations between the public and private sectors. This cooperation can use the advantages of each sector, merging public supervision with private sector effectiveness and ingenuity.

11.5.5 International Cooperation and Frameworks

- Global Collaboration for Technology Transfer
- Developing nations frequently face a dearth of resources and technological capabilities to facilitate the shift towards sustainable energy. Developed nations have the ability to significantly aid in the process of transferring technology by participating in international accords and organizations.
- Harmonizing Standards and Practices
- In order to promote the development of a worldwide market for sustainable technology, it is essential to establish standardized criteria. This guarantees that products and services are harmonious and adhere to specific environmental and performance standards, hence promoting international commerce.
- Financing and Support for Developing Countries
- Developing countries rely on international financial institutions and wealthier nations to play a crucial role in offering financial and technical assistance. This assistance is vital for these nations to establish the necessary infrastructure and capability required for the transition.

11.5.6 Role of Governance and Institutional Capacity

- Strengthening Institutional Capacities
- Robust institutions are essential for the successful implementation of policies. This entails not only establishing the requisite legal frameworks but also enhancing the capabilities of government authorities to enforce and oversee these rules.
- Community Involvement and Local Governance
- Community involvement is crucial for the successful implementation of policies. Local governance models can guarantee the customization of policies to suit local situations and the inclusion of community demands and preferences.
- Transparency and Accountability

  Transparent governance and accountability mechanisms are crucial for building public trust and ensuring effective policy implementation. This includes regular reporting on progress, stakeholder engagement, and independent auditing and review processes.

Policy and regulatory frameworks are foundational to the transition towards sustainable energy and circular economies. While significant progress has been made, continuous adaptation and refinement of these policies are essential. The collaboration of all stakeholders—governments, businesses, communities, and international bodies—is vital in steering the world towards a more sustainable and resilient future.

11.6 Business Models and Market Dynamics

11.6.1 Emerging Business Models in Sustainable Energy

- **Energy as a Service (EaaS)**: This model is transforming the energy sector. Companies provide energy solutions, such as solar panels or energy efficiency services, through subscription or pay-per-use structures, departing from conventional ownership patterns. This strategy reduces the obstacles for the adoption of renewable energy and promotes the use of energy in a more efficient manner.
- **Decentralized Energy Systems**: Decentralized approaches, such as microgrids and community-based renewable energy projects, provide local communities and businesses with the ability to exercise control and authority. Their primary objective is to advocate for energy autonomy and robustness, particularly in distant or underserved regions.
- **Green Financing and Investment**: Green bonds and impact investment, which are innovative financing strategies, are becoming increasingly popular. These models prioritize investments in sustainable energy projects, providing investors with both financial profits and environmental advantages.

11.6.2 Circular Economy Business Models

- **Product as a Service (PaaS)**: Under this paradigm, firms maintain ownership of their products and provide them as a service. This method provides firms with incentives to produce long-lasting, easily recyclable items, so minimizing waste and encouraging the efficient use of resources.
- **Recycling and Upcycling Ventures**: Businesses that prioritize recycling and upcycling are transforming discarded materials into lucrative commodities. These models not only minimize waste but also provide new market prospects in the circular economy.
- **Circular Supply Chains**: Companies are restructuring their supply chains to eliminate inefficiency and optimize the utilization and repurposing of resources. This transition not only decreases the negative effects on the environment but also improves the ability of the supply chain to withstand and maintain sustainability.

11.6.3 Market Trends and Investment Opportunities

- **Renewable Energy Market Growth**: The renewable energy sector is seeing a significant expansion, propelled by technological innovations and declining costs of renewable technologies like as solar and wind power.
- **Shift in Consumer Preferences**: Consumer demand for sustainable products and services is increasing. The current trend is exerting an impact on market dynamics, leading to an increased number of enterprises embracing sustainable practices in order to fulfill customer expectations.
- **Government Incentives and Policies**: Government policies, such as providing financial support for renewable energy and implementing legislation that encourage the adoption of circular economy practices, are influencing the way the market operates. These policies are generating fresh investment prospects in environmentally friendly technologies and sustainable business frameworks.

11.6.4 Role of Private Sector and Public-Private Partnerships

- **Corporate Sustainability Initiatives**: Many organizations are making firm commitments to achieve sustainability objectives, allocating resources towards renewable energy, and embracing circular economy techniques. These initiatives are motivated not only by environmental concern but also by the acknowledgment of enduring economic advantages.
- **Public-Private Partnerships (PPPs)**: Public-Private Partnerships (PPPs) play a vital role in facilitating the shift towards sustainable energy and circular economies. These collaborations utilize the respective advantages of both sectors - the private sector's efficiency and creativity, and the public sector's regulatory authority and public responsibility.
- **Innovation and Research Collaboration**: Partnerships among corporations, governmental bodies, and educational establishments are promoting the development of inventive sustainable technology and commercial strategies. These partnerships are essential for the development and expansion of innovative solutions.

The transition towards sustainable energy and circular economies is fundamentally transforming the structure of business models and market dynamics. These emerging approaches, such as energy as a service and circular supply chains, are simultaneously generating economic prospects and tackling environmental issues. Nevertheless, the effective execution of these models necessitates favorable regulations, inventive funding, and a dedication to sustainability from all participants in the market. In light of changing market dynamics, it is imperative for enterprises and governments to adjust and cooperate in order to fully utilize the opportunities presented by these emerging models. In the future, there is a high probability that we will witness an increase in the adoption of integrated and sustainable business practices. This will be primarily driven by technical advancements, growing consumer demand, and the shared obligation to construct a more sustainable global environment.

11.7 Social Implications and Community Engagement

Understanding the Social Context

- **Inequity in Energy Access**: Globally, there's a significant disparity in energy access. While developed countries enjoy stable and continuous energy supply, many developing nations grapple with energy poverty. This inequity extends to the availability of clean and renewable energy sources, impacting health, education, and economic opportunities in underserved communities.
- **Local Environmental Justice Issues**: Often, the environmental burden of traditional energy production disproportionately affects marginalized communities. For instance, fossil fuel plants are commonly located in low-income areas, leading to higher pollution levels and associated health risks. Recognizing and addressing these injustices is crucial in the transition process.

Community Engagement in Energy Transition

- **Community-Based Renewable Energy Projects**: Engaging communities in the planning and implementation of local renewable energy projects can be empowering. These projects, like community solar or wind farms, not only provide sustainable energy but also strengthen community ties and promote local ownership and decision-making.
- **Educational Programs and Awareness**: Education and awareness campaigns play a pivotal role in building community support for sustainable energy and circular economy initiatives. Informing people about the benefits of these transitions, and how they can contribute, is essential for widespread acceptance and participation.

Role of Social Innovation

- **Innovative Business Models**: Social enterprises are emerging as key players in promoting sustainable energy and circular economy practices. These businesses often focus on creating social and environmental value, offering innovative solutions like affordable solar lighting in off-grid communities or recycling programs that provide employment opportunities.
- **Collaboration and Partnerships**: Partnerships between various stakeholders – governments, NGOs, businesses, and communities – are vital for successful social innovation in this field. These collaborations can lead to more effective and inclusive policies and programs.

Impact of Policy and Regulation on Communities

- **Policy Inclusivity and Equity**: Policies are essential in guaranteeing that the shift towards sustainable energy and circular economies is fair and accessible to all. In order to prevent the worsening of current inequalities, regulations must take into account the effects on all sectors of society, particularly those that are more susceptible.
- **Incentives for Community Participation**: Governments and organizations have the ability to offer incentives to encourage community engagement in sustainable activities, such as providing financial assistance for the adoption of renewable energy or offering rewards for active involvement in recycling programs.

11.8 Challenges and Opportunities

- **Addressing Social Resistance**: The obstacle of resistance to change is a substantial task. This can arise from a deficiency in comprehension, apprehension of economic repercussions (such as employment reductions in conventional energy industries), or adherence to existing methodologies. Resolving these concerns by means of open discussion and imparting knowledge is crucial.
- **Leveraging Community Knowledge and Practices**: Communities frequently possess distinctive perspectives and customary methods that can be quite beneficial in formulating sustainable solutions. Incorporating and incorporating this indigenous knowledge can augment the efficiency and reception of sustainable projects.

The transition towards sustainable energy and circular economy models has significant social ramifications. Efficient community involvement entails more than just obtaining public approval; it also involves guaranteeing that this shift is fair and advantageous for everyone. By prioritizing inclusivity, education, social innovation, and good government, we can guarantee that communities not only take part in this change but also reap its benefits. The journey towards a sustainable future is marked by obstacles, but through cooperative and all-encompassing endeavors, it also offers significant prospects for societal advancement and progress.

11.9 Future Outlook and Pathways

Scenarios for the Future of Sustainable Energy and Circular Economy

- **Increased Renewable Energy Adoption**: Forecasts indicate a persistent increase in the utilization of sustainable energy sources. The utilization of solar and wind energy is anticipated to increase significantly due to developments in technology and the reduction in costs.

- **Advancements in Energy Storage**: It is quite probable that the future will witness substantial progress in energy storage technology, particularly in batteries, which play a vital role in resolving the problem of intermittent renewable energy.
- **Expansion of Circular Economy Principles**: The use of circular economy ideas is anticipated to extend beyond the realms of recycling and waste management to embrace wider sectors of the economy, such as product design, business models, and consumer behavior.

Potential Impacts of Climate Change and Other Global Trends

- **Accelerated Climate Change Effects**: With the increasing severity of climate change effects, there will be a growing need to shift towards sustainable methods. This encompasses heightened frequency and intensity of extreme weather phenomena, elevation of sea levels, and repercussions on biodiversity and agriculture.
- **Globalization and Technological Advancements**: The interaction between globalization and rapid technical breakthroughs will persistently influence the development of sustainable practices. The utilization of digital technologies, such as the Internet of Things (IoT) and Artificial Intelligence (AI), will have a crucial impact on improving the allocation of resources and increasing energy efficiency.
- **Socio-Economic Shifts**: The transition to sustainable energy and circular economy models will be influenced by socio-economic transformations, such as urbanization and changing consumer preferences. Consumer's increasing environmental consciousness is anticipated to stimulate the demand for sustainable products and services.

Strategies for Resilience and Adaptability

- **Building Resilient Infrastructure**: It is imperative to allocate resources towards building robust infrastructure, namely in energy and transportation sectors, in order to effectively endure and adjust to the consequences of climate change.
- **Policy Adaptability and Innovation**: Policymakers must maintain flexibility, continuously revising and creating policies to address changing environmental, technical, and socio-economic circumstances.
- **Strengthening International Cooperation**: Global concerns such as climate change necessitate international cooperation. This encompasses the dissemination of knowledge, technology, and optimal methodologies, along with offering assistance to emerging nations in their endeavors to transition.

The future prospects for sustainable energy and circular economy entail a combination of obstacles and possibilities. The escalating ramifications of climate change, along with global phenomena like as urbanization and technological progress, will determine the trajectory of this shift. The success of this shift hinges on a cooperative strategy, encompassing governments, corporations, communities, and international organizations. It necessitates constructing robust and flexible systems, promoting creativity, and upholding a dedication to sustainable development objectives. As we contemplate the future, it is evident that the choices and behaviors made at present will significantly influence the long-term viability of our planet. The pursuit of a sustainable and circular economy is not only imperative, but also presents a chance to establish a more resilient, fair, and affluent society for future generations.

11.10 Conclusion

This chapter is around acknowledging the significant capacity for sustainable energy and circular economy to bring about positive changes for both our planet and societies. The shift towards sustainable energy is not only an ecological obligation but also a socio-economic imperative. The imperative to transition to renewable energy sources and adopt more efficient and cleaner technologies is crucial due to the imminent danger of climate change and the limited availability of fossil fuels. Nevertheless, this transformation is accompanied by other difficulties, including technological obstacles, economic issues, and social opposition. However, these problems also present significant prospects. Progress in renewable energy technology, breakthroughs in energy storage and grid management, and the emergence of business models such as Energy as a Service (EaaS) are facilitating the transition towards a more environmentally friendly energy future.

Similarly, the concept of the circular economy arises as a potent remedy to the constraints of the prevailing linear paradigm. The circular economy concept presents a means to achieve substantial environmental advantages, economic robustness, and societal welfare by rethinking the utilization and reutilization of resources. The practicality and usefulness of waste management strategies, such as industrial symbiosis and consumer goods recycling, are demonstrated by their successful implementations. The pursuit of sustainable energy and a circular economy is intricate and diverse. It necessitates collaborative endeavors from governments, corporations, and communities. Policymakers have a vital responsibility to develop adaptable and robust policies that promote innovation while also guaranteeing inclusivity and equity. Global environmental concerns necessitate international cooperation to enable technological transfer and financial assistance to underdeveloped nations.

As we contemplate the future, the routes towards achieving sustainable energy transition and circular economy are characterized by a combination of obstacles and prospects. The themes described in this chapter, including technological advancements, policy dynamics, and market movements, are not merely theoretical but practical pathways that can effectively bring about the desired change. The shift is a continuous process, more like a voyage than a final point, necessitating flexible tactics and cooperative endeavors. The shift towards sustainable energy and the adoption of a circular economy are not optional decisions, but rather essential requirements for ensuring a sustainable future. Embracing these changes holds the potential for a future that is both environmentally sustainable and economically prosperous, while also being socially inclusive. As we progress, it is evident that the decisions we make now will have a lasting impact on future generations. The road is intricate, yet the trajectory is unambiguous - towards a sustainable, circular, and resilient future.

References

Hegab, H., Shaban, I., Jamil, M., & Khanna, N. (2023). Toward sustainable future: Strategies, indicators, and challenges for implementing sustainable production systems. *Sustainable Materials and Technologies, 36*, e00617. https://doi.org/10.1016/j.susmat.2023.e00617

IEA. (2020). Energy Efficiency 2020. In *Energy Efficiency 2020*. https://www.oecd-ilibrary.org/energy/energy-efficiency-2020_dfd85134-en

Rockström, J., Steffen, W., Noone, K., Persson, Å., Chapin, F. S., Lambin, E. F., Lenton, T. M., Scheffer, M., Folke, C., Schellnhuber, H. J., Nykvist, B., de Wit, C. A., Hughes, T., van der Leeuw, S., Rodhe, H., Sörlin, S., Snyder, P. K., Costanza, R., Svedin, U., et al. (2009). A safe operating space for humanity. *Nature, 461*(7263), 472–475. https://doi.org/10.1038/461472a

Zakaria, Z., Kamarudin, S. K., Ikhmal Salehmin, M. N., Rosyadah Ahmad, N. N., Aminuddin, M. A., Hanapi, I. H., Osman, S. H., & Mohamad, A. A. (2023). Energy scenario in Malaysia: Embarking on the potential use of hydrogen energy. *International Journal of Hydrogen Energy, 48*(91), 35685–35707. https://doi.org/10.1016/j.ijhydene.2023.05.358

SPRINGER NATURE

GPSR Compliance

The European Union's (EU) General Product Safety Regulation (GPSR) is a set of rules that requires consumer products to be safe and our obligations to ensure this.

If you have any concerns about our products, you can contact us on ProductSafety@springernature.com

In case Publisher is established outside the EU, the EU authorized representative is:

Springer Nature Customer Service Center GmbH
Europaplatz 3
69115 Heidelberg, Germany

The manufacturer's authorised representative in the EU is Springer Nature Customer Service Centre GmbH, Europaplatz 3, 69115 Heidelberg, Germany. If you have any concerns regarding our products, please contact ProductSafety@springernature.com

Printed and bound by CPI Group (UK) Ltd, Croydon, CR0 4YY
01/10/2026
02225896-0014